ROBOT PROGRAMMER'S BONANZA

ROBOT PROGRAMMER'S BONANZA

JOHN BLANKENSHIP

SAMUEL MISHAL

New York Chicago San Francisco Lisbon London Madrid
Mexico City Milan New Delhi San Juan Seoul
Singapore Sydney Toronto

The McGraw-Hill Companies

Library of Congress Cataloging-in-Publication Data

Blankenship, John, date.
Robot programmer's bonanza / John Blankenship, Samuel Mishal.
p. cm.
ISBN 978-0-07-154797-0 (alk. paper)
1. Robotics. I. Mishal, Samuel. II. Title.
TJ211.B565 2008
629.8'9251—dc22 2008010227

1 2 3 4 5 6 7 8 9 0 DOC/DOC 0 1 4 3 2 1 0 9 8

ISBN 978-0-07-154797-0
MHID 0-07-154797-5

Printed and bound by RR Donnelley.

This book is printed on acid-free paper.

Sponsoring Editor
Judy Bass

Production Supervisor
Richard C. Ruzycka

Editorial Supervisor
Stephen M. Smith

Project Manager
Vastavikta Sharma, International Typesetting and Composition

Copy Editor
Priyanka Sinha, International Typesetting and Composition

Proofreader
Divya Kapoor, International Typesetting and Composition

Art Director, Cover
Jeff Weeks

Composition
International Typesetting and Composition

To my wife Sharon for putting up with all the hours I spent at the computer. A special thanks to Sam Mishal for the countless hours he spent developing RobotBASIC. Our constant debates about how this book should be written have made it better than either of us ever envisioned. In many ways, this book is far more his than mine.

JOHN BLANKENSHIP

To my sister May for always being there for me and for setting an example of excellence. To my nephew Rany just because I love him. To my good friend Tom Emch for all the interesting discussions we had over the years and for editing and reviewing this book. To my good friend Ted Lewis for all his psychological support. A special dedication to John Blankenship for having been an inspiration to me in many aspects and during the writing of this (my first) book.

SAMUEL MISHAL

ABOUT THE AUTHORS

JOHN BLANKENSHIP taught computer and electronic technology for 33 years at the college level. He has also worked as an engineer and as an independent consultant. He received a B.S. in electrical engineering from Virginia Tech, a masters in electronic engineering technology from Southern Polytechnic State University, and an M.B.A. from Georgia State University. This is his sixth book.

SAMUEL MISHAL is a software engineer and systems analyst. He worked as a consultant for major government departments and businesses around the world. He taught mathematics and computing at the college level. He received a B.S. in electronics engineering technology from DeVry University, a bachelors in computer science from the University of Western Australia, a masters in engineering science from Oxford University, and a masters in structural engineering from Imperial College London.

CONTENTS AT A GLANCE

CONTENTS

PREFACE

The field of hobby robotics has many parallels to personal computing. If you wanted to own a computer in the 1970s, you had to build it yourself. Less than a decade later, you could buy a fully assembled computer and people quickly discovered that programming a computer led to far more enjoyment, satisfaction, and productivity than constructing one.

In the 1980s robot hobbyists spent most of their time building robots from wood and sheet metal. They powered their creations with surplus parts like windshield wiper motors salvaged from car junkyards. So much time was spent in the construction phase that minimal thought was given to the electronic aspects of the project—many of the early robots were controlled with doorbell buttons and relays.

As the personal computer became more powerful a more sophisticated robotics hobbyist began to evolve. They learned more about electronics and started building crude sensors and motor control circuitry that, along with a personal computer, gave their robots, at least, the potential to interact with their environments. These new hobbyists renewed the dream that intelligent robots could actually be built. Unfortunately, most of the people interested in robotics still lacked the required electronics skills and knowledge.

In the years that followed, many books and magazines were published that promised to help robot enthusiasts create circuitry to give their robots more intelligence. However, often, due to complexity and lack of experience, many people had trouble duplicating the authors' works.

Despite all these difficulties, the desire to build personal robots did not diminish. New companies emerged offering robot kits that required minimal experience to build and actuate. These early kits were not programmable, and thus did not satisfy the hobbyists' desire to create intelligent machines. Nowadays there are many companies that offer sophisticated sensors and embedded computers that make it possible to build intelligent, capable and useful robots.

Today, you can buy electronic compasses, ultrasonic rangefinders, GPS systems, infrared perimeter sensors, line and drop-off detectors, color detectors, electronic accelerometers, and even cameras. Reasonable knowledge and often a lot of time are still required to interface these devices to a robot's microcontroller, but the abundance of manuals and books make details available to any hobbyist willing to expend the effort. With sophisticated hardware available to everyone, hobby robotics is now able to turn its attention to programming, finally making it possible to create truly intelligent machines.

Considering these developments, it is easy to feel like all the hard work has been done, when in fact, the real work is just beginning. Remember, personal computers were just a curiosity until the emphasis shifted from building them to programming them. This *paradigm shift* enabled innovative hobbyists and entrepreneurs to create word processors, spreadsheets, and graphical user interfaces (GUIs) that changed the world. The world of hobby robotics is now entering such an era. Today's robot enthusiasts no longer need a degree in electronics and a machine shop in their garage to create robots that are ready to be programmed. They do, however, need to understand programming, because it is software that truly creates a useful robot.

Sophisticated kits and fully assembled robots are available from many vendors. Numerous companies offer off-the-shelf hardware modules that enable a typical hobbyist to *assemble* a custom robot with capabilities that were only a dream a few years ago. A hobbyist that understands the concepts of robot programming can use these new platforms to create the projects robot builders have been seeking for years.

Unfortunately learning to program a robot can be very frustrating, even if you have the appropriate hardware. Sensors often need adjusting and realigning and batteries always seem to need recharging. When the robot fails to respond properly you run the risk of damaging it or even your home or furniture. Because you can't *see why* the robot is failing, the task of debugging the code can often be exasperating. With the world of robotics entering its new era, there has to be a better way for hobbyists to learn about *programming* their machines.

This book is aimed at the new hobbyist who is interested in *programming* robots. Today there are numerous microcontrollers that can be used to control robots. These controllers can be programmed using a variety of programming languages (Assembly, C, BASIC, and others). This lack of homogeneity in hardware and software tools make it hard to learn how to program a robot, even if you have previous programming experience.

In reality, the details of the implementation using a specific combination of software and hardware are of secondary concern. What is important in programming a robot to do useful tasks is the *algorithm* that achieves the desired logic. Once the algorithm is determined it can be easily *translated* into any programming language to work on any appropriate microcontroller.

RobotBASIC is a full-featured, interpreted programming language with an integrated robot simulator that can be used to *prototype* projects. The simulator allows you to research various combinations of sensors and environments. You can change the types and arrangements of sensors in seconds, making it possible to experiment with numerous software ideas. You can test your algorithms in environments that would be impractical to create in real life.

The simulated mobile robot is two-dimensional, but programming it lets you learn how to use all the sensors you would expect to find installed on robots costing hundreds if not thousands of dollars. And you will soon discover that programming the simulation is so much like programming the real thing (less all the frustrating aspects) that you will soon forget it is just a circle moving on your screen.

RobotBASIC has capabilities far beyond the robot simulator. It is a powerful programming language with functions that support graphics, animation, advanced mathematics, and access to everything from I/O ports to Bluetooth communication so that you can even use it to control a real-world robot if you choose. When you learn about robot programming with RobotBASIC you won't have to spend months building a robot. You will be able to start programming immediately and never have to worry about charging batteries or damaging furniture, although you can simulate those events too.

The book is divided into four parts. Part 1 explores the advantages of using a simulator and teaches how to use the simulated robot and its sensors. It also introduces the RobotBASIC language and programming concepts in general. By the time you finish Part 1, you will be able to write and debug simple programs that move the robot around a simulated environment while avoiding objects that block its path.

Part 2 examines everything you typically find hobbyists doing at robot clubs. You will learn ways to make the robot follow a line on the floor, hug a wall, or stay away from a drop-off such as a stairway. All of these topics (and more) are examined with simple easy-to-understand approaches. The simulation is then used to expose problems and deficiencies with the initial approaches. New and better algorithms are then developed and explained. Learning about robotics using this building blocks approach can be very motivational because it is exciting and relevant. As you proceed through the book you will gain more knowledge about programming and problem solving principles. This makes RobotBASIC an ideal first language for teaching students about programming, mathematics, logical thinking, and robotics.

The chapters in Part 3 combine the behaviors developed in Part 2 into compound complex behaviors, that enable the robot to solve real-world projects such as charging the robot's battery, mowing a lawn, solving a maze, locating a goal, and negotiating a home or office environment. As in Part 2, the projects are first explored with simple approaches before introducing more complex concepts. The advanced reader will find this part of the book interesting because many behaviors are evolved using mathematics and computer science topics.

Part 4 explores advanced topics such as adaptive behavior and how RobotBASIC programs can be used to control real-world robots using wireless links. Additionally, ideas are forwarded for why RobotBASIC can be useful in robotic contests and as a teaching tool in the classroom.

The RobotBASIC program along with all the programs in this book can be downloaded from www.RobotBASIC.com. The language is subject to change as alterations and upgrades are implemented. The help files accessible from the latest IDE will have the most valid up-to-date descriptions of all the functionalities of the language. Make sure to always download the latest version and to consult the help files for any new and modified features. Also make sure to check the site for:

- Updated listings of all the programs in the book.
- Solutions for some of the exercises in the book.
- Any corrections to errors that may have slipped into the book.
- Other information and news.

ACKNOWLEDGMENTS

We thank William Linne and Thomas Emch whose suggestions and comments have added greatly to the final text. We also thank Stephanie Lindsay at Parallax, Inc. for her support and contributions. A special thanks to everyone at McGraw-Hill, especially Judy Bass, for making the huge task of writing this book an enjoyable experience.

ROBOT PROGRAMMER'S BONANZA

PART 1

BUILDING BLOCKS

In Part 1, besides exploring the advantages and utility of simulators, we introduce the RobotBASIC IDE (integrated development environment) and language along with the robot simulator. Initially we develop simple programs to illustrate the mechanisms for creating and animating a robot. Later chapters introduce the available sensory systems and show how to use them to avoid obstacles while the robot is roaming around its environment. The RobotBASIC programming language is introduced in stages in Chaps. 2 to 5. Flow-control statements, conditional execution, binary math, bitwise operators, and subroutines are introduced with application to the simulator. Many commands, along with some mathematical functions and concepts, are introduced while writing programs to control the robot.

Each chapter introduces pertinent new skills while building upon previous knowledge to accumulate the expertise necessary for building the toolbox of behaviors that will be developed in Part 2.

Upon completing Part 1 you will be able to:

- Create, edit, open, and save programs using the IDE.
- Write programs using the language to a good level of proficiency:
    - Get input from a user using the mouse and keyboard.
    - Display output and graphics on the screen.
    - Do conditional execution.
    - Use looping constructs.
    - Understand and utilize commands and functions.
    - Use binary numbers and bitwise operations.
    - Apply modularity and utilize subroutines.
- Manipulate the robot and utilize most of its sensory systems:
    - Move the robot in a simulated environment.
    - Interrogate and interpret the infrared and bumper sensors.
    - Be aware of other sensors and instrumentation.
- Use the Debugger to debug programs.

CHAPTER 1

WHY SIMULATIONS

Since you are reading this book, you must be interested in robotics to a certain extent. Perhaps you are a member of a robot club or attend a technical school and have a little experience building your own robots. Maybe you have purchased a robot kit and want to learn how to customize it. Maybe you want to learn about robotics but don't have the funds to buy or build a robot of your own. If you fall into any of these categories, a robot simulator is a very effective way to learn about robotics and robotic algorithms. A robot simulator is also a valuable tool for experimenting with various possibilities and combinations of hardware and software arrangements without the time delay and expense incurred when building an actual robot.

1.1 What Is RobotBASIC?

In general, this book is about a computer language called RobotBASIC. More specifically, this book is about how you can use RobotBASIC to prototype algorithms that enable a robot to interact with its environment. The advantage of a simulator is that you can do this without having to buy or build an actual robot.

RobotBASIC allows you to create a *simulated* robot on your computer screen. As we progress through the algorithms in this book you will find that the simulated robot is very much like the real thing. It can be placed in rooms with furniture, or outside so that it can mow a yard. You can program the simulator to do nearly anything a real robot can do.

After studying this book you will be able to program a robot to, for example, navigate throughout the rooms in your home to find and plug itself into a battery charging station.

That last statement was very important. Notice that we did not say that you would be able to program the simulated robot—We said you would be able to program *a* robot. The robot in RobotBASIC is so realistic and accurate in its ability to mimic a real robot, that the very same algorithms and principles you use to program the simulated robot can be used to control a real one. Chapter 17 shows how to build a real world equivalent of the robot simulated in RobotBASIC and shows how you can utilize the algorithms developed in this book to program an actual robot.

1.2 Flight Simulators

The fact that a simulation can truly mimic the real world may be unfamiliar to you if you are not acquainted with how simulations are used nowadays. Pilots, for example, are trained on flight simulators that are so accurate and realistic that they can be used for certification purposes. Simulators have economic advantages over using a real airplane for training purposes, but there are other advantages too. A simulator allows situations to be tested that would otherwise be difficult or dangerous to implement. We want, for example, commercial pilots to be able to land a plane even if one engine fails because several geese were sucked into it during approach to the runway. Simulating such an emergency on a real airplane by shutting down one of the engines is dangerous and expensive. Using a realistic simulator would be much safer and cost efficient.

Obviously, if flight simulators are going to be effective they have to feel real to the pilot being trained. They have to respond to the pilot's commands exactly like the real airplane would. In order to be useful, they have to make the pilot *forget* the fact that he is commanding a simulator. Flight simulators today have cockpits mounted on hydraulic actuators where the windows are actually computer screens that display what would be seen out of a real window. It is not unusual for the simulation to be so detailed that you can feel the plane bump as it rolls over the tar-filled cracks on the runway.

1.3 Comparing RobotBASIC with Other Simulators

If you search today you can find programs that allow you to create simulated robots of various shapes and sizes with sensors tailored to your specifications. Some simulators will display your creations in three dimensions on your computer screen, perhaps even complete with the appropriate shading and shadows. Unfortunately such programs are often expensive, complex to learn and use, and slow if not being run on a very fast system.

RobotBASIC was developed to address all these issues. RobotBASIC is free for everyone to use. This includes individuals, clubs, schools, or any other organization. Give it to your friends, distribute it to your students, tell your club members to download it—our aim is for RobotBASIC to be of utility to people of various skills and ages. The only thing you are not allowed to do with RobotBASIC is sell it.

RobotBASIC does not display the simulated robot in three-dimensional graphics, however, you will find that the robot has all of the sensors you would expect to find on a hobby robot as well as a few that most people wish they had the means to implement. Other simulators may have sophisticated graphics but displaying the robot in three dimensions does not enhance the *functionality* of simulations for a robot that moves in two dimensions.

RobotBASIC is easy to use. It is a BASIC-like language that is easy to learn, even for people who have never programmed before. A teacher can utilize RobotBASIC to make even sixth graders excited and productive in only a few hours, and they won't just be learning to play with a robot, they will be developing significant problem solving skills and learning the principles needed to program a computer in any language. RobotBASIC can be used to create challenges appropriate to various age groups.

Even though RobotBASIC is easy enough for beginners, you will find it is also powerful enough to be used by sophisticated hobbyists and experienced programmers. It has all the standard flow-control structures and a virtually unlimited space for variables and arrays. As a RobotBASIC programmer you have a full complement of graphics commands and functions for manipulating strings. The mathematic functions available include the ones you would expect in any powerful scientific calculator, but you will also find matrix operations seldom found in any language.

1.4 Developing Robot Behaviors

The debugging tools in RobotBASIC are both powerful and easy to use. They let you watch the value of variables in your program while you observe the robot's behavior. You can even see the areas around the robot's perimeter where the infrared sensors are checking for objects. These features help you understand how your robot is *seeing* its environment, which in turn helps you develop algorithms that give your robot intelligent behaviors.

RobotBASIC lets you easily and quickly simulate a wide variety of environments and situations for testing your algorithms. Testing a real robot can often be extremely time-consuming. Typically, when programming a real robot, you have to edit a file, compile it, plug the robot into the computer, download the program to the robot's memory, unplug the robot, position the robot in the testing environment, switch it on, and then observe its behavior while making sure it does not damage itself or the environment. It is often difficult if not impossible to *see* why the robot is not responding as you expected. You often have to repeat this cycle many times until you get the required result. The inconvenience of this iterative process can lead you to compromise and accept a working algorithm rather than an optimal one you could have developed had you persevered in trying to optimize your algorithm.

With the simulator, you can make changes in seconds, not only to your algorithm, but to the environment as well. And during testing, you don't have to guess what your robot is seeing. With the debugging tools you can step through sections of your code, watching exactly what the robot is detecting and how it is reacting to obstacles in its path. We can't emphasize enough how important this ability is. When you develop an algorithm to control your robot's behavior it is crucial to be able to view the environment from the robot's perspective. A simulator is by far, the best way to achieve this.

1.5 Simulation Can Improve Hardware Choices

When you design a robot, you need to make many decisions. What type of sensors should it have, how many of each should there be, and how should they be mounted. For example, you might want to have infrared sensors around the perimeter of your robot so that it can detect objects before bumping into them. (Infrared sensors work by emitting infrared light and detecting if any of that light is reflected back to the robot.) You may choose to have only one sensor facing the front of your robot, or you might want one on each side in addition to the front one. The correct choice will be influenced by the type of environment in which you expect your robot to operate.

RobotBASIC's robot has five infrared sensors, one directly in the front, two more offset 45° to the sides, and two more directly to the left and right of the robot. When programming the simulator you may use any or all of these sensors. You also have the capability of creating as many custom sensors as you might need for special situations (see Chap. 9). Imagine how this can help in designing your robot.

Without a simulator you would have to mount and remount your sensors while going through numerous programming alterations and tests to see how your robot would react to your choices. With the simulator you can do all of this in a fraction of the time. The simulator also lets you easily test your sensor placements and programming algorithms under a wide range of conditions, such as extremely crowded environments or objects with sharp points and so on.

If you use a simulator to test your ideas you can make decisions about what sensors your robot should have and how they should be placed *before* you actually construct the robot.

1.6 Robots Are Not Just Hardware

Many people may feel discouraged by the previous discussion because it means they have to do a lot of programming. Some may say: "I just want to build a robot—I don't want to sit and program all day". Without software and sensors a robot is nothing more than a motorized toy. An autonomous mobile robot needs to be able to make *its own* decisions about how to react to its environment. Autonomous robots are more challenging to design, but are much more versatile and useful.

Imagine if the Mars Rover was not autonomous. Controllers on Earth trying to manipulate it would be very frustrated due to the fact that signals from Earth take nearly 10 minutes (depending on orbital positions) to reach Mars and vice versa. So a human trying to remote control the robot would have to wait a considerable time to see the results of the most recent control input and a considerable time to be able to command a correction. The robot can fall off a ledge, or collide with a rock by the time a corrective command reaches it. The only way to have an effective Mars Rover is to build it with a collection of intelligent algorithms to autonomously achieve the desired tasks.

An algorithm that controls a robot's behavior is basically a set of rules that tell it how to respond to various situations as defined by the state of its sensors. As these rules become more numerous and more complex you will start to see the robot behave in ways you never expected. The robot may appear to deal intelligently with situations you never even considered when you wrote the program. At the other extreme, your robot might look really unintelligent when it encounters some situations.

Programming your robot, or your simulator, is how you give it life. It is how you create its personality and how you determine its behavior. Once you appreciate this concept your experience with building robots will be enhanced and enriched. The RobotBASIC simulator will help you learn to program a real robot, and you will soon find that it can be just as challenging as programming the real thing. You may also be surprised to find that it can be just as exciting and rewarding too. You may not believe that a simulator can make you feel this way, but trust us, RobotBASIC can.

1.7 RobotBASIC Teaches Programming

Novice programmers learn programming much faster when they are writing programs to solve real-world problems (like programming a robot). A simulator helps them see flaws in their programs because they get immediate and useful feedback on the effectiveness of their algorithms. This feedback alone is a compelling reason for using a language such as RobotBASIC to teach programming, but there are additional advantages.

Typically, students in a programming class write small programs that only demonstrate some concept or syntax. Unfortunately, these initial programs are often extremely boring to students because there is little relevance to real-world problems.

It has been our experience that programming a robot is a valuable teaching tool for everyone from young children to college students. When introduced to the robot properly, students find controlling it enjoyably challenging and viewing its responses helpful in their understanding of programming principles. Furthermore, since the programs being written address real situations, the students learn *problem-solving skills* that are hard to obtain by other means.

Above all, students who learn programming with a simulator have fun. They enjoy learning how to make their creation smarter. They *want* to learn about new concepts, new syntax, and new techniques to improve their programs. Teachers know this makes a big difference.

1.8 Summary

In this chapter you have learned that:

- RobotBASIC is a programming language that allows you to simulate a robot with realistic behavior.
- Simulators are used in many fields, and are a valuable training and prototyping tool.
- RobotBASIC is easy to use yet full of powerful features. Both the novice and the experienced programmer can create realistic, enjoyable, and effective simulations.
- RobotBASIC's debugger gives you insight into the robot's view of the environment, which aids in developing more effective algorithms.
- Building simulations with RobotBASIC enables you to make better choices when it is time to design and build a real robot.
- Robots without a well-designed controller program are no more than a toy.
- Learning to program with RobotBASIC is more fun and more effective than traditional methods.

CHAPTER 2

INTRODUCTION TO ROBOTBASIC

RobotBASIC is a fully featured programming language similar to the standard BASIC language, but with major enhancements, additional flow-control structures and other features; all of which help you create powerful structured programs with ease.

RobotBASIC has an integrated development environment (IDE) that enables you to create and edit programs and then run them instantly on a terminal screen. The IDE will indicate any syntactical errors in your program and point out the nature and location of the error. Additionally, there is a debugger that can help in figuring out logical errors that might otherwise be hard to locate.

RobotBASIC has tools, commands, and functions to help you write programs that:

- Create realistic and effective robot simulations.
- Create graphical displays.
- Interact with the user with input and output commands.
- Perform mathematical, trigonometrical, and statistical calculations.
- Create and manipulate strings, and convert between strings and numbers.
- Create and manipulate matrices with a set of functions and commands that allow for many of the matrix operations that are encountered in advanced mathematical courses.

Most of the features above will be discussed as the need arises in later chapters. This chapter will show you how to download and run RobotBASIC including how to create, save, load, edit, and run programs. You will also be introduced to the robot simulator, where you will write simple simulations that make a robot come to life.

2.1 Running RobotBASIC

You can download a zip file that has RobotBASIC.exe and all the programs in this book from www.RobotBASIC.com. Open the zip file using Windows Explorer and drag-and-drop the RobotBASIC folder onto your desktop. You can now close the zip folder and open the newly created RobotBASIC folder. This folder contains the RobotBAISC.exe and a subfolder called RobotProgrammersBonanza. This subfolder contains subfolders for each chapter in the book that has programs. There are also subfolders for other demo programs. If you wish, you can create a shortcut to the RobotBASIC.exe on your desktop. This makes it easier to run the interpreter on a regular basis.

You will now be able to run RobotBASIC and execute programs from the RobotProgramersBonanza subfolder. If you create new programs you can save them in this subfolder or you may create another folder from within the IDE.

NOTE: The language is subject to change as alterations and upgrades are implemented. The help files accessible from the latest IDE will have the most valid up-to-date descriptions of all the functionalities of the language. Make sure to always download the latest version and to consult the help files for any new and modified features. Also make sure to check the site for updated listings of all the programs in the book, solutions for some of the exercises in the book, corrections to errors that may have slipped into the book, and any other information and news.

2.2 The RobotBASIC IDE

The RobotBASIC IDE consists of an Editor Screen, a Terminal Screen, a Help Screen, and a Debugger Screen. Each screen has various buttons and menus that facilitate the numerous actions required in each one. This section will discuss the Editor Screen, Terminal Screen, and Help Screen. The Debugger Screen will be discussed in Chap. 6. Only the features required in this chapter will be described for each screen. For a more detailed description of all the actions available refer to App. A.

2.2.1 THE EDITOR SCREEN

The Editor Screen (Fig. 2.1) has a number of buttons and menu items that facilitate the creation, editing, and running of programs. If you place the mouse cursor on a button and wait for a second, a description will pop-up showing the button's intended action (Fig. 2.2). In addition, each button has an icon that is helpful in remembering the button's functionality. It is also possible to achieve all the buttons' functionalities by using drop-down menus or keyboard shortcuts.

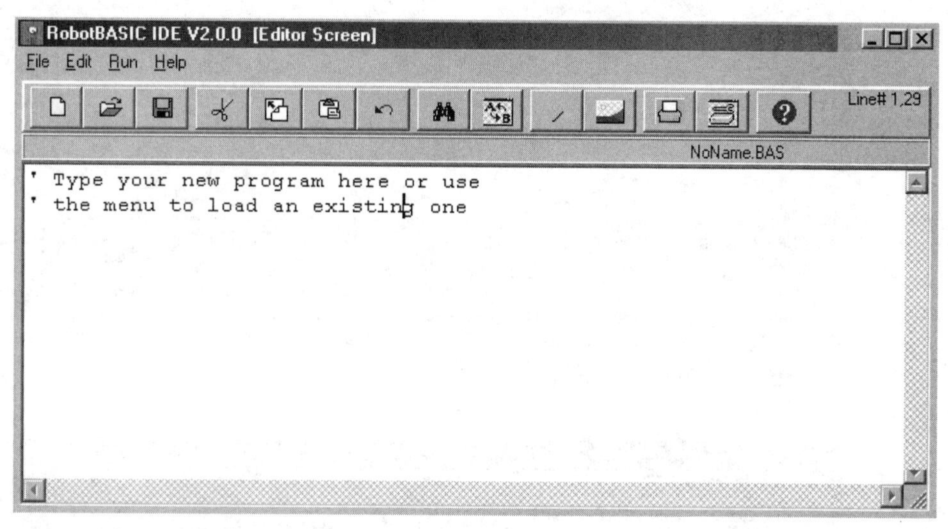

FIGURE 2.1 The Editor Screen.

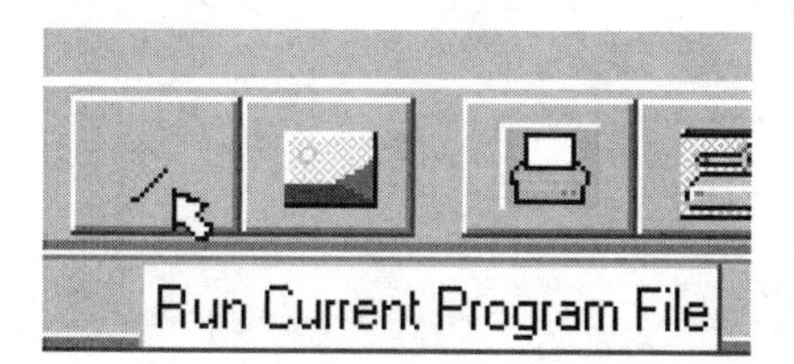

FIGURE 2.2 Button hints.

To run the program currently being edited either, click the *Run* menu and the *Run Program* submenu, or press the button, or use the *Ctrl+R* key combination on the keyboard. Running a program will open the Terminal Screen window and display any program interaction on this screen.

2.2.2 THE TERMINAL SCREEN

The Terminal Screen (Fig. 2.3) is where the program's input and output take place. This screen has many features. For complete details on these features and how to utilize them, refer to App. A.

2.2.3 THE HELP SCREEN

The Help Screen (Fig. 2.4) provides explanations and details of the RobotBASIC language and other aspects of the entire system. The screen has a drop-down combo-box that allows you to choose the desired Help Screen from a list of topics. Information given in this screen is discussed in Apps. A, B, C, and D. Having all the information available on this screen is convenient while writing programs and provides the most up-to-date details.

Any help text can be selected and copied to the Windows Clipboard using the button or *Ctrl+C* key combination. The button or *Ctrl+F* allows you to search the text in the currently displayed section for easy location of the topics relating to your query.

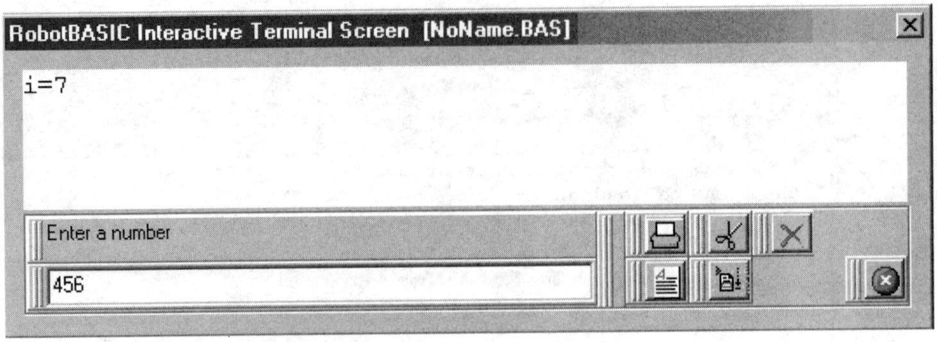

FIGURE 2.3 The Terminal Screen.

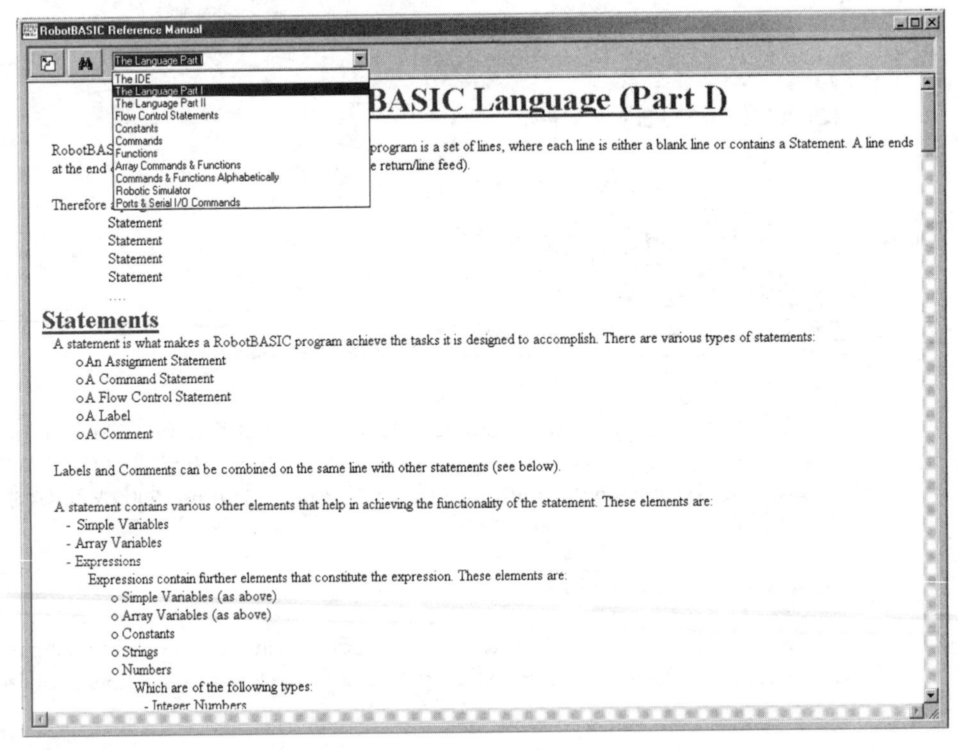

FIGURE 2.4 The Help Screen.

2.3 Creating, Running, and Saving a Program

The Editor Screen (Fig. 2.1) is where you create your programs. The editor is very similar to the Windows Notepad program. You can type text, cut, paste, copy, search, search and replace, print, save to a file, and load from a file.

To create a new file, press the button. There is a button for each of the actions listed above. If your program has been previously created and saved you can load the program using the button, which will bring up a dialog box that allows you to select the file required. Pressing the button brings up another dialog box that allows you to save the text currently in the editor to any file you name, or overwrite an existing file if required.

Once you are ready to test your program press the button to run the program currently in the text editor. This will show the Terminal Screen (Fig. 2.3) and the program's output will be displayed on this screen.

2.4 The Robot Simulator

RobotBASIC makes it easy to simulate a robot on the Terminal Screen. There are many aspects to the simulated robot that will be described in later chapters. Here we will show you how to create a robot and make it move around the screen.

The Terminal Screen simulates a room with four walls that normally measures 800×600 pixels. The robot's world is limited by the confines of this room. Given a robot diameter of 40 pixels, we can get a feel for the scale of things. Assuming a real robot of 12-in diameter we can calculate the room dimensions to be $800 \times 12/40 = 240$ in, that is, 20 ft and $600 \times 12/40 = 180$ in, that is, 15 ft. So the default simulated robot represents a 1-ft diameter robot in a room measuring 20×15 ft. These proportions can be altered, if needed, by changing the size of the robot. The room can be empty or filled with objects like sofas, tables, chairs, toys, and so on. You can even divide it up into further rooms or partitions such as in an office environment.

For some simulations, discussed in the coming chapters, you will need to draw lines on the floor and hang lights from the ceiling to act as homing beacons. RobotBASIC has many commands for drawing graphics on the screen that can be used to simulate all of the above. See Sec. C.7 for details on these drawing commands.

2.4.1 INITIALIZING THE ROBOT

Before you can use the robot in any simulations you must initialize the robot and place it in the environment. The environment has to be created before placing the robot in it. The command to initialize and place the robot on the screen is:

`rLocate` *X,Y,Heading,Size,Color*

X and *Y* are required parameters that define the position on the screen to place the robot. Both *X* and *Y* have to be whole numbers and must be within the limits of the screen (800×600 pixels). If you try to place the robot off the screen it will be placed at the limit of the screen.

Heading is optional and if it is not specified, 0 will be the default. *Heading* specifies the direction the robot will be facing (0° to 359°) 0° is north, 90° is east, 180° is south and 270° is west. Intermediate headings like northwest would be 315°, and so on. If you need to specify the next parameter *Size* you must also specify the *Heading*.

Size is optional and if it is not specified 20 pixels will be the default. You can specify a maximum of 50 pixels and a minimum of 5 pixels. If you try to specify a number for *Size* outside these limits the closest limit will be assumed. You must specify *Heading* and *Size* if you want to specify the next parameter *Color*.

```
rLocate 300,300,45,40,Red
End
```

FIGURE 2.5 Program to initialize the robot.

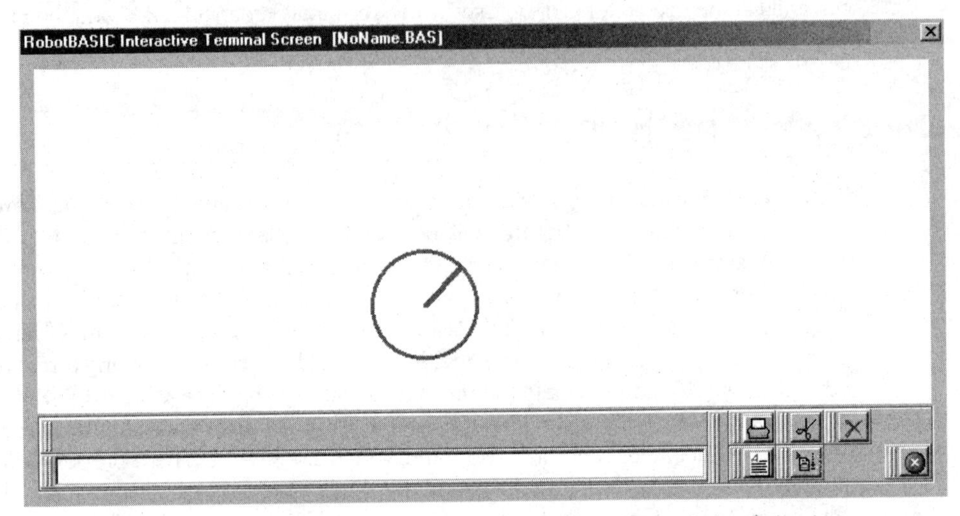

FIGURE 2.6 Locating the robot. (*Note:* The screen has been rescaled to fit here.)

Color is also optional and if it is not defined the color blue will be the default. You can specify any of the colors listed in Sec. B.7.6. When you specify a color consider the floor color the robot is being drawn over. If you specify the same color as the floor color RobotBASIC will select the next color up to avoid making the robot invisible.

Let us write a program to place the robot on the screen. The room will be empty. Type the lines of code shown in Fig. 2.5 in a new editor screen and then press the run button to execute the program. You will see the screen in Fig. 2.6 (notice the color, heading, and size of the robot).

NOTE: The rLocate command and the End statement use capitalization to make them easier to read. RobotBASIC does not care what combination of lower and uppercase lettering you use in writing the commands. However, there are situations where the combination matters. These will be detailed in the appropriate sections.

That's all; you have just created a program to create a robot. But you will, of course, need to make the robot move and turn. Remember that you need to always `rLocate` the robot before you do any further robot manipulations. If you do not do so, an error will be issued and the program will be halted.

2.4.2 ANIMATING THE ROBOT

There are two commands to make the robot move around:

`rForward` *nPixels*

This command makes the robot move *nPixels* forward or backward in the direction it is facing. The parameter *nPixels* is a positive or negative whole number. If *nPixels* is positive the robot will move forward, if it is negative the robot will move backward, maintaining the same heading.

`rTurn` *nDegrees*

This command will make the robot turn *nDegrees* clockwise or counter-clockwise. *nDegrees* is a whole number. If *nDegrees* is negative the robot will turn counter-clockwise. If it is positive it will turn clockwise. If the number is 0 no turning will happen. Turning occurs around the center of the robot, so no forward or backward motion will occur while turning.

NOTE: All the commands and functions that relate to the robot in the RobotBASIC language start with an "`r`." See Sec. C.9 for a list of the commands and functions relating to the robot simulator.

Let us write a program to make the robot move around. Type the lines of code in Fig. 2.7 in a new editor file and press the *Run* button. This program causes the robot to move and turn. This shows how easy it is to animate the robot.

You might wonder what would happen if the robot tries to move beyond the room's boundaries (run into walls), or what if there were objects in the room. Type the program in Fig. 2.8 and run it.

The following screen will be the result:

```
rLocate 100,100
rTurn 90
rForward 300
rTurn 45
rForward 50
rTurn  -90
rForward -200
End
```

FIGURE 2.7 Program to make the robot move around.

```
rLocate 100,100
rTurn 90
rForward 300
rTurn 45
rForward -50
rTurn  -90
rForward 200
End
```

FIGURE 2.8 Program that causes the robot to crash into a wall.

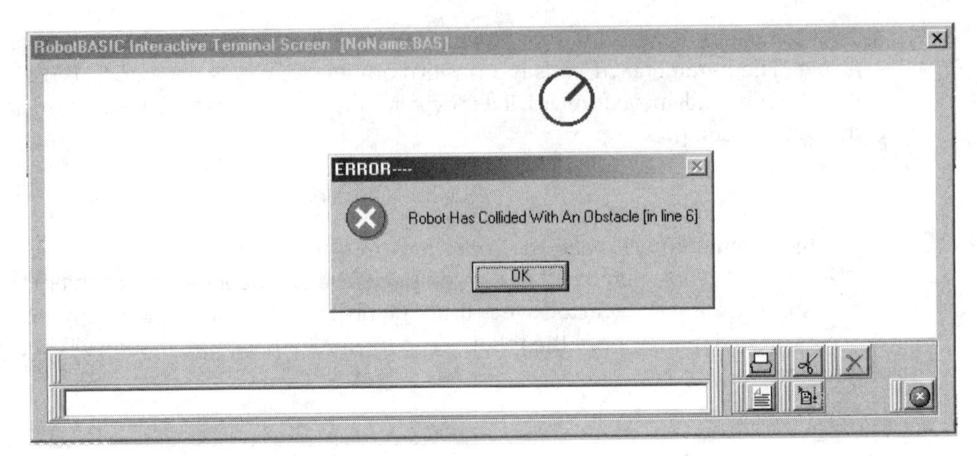

FIGURE 2.9 Robot crash.

The program in Fig. 2.8 causes the robot to crash into the north wall and Fig. 2.9 is displayed. The error message indicates this fact, after which the program is halted. A similar situation occurs if the robot collides with an object in the room.

Perhaps you are wondering how we can make the robot avoid crashing into objects and walls. In order for the robot to avoid obstacles it has to be able to detect them. This is achieved by giving the robot the ability to *sense* objects in the environment. We will study various sensory systems in Chap. 3, but for now, we will make the robot avoid objects manually.

2.4.3 MOVING AROUND OBSTACLES

Let us place some objects in the room and see if we can make the robot move around them. We will do this by *telling* the robot how to move. This is not the most effective way, since objects in the room can change position. If we build into the robot how to avoid objects and make assumptions about where these objects are, then, when the environment changes, the robot may crash because it does not have an up-to-date plan.

A better method would be to have the robot avoid any objects it encounters automatically by *sensing* its way around the environment. We will learn how to do this in Chap. 5 and other chapters. For now we will only use the commands we have learned so far, albeit the robot won't be as intelligent as it could be if it had *senses* and could decide on its own how to move around and avoid objects. Without *autonomous decision-making,* a robot is really just a remote controlled vehicle.

To simulate objects in the room we are going to use RobotBASIC's graphics commands to draw on the screen circles and rectangles. The two commands are:

```
Circle X1,Y1,X2,Y2,PenColor,FillColor
Rectangle X1,Y1,X2,Y2,PenColor,FillColor
```

These commands will draw a circle or rectangle bounded by the coordinates *X1, Y1, X2,* and *Y2* with the outline being *PenColor* and the inside filled with the *FillColor*. For more detailed information on these and other graphics commands see Sec. C.7. Type the

```
01 rectangle 300,300,500,500,red,red
02 circle 100,100,200,200,blue,blue
03 circle 600,500,700,550,magenta,magenta
04
05 rlocate 50,50
06 rturn 90
07 rforward 700
08 rturn 90
09 rforward 500
10 End
```

FIGURE 2.10 Program to manually negotiate around obstacles.

program in Fig. 2.10 and run it. The line numbers are not needed they are in the figure for the purpose of the following discussion.

In Lines 01 to 03 we create the obstacles. In Line 05 we locate the robot at the top left-hand corner. We aim to make the robot reach the bottom right-hand corner. Notice how the commands in Lines 06 to 09 achieve this.

What will happen if we change the number 700 in Line 03 to 770? Change the number and see the result. You can now appreciate the problem of telling the robot how to move. It is not as versatile as automatically deciding on a moving strategy. A program that tells the robot how to move to get from one place to another will have to be modified every time the environment changes. Imagine if we could write a program that enables the robot to move around regardless of the details of the environment. This is what autonomous mobile robot programming is all about, and RobotBASIC simulations help you develop algorithms that achieve this goal. We will see many examples of this in later chapters.

2.5 Summary

In this chapter you have learned:

- How to obtain a copy of RobotBASIC and how to install and run it.
- About the various IDE screens and their functionalities.
- How to create, save, edit, and run programs.
- How to initialize a robot simulation and locate the robot on the screen.
- How to move the robot around the screen.
- What happens if the robot crashes into walls or objects.
- How to draw graphics on the screen to simulate objects in a room.

Now, try to do the exercises in the next section. If you have difficulty read the hints.

2.6 Exercises

1. Use Lines 01 to 03 from the program in Fig. 2.10 and then add your own lines to locate the robot at position (250, 250). Make the robot move all the way around the red rectangle and back to where it was but facing north-west (315°).

HINT: Do four sets of turning 90° and forwarding 300 and then turn −45°.

2. From where you ended up in the previous exercise, what would happen if you add one more line with the command `rForward 100`?

HINT: There is an obstacle in the robot's path, will it crash?

3. Create a program (no obstacles) that makes the robot move from location 100, 100 to location 300, 300, then location 500, 100 then back to 100, 100.

HINT: Locate the robot at 100, 100 facing 135°, then forward 283, turn −90°, forward 283, turn −135° and finally forward 400. Can you explain the numbers?

CHAPTER 3

ROBOTBASIC SENSORS

In Chap. 2 we made the robot move around the screen but we had to be careful when specifying the commands to avoid making the robot crash into walls or objects in the room. This method of making the robot move around is not very effective when:

- The robot must be able to function in various environments.
- The positions and shapes of obstacles are not known in advance.
- The environment changes dynamically.

The robot in RobotBASIC has a collection of sensors that enable it to *feel* and *see* its environment. Algorithms use sensors to analyze the environment and then allow the robot to take action to avoid crashing into objects and to be able to find and locate objectives. In this chapter we will examine some of the sensors on the robot and explore how we can use data from these sensors to program effective behaviors for the robot. The objective is to introduce the standard sensors and explain how to gather information from them. Later chapters will use the sensors in simulations to do useful and interesting work and will show how to use customizable sensors.

3.1 Some Programming Constructs

Many programming constructs will be introduced throughout Part 1 as the need for them arises. These constructs are necessary to be able to create useful simulations using the Robot Simulator and the RobotBASIC language.

NOTE: In general, RobotBASIC *is not* case-sensitive. You can write most of the constructs in the language using any upper- and lower-case letter combinations. So `IF`, `if`, and `If` are all the same. There are three constructs where RobotBASIC *is* case sensitive. These are variable names, array names, and labels and will be made clear when we discuss them later.

3.1.1 COMMENTS

Comments are an indispensable programming construct. They are used to annotate and document a program with information to readers of the code who may find it hard to understand exactly what the code achieves.

Comments are also used to make the code easy to scan so a reader can quickly pick out pertinent sections. Even the writer of the code may appreciate her/his own comments. When you go back to read your code, after some time has passed since you have written it, you will appreciate the fact that you have a reminder of the intent of certain sections of code with explanations of the harder to grasp aspects of the algorithm and other details.

Comments are not executable code and RobotBASIC ignores them. They are there only for human readers of the code. A comment in RobotBASIC is designated as such with a // which makes any text that follows, including the // itself, a comment. You can put comments on a separate line or on a line following an executable statement. Anything on the line after the // becomes a comment.

You may also want to make certain parts of your code not execute to test something or another. Rather than actually deleting the lines of code, you can *comment them out* by putting // before each line. If you later determine that you actually need the code simply remove the // to make the code executable again.

You will see examples of comments in the programs throughout the book. (Refer to Sec. B.2 for more details.)

3.1.2 CONDITIONAL STATEMENTS

It is often necessary to perform certain actions only *if* a condition is *true*. Sometimes you need to perform a set of actions *if* a condition is *true* but *if* it is *not true* perform other actions.

This is achieved by using the `if-then` and the `if-else-endif` programming constructs. `if-then` is used when you need to do *one* action only if a condition is true. `if-else-endif` allows you to do as *many* actions as needed, and also allows for doing *other* actions if the condition is not true.

The first construct looks like this:

```
if some condition then do an action
```

Only one action is allowed after the `then`, which will be executed only if the condition is true. If the condition is false the program will skip the action after the `then` and proceed to the next line.

The second construct can be used like this:

```
if some condition
   Do some action
```

```
        Do another
        Do yet another
        And so on
    endif
```

Notice here we do not use the `then`. The statements between the `if` and the `endif` will be executed if the condition is true. If the condition is false the program will skip them and go on to the statement right after the `endif`.

Another way to use this construct is:

```
if some condition
    Do some action
    Do another
    And so on
else
    Do some action
    Do another
    And so on
endif
```

In this construct the statements between the `if` and `else` will be executed if the condition is true but not the statements between the `else` and `endif`. If the condition is false the statements between the `else` and `endif` will be executed but not the ones between the `if` and `else`.

You will see examples of these three constructs in programs throughout the book. Refer to Apps. B.6 and C.6 for more details and additional ways to use this construct.

3.1.3 COMPARISON OPERATORS

In RobotBASIC you can compare if something is greater than another (>), is equal to another (=), is less than another (<), if it is less than or equal to another (<=), if it is greater than or equal to another (>=), and finally if it is not equal to another (< >).

All these operations are achieved with comparison operators. In the above section we test for conditions using these operators. See Sec. B.7.5 for further details.

3.1.4 LOOPS

It is often necessary to *repeat* a section of code a certain *number of times* or *while* a certain condition is true. We will discuss these looping constructs in detail in Chap. 4. In this chapter we will use this construct in a simple way.

The `for-next` and `while-wend` looping constructs are used here to move the robot forward a fixed number of steps in the first example and while it is not bumping into objects in the second example. For now, study the use of these constructs in the light of the programs given.

3.1.5 BINARY NUMBERS

In order to understand how most of the sensory data is organized you will need a basic knowledge of binary numbers.

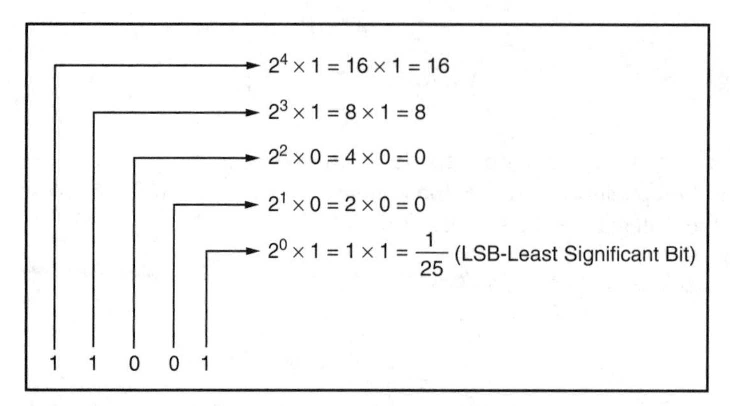

FIGURE 3.1 The value for each digit in a binary number is a power of 2.

In a decimal number like 234 the convention is that the first digit (going right to left) is the ones digit, the second is the tens digit, the third is the hundreds digit, and so on (1000, 10000, etc.). You will notice this is the same as saying 1, 10, 10×10, $10 \times 10 \times 10$, and so on or in more mathematical language 10^0, 10^1, 10^2, 10^3, and so on. So the number 234 can be understood to mean $2 \times 10^2 + 3 \times 10^1 + 4 \times 10^0 = 200 + 30 + 4 = 234$. Notice that we have ten digits 0 to 9. We do not have a symbol for ten. Since ten is 10, that is 1 in the 10^1 place and 0 in the 10^0 place which means $1 \times 10 + 0 \times 1 = 10 + 0 = 10$. The decimal system is referred to as base-10.

Computers are made up of switches that can be either on or off. We can represent the on state by a 1 and the off state by a 0. This means that computers are binary systems (binary means two). This means that there are only two possible numbers 0 and 1. Just as each digit in a base-10 number is based on ten raised to a power, a binary or base-2 system is based on two raised to a power. So the number 1010 in base-2 is $1 \times 2^3 + 0 \times 2^2 + 1 \times 2^1 + 0 \times 2^0 = 1 \times 8 + 0 \times 4 + 1 \times 2 + 0 \times 1 = 10$ (in base-10).

The binary (base-2) system is how numbers are represented in computers. If we put a set of five switches in a row we can represent numbers from 0 to 31. The maximum value of the number can be made up of the sum of the numbers 16, 8, 4, 2, and 1. Look at the example 5-bit binary number (11001) in Fig. 3.1. Only three positions in the original number have 1's in them. The weights of these positions are 16, 8, and 1. The sum of these weights is 25 thus 25 base-10 is the same as 11001 base-2 (the number 25 is 11001 in binary).

Many of the sensors on the robot are made up of switches arranged in groups as described above. These groups can be read as numbers in base-10 or we can examine them a bit at a time. As you use the sensors available in RobotBASIC, you will see why binary numbers are important.

3.2 Avoiding Collisions Using Bumpers

The first type of sensor we will consider is a set of collision detectors around the perimeter of the robot. In the real world these sensors could be bumpers mounted on simple leaf-switches. When the robot collides with an object, the pressure causes one or more leaf-switches to close. The electronics of a real bumper system sends a logical 1 (collision detected) or 0 (no collision) for each sensor to its corresponding bits on a computer input port. The combination of these 1s and 0s form a binary number that indicates the state of the bumpers. This number can be obtained by using a function in the programming language controlling the robot and can be analyzed as a binary number or its equivalent in decimal to determine which bumpers have been activated.

3.2.1 BUMPER SENSORS

The robot in RobotBASIC has four bumpers of the type described above. The front and rear bumpers each compose a 130° arc making them larger than the side bumpers, which are only 50°. Figure 3.2 shows how the bumpers are arranged.

The number indicating the status of all four of the bumpers can be obtained using the function `rBumper()`. As you know from Chap. 2, all robot-related statements in RobotBASIC start with the letter "`r`". Each of the four bits in the number obtained represents the state of one of the bumpers as indicated in Fig. 3.3.

If, for example, the robot bumped into something directly ahead of it (pressing the front bumper) the binary number generated would be 0100 or 4 in base-10. If the robot was backing up and wedged itself into a corner where the back bumper and the left bumper were both pressed, then the number formed would be 1001, or 9 in base-10.

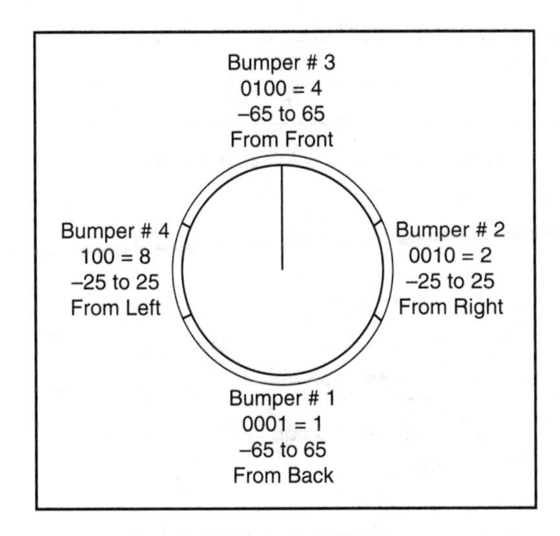

FIGURE 3.2 Four perimeter bumpers are used to detect collisions.

Bumper	Bit position	Value
Rear bumper	2^0 (LSB)	1
Right bumper	2^1	2
Front bumper	2^2	4
Left bumper	2^3	8

FIGURE 3.3 The conditions of the robot's bumpers form a binary number.

3.2.2 AVOIDING COLLISIONS

Let's see how you can use this information to control the behavior of the robot. We will start by locating the robot near the center of the screen and making it move upward (north) using the program in Fig. 3.4.

Since the robot will be pointed north when it is created, this program will make it move forward until it hits the north wall and causes an error. One way to avoid this error is to monitor the bumpers and stop moving the robot forward when they indicate that an object has been touched. The program in Fig. 3.5 shows how this can be done. If you are unfamiliar with any of the programming statements used here, refer to Sec. C.9.

Instead of just telling the robot to move forward 500 times (as in Fig. 3.4), the program of Fig. 3.5 uses a `for-next` loop to make the robot *consider* moving forward 500 times. The `if-then` statement inside the loop checks the bumpers and if none of them are on (the value returned is 0) then the robot moves forward one position. Notice that when using a program to move the robot you will usually move the robot only one position at a time so we can monitor the environment before moving again.

Figure 3.6 shows two example programs that perform similar actions to the program in Fig. 3.5, but using different RobotBASIC statements.

```
rLocate 400,300 //position the robot on the screen
rForward  500 //--make the robot go forward 500 pixels
End
```

FIGURE 3.4 This short program will cause a collision with the north wall.

```
rLocate 400,300
for a = 1 to 500
    //--only go forward if bumpers are free
    if rBumper() = 0 then rForward 1
next
End
```

FIGURE 3.5 This program checks 500 times to see if it can move forward and only moves if nothing is in the way.

```
rLocate 400,300
for a = 1 to 500
   if rBumper() = 0
      rForward 1
      // more statements
      // can be placed here
   endif
next
End
```

```
rLocate 400,300
while rBumper() = 0
  rForward 1
  //more statements can
  // be placed here too
wend
End
```

FIGURE 3.6 These two programs perform similar functions to the one in Fig. 3.5.

The program on the left in Fig. 3.6 still uses a `for-next` loop, but it shows how to use an `if-endif` statement. The `if-endif` should be used when there are several things that need to be done when the `if`-condition is true.

3.2.3 IMPROVING EFFICIENCY

The program on the right side of Fig. 3.6 does not use an `if`-statement at all. Instead it uses a `while-wend` loop that executes all of the statements inside the loop as long as the condition specified is true. Notice that the program in Fig. 3.5 and the one on the left of Fig. 3.6 both continue to attempt to move the robot even after a bumper has closed. The program's logic will not move the robot if the bumpers are closed but it will continue to try to do so 500 times. The program on the right of Fig. 3.6, however, will stop attempting to move the robot as soon as any bumper is closed. This implies that the algorithm on the right of Fig. 3.6 is more efficient than the other two (on the left of Fig. 3.6 and in Fig. 3.5).

These example programs bring up an important point that is especially pertinent to novice programmers. There is no *right* way to create a program. If you ask ten people to write a story about a particular incident, they might all tell the same story but each would have their own style and would use their own words. Programming is the same. Different people will use different statements and different approaches to solving the same problem. You could argue that some approaches may be more efficient (such as the program on the right side of Fig. 3.6) but if a program accomplishes its goal, you can't say it is wrong. Of course you should always strive to design programs that are as efficient as possible. However, sometimes you may have to compromise to make the program faster or simpler or even, easier to read and maintain.

3.2.4 MAKING BETTER DECISIONS

In all the example programs above, a decision was made about what to do based on the value of the bumpers being 0, meaning none of them was pressed. In more realistic programming, we might want to do different things depending on *which* bumpers are pressed. For example, if we know that the left bumper is being pressed we might want our robot to turn right to avoid the obstacle. Figure 3.7 shows some example expressions that can help analyze what the bumper data are indicating.

All the expressions in Fig. 3.7 can be used as conditions for `if` and `while` statements. In the chapters that follow, you will learn more about how to write programs that analyze sensor information and how to use the information to control the robot.

Expression	Situation that makes it true
rBumper( ) = 0	if all bumpers are not pressed
rBumper( ) =15	if all bumpers are pressed
rBumper( )	if any bumper is pressed
rBumper( ) = 4	if only the front bumper is pressed
rBumper( ) = 12	if both the front and the left are pressed

FIGURE 3.7 Example expressions for testing bumper conditions.

3.3 Other Sensors for Object Detection

In the previous examples we used the robot's bumpers to avoid collisions, however, it took a collision (although a very minor one) to activate one of the bumpers. Bumpers are very important because they are a reliable means of making sure the robot does not try to push furniture around the room. Nevertheless, it would be better if the robot could detect an object in its path before actually touching the object.

3.3.1 INFRARED SENSORS

One method for enabling the robot to detect obstacles without touching them is to use infrared sensors. The principle is to use an infrared LED (light emitting diode) to shine light away from the robot. A phototransistor circuit detects if that light is reflected back to the robot. If the light is reflected back then we can assume that some object is close by. The robot in RobotBASIC has five infrared sensors mounted 45° apart as shown in the Fig. 3.8.

As with the bumper sensors, the state of the infrared sensors is encoded into a number that can be obtained using the function `rFeel()`. The sensor on the right side of the robot is the least significant binary (LSB) position in the number. Each sensor, moving counter-clockwise, corresponds with the next bit position. The information obtained from `rFeel()` can be used in a similar manner to that from `rBumper()`.

The program in Fig. 3.9 is very similar to the one on the right of Fig. 3.6 but it uses `rFeel()` in place of `rBumper()`. Run this program and compare where the robot stops in comparison to the one in Fig. 3.6.

In general, it is better to detect objects with `rFeel()` rather than `rBumper()` because it is best not to have any collision, no matter how small. The disadvantage of infrared

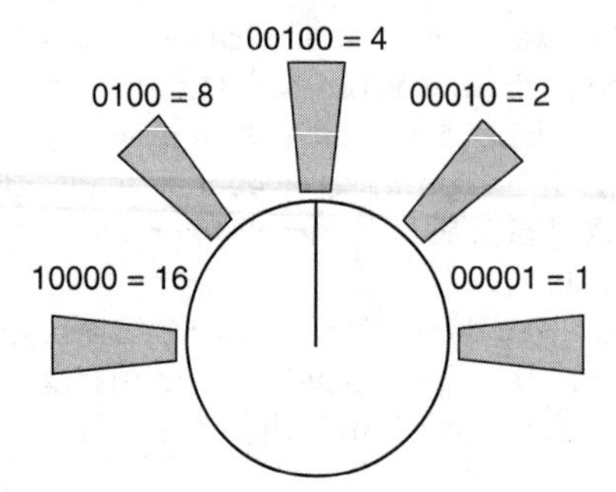

FIGURE 3.8 The robot can *feel* objects without touching them using five infrared sensors.

```
rLocate 400,300
while rFeel() = 0
  rForward 1
wend
End
```

FIGURE 3.9 This program uses `rFeel()` to detect an obstacle.

sensors is that it is possible for a small object (or perhaps the corner of a large object) to slip between the sensors and cause a collision (refer to Fig. 3.8). For this reason it is recommended that you analyze the data from both the infrared sensors and the bumpers when trying to avoid a collision. This principle will be discussed in detail in later chapters.

3.3.2 ULTRASONIC AND INFRARED RANGING

One limitation of the infrared and bumper sensors is that they only detect objects that are very close to the robot. It may be advantageous for the robot to detect distant objects along its path so it could take action before it becomes too late to act. You can buy sensors that report not only the presence of objects in the path, but also the distance to the objects. Some of these sensors use ultrasonic technology (sound waves) and others use infrared or laser.

Our robot has a single ranging sensor mounted so that it faces in the same direction as the robot. You can get the data from that sensor using the function `rRange()`. If, for example, `rRange()` returns a value of 27 it is telling you there is some object 27 pixels away. The `rRange()` function simulates laser technology, which makes it very directional.

The program in Fig. 3.10 makes the robot approach the north wall stopping 20 pixels away from it.

3.3.3 ROBOT VISION

Another sensor that the robot can use to detect objects at a distance is a camera pointed in the direction the robot is facing. This camera is not intended to provide full pictures to analyze, which is the subject of an interesting field in robotics called *robotic vision*. Rather, the RobotBASIC camera returns a number to indicate what color it is seeing. The function for the camera is `rLook()`. The program in Fig. 3.11 shows how the robot can use the camera to determine when it is facing an object of a particular color. In this case, the robot will turn until it sees the red circle.

```
rLocate 400,300
while rRange() > 20
    rForward 1
wend
End
```

FIGURE 3.10 The function `rRange()` allows the robot to determine how far objects are in front of it.

```
Circle 600,500,620,520,red,red   // draw a red circle
rLocate 400,300
while rLook() < > RED   // turn until red is seen
    rTurn 1
wend
End
```

FIGURE 3.11 The function `rLook()` allows the robot to determine what color is seen straight ahead.

```
Circle 600,500,620,520,red,red  // draw a red circle
rLocate 400,300
while rBeacon(RED) = false  //while the beacon is not seen
   rTurn 1
wend
End
```

FIGURE 3.12 The function `rBeacon` (*Color*) allows the robot to determine when a specified beacon is ahead, even if the path is blocked.

3.3.4 BEACON DETECTION

One way of locating a desired location is to hang a sign above it indicating the location below the sign. Before electronic compasses and global positioning systems (GPS) were available at affordable prices lighthouses served as beacons for ships at sea and radio automatic direction finder (ADF) beacons provided navigational data to aircrafts. These systems are still in use today, though they are being gradually replaced by newer technologies.

If we want the robot to find a location in a room, we could hang a flashing light (either visible or infrared) above the location. Since this flashing "beacon" is high in the room it can be seen at all times even if other objects are in the way between the robot and the location. Like the camera and ranger sensors, the beacon detection sensor faces directly ahead of the robot.

The function `rBeacon` (*Color*) returns a non-zero value (true) which indicates that the robot is facing a beacon of that color, or zero (false) which indicates that the robot is not facing the beacon. If the number returned by the function is not zero then there is a beacon ahead of the robot but this number is actually the distance in pixels to the beacon. This functionality simulates more complex beacon detection. If you do not wish to use the distance data then just use the returned number as a *true* or *false* indicator. The program in Fig. 3.12 shows how the robot can turn to face a beacon. We must tell the function what color beacon to search for by passing it the color of the beacon. This function is very similar to the camera function, but the beacon function can see over objects that might be in the way (because the beacon is assumed to be hanging high in the air).

3.3.5 CUSTOMIZABLE SENSORS

There are several other sensors available on the robot, some of which can be customized so that you can create the exact type and configuration of sensors you need to allow your robot to achieve a desired behavior. Some of the sensors described above can be configured in other ways that will be discussed in subsequent chapters. Additionally, there are alternate ways of interrogating the sensory data as will be described in Chap. 5. Refer to Sec. C.9 for more information.

3.4 Other Instruments

The robot in RobotBASIC has navigational instruments that enable it to determine its position and orientation. There is also a self-diagnosis instrument that enables it to check the condition of its battery.

```
rLocate 400,300
while rCompass() < > 90 //east is 90 degrees
   rTurn 1
wend
End
```

FIGURE 3.13 The function `rCompass()` allows the robot to determine what direction it is facing.

3.4.1 COMPASS

RobotBASIC has a compass function, `rCompass()`, that returns the current direction, in degrees, the robot is facing. In the chapters that follow you will see how this function can be used to help our robot make better decisions about where it is and how it should move to get to a desired location. The program in Fig. 3.13 uses the compass to make the robot turn due east. Remember that north is up on the screen, south is down, east is to the right, and west is to the left.

The compass in RobotBASIC is accurate to 1°. Inexpensive electronic compasses can rarely be this accurate. If you wish to simulate a compass that is accurate to 3°, for example, you can divide the value returned by `rCompass()` with the number 3 (forcing an integer divide) and then multiply the result by three as in the formula:

3*(`rCompass()`/3)

3.4.2 GLOBAL POSITIONING

Nearly everyone nowadays is familiar with GPS systems in vehicles that display exactly where you are on a map. Our robot has two GPS functions, `rGpsX()` and `rGpsY()` that return the *x* and *y* values for the robot's position. The GPS in RobotBASIC is accurate to a single pixel. Standard real-world GPS systems are not this accurate, but the later chapters will discuss a variety of ways to circumvent this limitation. You can simulate a less accurate GPS system in the same way described to simulate a less accurate compass.

The program in Fig. 3.14 shows how the robot can avoid the north wall by keeping track of its position and stopping when it is 10 pixels from the wall. It uses the function `rGpsY()` to find its position on the screen (an *x*, *y* of 0, 0 is the upper-left corner). It moves while its *y*-coordinate is greater than 30. Remember the robot's default radius is 20 pixels; also the GPS functions report the position of the center of the robot; therefore we use the number 30 which means that the edge of the robot will be 10 pixels away from the north wall which has a *y*-coordinate of 0.

```
rLocate 400,300
while rGPSy() >30
    rForward 1
wend
End
```

FIGURE 3.14 The function `rGpsY()` allows the robot to determine its vertical position on the screen.

3.4.3 BATTERY CHARGE LEVEL

A reasonable requirement of any real mobile robot is that it should be able to monitor its battery condition and determine when a recharge is required. The function `rChargeLevel()` returns the percentage of battery life left. In Chap. 13 we will use this function as well as other sensors to teach our robot how to find and utilize a charging station when the battery charge level is low.

3.5 Summary

In this chapter you were introduced to:

- Different programming structures (`if-then`, `if-else-endif`, `while-wend`, and `for-next`) and how they can be used to control the robot more effectively.
- Binary numbers in preparation for more powerful sensor manipulation.
- `rBumper()` and `rFeel()` and how they can be used to avoid crashing into objects in the robot's environment.
- Detecting objects at a distance with `rRange()`, `rLook()`, and `rBeacon()`.
- Navigational instruments with `rCompass()`, `rGpsX()`, and `rGpsY()`.
- Battery charge level information with `rChargeLevel()`.

In subsequent chapters we will explore how to use sensors to solve realistic problems. For now try to solve the exercises in the next section. Try to do so without reading the hints, but by all means use the hints if you need to.

3.6 Exercises

1. Write a program to place a gray object at position 100, 200 on the screen and the robot at 400, 300. The program should then make the robot face that object and report the distance to it. Can you predict what the value will be? How accurate was your prediction? Can you explain the difference?

HINT: Draw a circle to simulate the object and then use `rBeacon()` or `rLook()` in a loop to face the object. Use `rRange()` to find the distance. Use the *Print* command to report the distance (see Sec. C.7).

2. Enhance the program in Exercise 1 to make the robot go to the object.

HINT: Use `rRange()` and a `while-wend` loop to go to the object. Did the robot crash into the object? Can you make it not do so? Use `rBumper()` or `rFeel()` and an `if-endif` to avoid the object.

3. Write a program that places the robot at 400, 300, then make the robot move to a point that is in the direction 135° and 350 pixels away. Can you predict the coordinates of this point? How accurate was your prediction?

HINT: Use `rCompass()` to face the robot then a `for-next` loop to move. Use `rGpsX()` and `rGpsY()` to get the position when on that point and use *Print* to report the values.

4. Modify the program in Exercise 3. At the top of the program, before the line that initializes the robot, place this line:

rectangle 450,400,500,500,black,black

Now run the program. What happens? Can you avoid this?

HINT: Use the same method to avoid the object as discussed in this chapter. Do not attempt to go round the object to continue reaching the goal. You will see how to do this in Chap. 12.

CHAPTER 4

REMOTE CONTROL ALGORITHMS

For many robot hobbyists, their first project is building a mobile platform that can be manipulated using some form of remote control. The ultimate goal, of course, is to create a robot that can make its own decisions on how to move around based on sensory data obtained from its environment. However, before we can make the robot decide on its own how to move and turn we need to gain some experience with programming it and controlling it. In subsequent chapters you will learn many methods for giving the robot the ability to think autonomously. In this chapter we will explore methods of moving the robot manually by remote control.

As you have seen in Chaps. 2 and 3 you can make the robot move and turn easily enough with a program that gives the robot a set of instructions on how to move and how much to move. If you want the robot to move in a different direction or distance you would have to reprogram it with the new data. This is not a convenient way of making the robot move wherever we want. A more efficient way is to have a program that can receive instructions from us on how to move and then execute the right commands to move the robot as we indicated.

There are a variety of ways to remote control a real-life robot. Whether you use a wired or wireless (radio or infrared) controller, the principle is the same: The controller sends signals to the robot to make it move or turn. There may also be other actions the robot can accomplish so there usually are additional buttons on the remote controller to tell the robot to perform the additional functions.

There are three general styles of remote control:

- As long as a button is pushed the robot will move. When the button is released the robot stops.
- You push the button to make the robot move and release it. The robot will continue moving until you push the button again to stop it. The button is used to toggle the action.
- Given an instruction, the robot executes certain actions to complete the instruction then waits for the next instruction.

We will develop algorithms for each of the above styles. To control the simulated robot we will use the keyboard and the mouse to simulate a remote controller. Sometimes it may be desirable to display information about the robot's condition, so we will explore some display commands.

The third style of control is a little more complex than the other two. To accomplish the necessary programming, some mathematics will be required. RobotBASIC has many mathematical functions that will help in designing this style of control.

4.1 Some Programming Constructs

The algorithms in this chapter will use programming constructs that allow for repeating a section of code many times. We will discuss the various constructs to achieve this. Also, RobotBASIC has commands to obtain input from the user and to display output back. The two devices for accepting input are the Keyboard and the Mouse. The device for displaying output is the Screen.

4.1.1 VARIABLES

A variable is a storage space for holding a number or a string (text). A variable name is assigned to the storage space for use in a program. A variable name must start with a letter followed by any combination of letters and numbers. A variable can be used anywhere a number (or string) is needed. Of course you must assign the variable a value before using it.

NOTE: Variable names are case sensitive, so *Distance, DISTANCE,* and *distance* are **not** the same variable. RobotBASIC is generally not case sensitive. Variable names, array names and labels are the exceptions to this rule.

In most computer languages variables have to be assigned a type and cannot be used to store values of different types. So, for example, in the standard BASIC language you have to name a variable with a $ at the end of the name to indicate that it is to hold a string value. If you try to store an integer in it you will get an error. The variables in RobotBASIC are more versatile than this. When you name a variable you are not restricted as to what to name it and you can store values of any type in it. Furthermore, you can change the type and value of the data stored in a variable at any time. This is a very powerful feature. You can read all about variables in Secs. B.7.3 and C.4. Also you can read further about data types in Secs. B.7.1 and B.7.2.

4.1.2 THE KEYBOARD

There are various commands to obtain input from the user using the keyboard. We will only use two of them here. See Sec. C.7 for details.

4.1.2.1 `GetKey` *Var* This command is useful in loops like `for`-loops, or `while`-loops (see later) where you want something done repeatedly but want the user to be able to affect the action by pressing a key. You do not want the repeated action to pause until the user presses a key, but you want it to change if she/he does press a key.

This command checks if any key is pressed on the keyboard. If there is a key pressed its code value is placed in the variable *Var*. If no key is pressed the value 0 is placed in the variable.

The program flow is not paused until the user presses a key. If the user presses a key when the command is executed the key will be reported, but if the user does not press any key by that time, the program will report a 0 and go on to the next command and proceed with the rest of the program. (See also the command `GetKeyE` *Var*.)

4.1.2.2 `WaitKey` *{ExprS}, Var* This command pauses the program flow until the user presses a key. This command is useful for allowing the user to press buttons on the keyboard to achieve various actions, where each action is assigned a key value.

Once the user presses a key, the key code is placed in the variable *Var* and then the program continues with the next command. Read Sec. C.7 for more information on this command.

The two commands above obtain input from the user but only one key press at a time. The information obtained is a number that represents the key the user pressed. This number is a standard code for computers called the ASCII code. You can convert this code back to a letter by using the `Char()` function. You can also convert a letter to its code by using the `Ascii()` function. So `Ascii`("A") gives the numeric code 65 and `Char` (66) gives " B". See Sec. C.8 for details of these functions.

There is another way to obtain input from the user. This way allows the user to input any combination of keystrokes to form a sentence or a number. The user then presses *Enter* to indicate completing the entry. The command is `Input` *ExprS, Var*. Read more about this command in Sec. C.7.

4.1.3 THE MOUSE

The Mouse is a very useful input device. In RobotBASIC there is a function that enables you to obtain the information a mouse provides. The command looks like this:

`ReadMouse` *Var1, Var2, Var3*

When the command is executed *Var1* and *Var2* are filled with the screen coordinates of where the mouse cursor is when the command is executed (0, 0 is the top left corner). *Var3* will be filled with a value that specifies which mouse button was pressed and in what combination with the *Shift, Alt,* or *Ctrl* keys. Read Sec. C.7 for details of these codes.

When the command executes it does not pause the program or wait for the user to press anything. If the user happens to be pressing the mouse buttons then *Var3* will be set to the code, if not then it is set to 0. If the cursor is inside the terminal screen then *Var1* and *Var2* will be filled with the cursor's position. If the cursor is outside the terminal screen when this command executes, the values in *Var1* and *Var2* will be the last valid values obtained from the mouse. This information can be useful for knowing how the mouse exited the screen.

4.1.4 OUTPUT TO THE SCREEN

There are many commands for sending output text and numbers to the user. In this section we will only be concerned with two of them (see Secs. B.7 and C.3 for details on expressions):

4.1.4.1 `Print` *{Expr,Expr;...}* This command writes out to the screen the results of the expressions. The first time you issue a `Print` the first line on the screen will be used, the next time will use the second line, and so on. Once the last line is reached the next time a `Print` command is executed the screen will scroll one line up and the data is printed on the last line. The comma (,) is used to display the output using no spaces between the expressions and the semicolon (;) puts a tab space between them. Read Sec. C.7 for more details.

4.1.4.2 `XYString` *X,Y, Expr{,Expr;...}* This command writes out to the screen the results of the expressions. Values *X*, *Y* are screen coordinates in pixels where the output will be printed. Read Sec. C.7 for more details. This command is just like `Print` but it puts the text on a particular screen coordinate. No scrolling occurs.

4.1.5 LOOPS

In programs it is often necessary to repeat execution of some lines of code a certain *number of times* or *while* a certain condition is true or *until* a certain condition becomes true. For example, if you want to print the numbers 1 through to 10, you can have 10 separate print statements like `Print 1`, then `Print 2`, and so on. Or you could write:

```
for I = 1 to 10
    Print I
next
```

In this method you have written 3 lines instead of 10. Imagine if you wanted to print 1 to 100. You can appreciate the savings in time and space.

Imagine you want to print a random number every time the user presses a key but if she/he presses the key "q" you want to stop. You cannot use the above since you do not know how many numbers the user needs. You can use this:

```
K = 0
while K < > Ascii("q")
    Print Random(1000)
    Waitkey K
wend
```

This way the program keeps repeating the printing and waiting for a key until the user presses the "q" button. The function `Random` (*n*) is used to generate a random number from

0 to $n-1$. Also the function `Ascii()` is used to get the code value of the letter "q" which is the value returned by the `WaitKey` command inside the variable *K* when the user presses the "q" button.

Notice how the variable *K* had to be initialized before entering the loop. This is because the condition for the loop checks to see if *K* is not equal to the code for "q," and if *K* has not been defined yet you would get an error. Another way to do exactly the same thing is:

```
repeat
   Print random(1000)
   WaitKey K
until K = Ascii("q")
```

Notice that *K* did not have to be initialized this time. This is because *K* is not used until after it has been assigned a value by the command `WaitKey`. Otherwise this **flow-control** structure is very similar to the one above. Notice the condition for the `while-wend` is exactly opposite to the one in the `repeat-until`.

Here is another way to do the same as above but this time instead of checking for the condition to exit out of the loop, in the loop itself we will use an `if`-statement to decide when to *break* out of the loop. This has an advantage if the condition for exiting the loop is one that is not easily testable in one place in the program or is not suitable to be tested only once at the top (or bottom) of the loop. Let's say we want the loop to finish if the user presses "Q" or "q." We can do this:

```
while True
   Print Random(1000)
   Waitkey K
   if K=Ascii("q") then break
  if K=Ascii("Q") then break
wend
```

```
repeat
   Print Random(1000)
   Waitkey K
   if K=Ascii("q") then break
   if K=Ascii("Q") then break
until False
```

The `Break` command causes the program flow to go to the line right after the `wend` (or `until`) statement, effectively ending the loop. Notice that you do not need to assign a value for *K* before entering the `while`-loop since you do not use the variable before it is defined. Also, notice that the condition for ending the loop is *True* (*False* for the `repeat-until`), which means that the loop will never end unless a `Break` is executed.

The above is just an example. A better way to accomplish the same action would be:

```
K = " "
while K<>"Q" AND K<>"q"
   Print Random(1000)
   Waitkey K
   K = char(K)
wend
```

```
repeat
   Print Random(1000)
   Waitkey K
   K = char(K)
until K="Q" OR K="q"
```

Notice the condition for the `while-wend` loop and the `repeat-until` loop. They are exactly opposite. As a matter of fact, in **Boolean algebra** (the mathematics of logic) we know that:

```
Not(X) AND Not(Y) = Not(X OR Y)
```

This might be confusing but in English the `while` condition in the above example is the equivalent to "keep looping while the user has not pressed the 'q' button *and* not pressed the 'Q' button." For the `repeat` loop the meaning is "keep looping until the user presses 'q' *or* 'Q'."

The method used for creating a loop depends on the logic of the algorithm you are using. You have seen, above, various methods, but there are many ways you can create a loop. It all depends on the logic you are trying to achieve. Refer to Secs. B.6 and C.6 for flow-control structures and Sec. B.7.5 for logical operators.

4.1.6 FUNCTIONS

There are two ways to obtain a value in RobotBASIC, commands and functions. Commands tell the system to perform some action and given a variable name, the command will assign the variable a value depending on the action of the command (as you have seen in the `WaitKey` command above).

Functions perform an action too, but after performing the action they act like a variable, taking on the value generated by the action. As you have seen in the discussion above the function `Ascii`("A") returns the value 65 and you can use this number as if you have typed 65 in the statement. You can say

```
y = Ascii("A")+3
```

this will cause the number 68 to be stored in y just as if you typed

```
y = 65+3.
```

In RobotBASIC there are functions to obtain the length of a string, to convert a number to a string, to get the sine of an angle, and more. There are math functions, string functions, functions relating to the robot, and so on. Read Sec. B.7.7 for details about functions and Sec. C.8 for a list of functions. Some functions will be used in this chapter and many more throughout the book.

4.2 Simple Remote Control

The first two styles discussed in the beginning of the chapter will be implemented below. The advantage of the first style is that you can easily control the robot accurately, but it is slow. The advantage of the second style is that the robot will move quickly and you do not have to keep the button pressed, but it is hard to control the robot with accuracy.

4.2.1 FIRST STYLE OF REMOTE CONTROL

In this style the user will press

"f" or "F" to go forward	"b" or "B" to go backward
"l" or "L" to turn left	"r" or "R" to turn right

The robot will move as required as long as the key is pressed. If the key is released the robot will stop moving. The robot will use data from its sensors and not go forward or backward if there are obstacles blocking the direction of travel even if you try to make it do so.

In order to display the robot's current position and heading we will use the GPS and compass instruments described in Chap. 3. See Sec. C.9 if you need more details on the `rGpsX()`, `rGpsY()`, and `rCompass()` functions. The algorithm is in Fig. 4.1 (don't type the line numbers; they are only there for the discussion that follows).

As you have seen from the previous section the `WaitKey` command is ideal here. We use the `XYString` command to display the data. We also use the drawing commands you saw in Chap. 2 to place some obstacles in the robot's environment.

The function `Char()` used on Line 10 converts the key code to a character so that we can compare it to the characters used to control the robot. Notice how the values returned by the functions `Char()` and `rBumper()` are stored and then used in the `if`-statements. This is more efficient than if we were to call the function in each if-statement by saying:

```
if Char(k) and not(rBumper() & 4) then rForward 1
```

Calling functions is a little slower than accessing a variable. We would be calling functions eight times, each time we loop, if we use the function in each statement directly. It is important to realize that we can only use the stored data because the robot is not moving after the `rBumper()` statement is executed. Lines 12 to 15 use `if`-statements to determine what key was pressed and execute the right action. Line 11 gets the state of the bumpers using `rBumper()` as in Chap. 3. We use this value (*B*) to test to see if the front bumper (Line 12) or the back bumper (Line 13) is pressed before moving forward or backward, respectively. You will learn more about this action in Chap. 5. In this chapter just accept that the statement `not(B & 4)` means that the front bumper is not pressed and `not(B & 1)` means the back bumper is not pressed [remember `B = rBumper()`].

```
01 rectangle 300,300,500,500,red,red
02 circle 100,100,200,200,blue,blue
03 circle 600,500,700,550,magenta,magenta
04 rectangle 0,0,130,22,blue,blue
05 rlocate 400,200,270
06 //--style 1
07 while true
08   XYString 2,2,rGpsX(),",",rGpsY(),",",rCompass(),"     "
09   waitkey "Press l,r,f, or b", k
10   C = char(k)
11   B = rBumper()
12   if (C="f" or C="F") and not(B & 4) then rForward 1
13   if (C="b" or C="B") and not(B & 1) then rForward -1
14   if (C="l" or C="L") then rTurn -1
15   if (C="r" or C="R") then rTurn 1
16 wend
17 End
```

FIGURE 4.1 First style of remote control.

In Line 08 the robot's position and heading are displayed. Notice the use of the commas to make the display look nice. The box around the text was drawn on Line 04. The box is needed to stop the robot from going into the text area.

4.2.2 SECOND STYLE OF REMOTE CONTROL

In this style we do similar actions as in the previous program. The difference is that we don't wait for the user to press a key. The algorithm was designed so that the last key pressed is saved and used to make the robot move continuously until the user presses another key. If the new key pressed is the same as the last one then the last movement is turned off. If it is a different command then the new command will be executed. The new algorithm is shown in Fig. 4.2 (do not type the line numbers).

In Line 07 the variable *LC* is initialized to 0. This variable will hold the value of the last command issued. In Lines 11 to 15 we check if a key is pressed, and if so, we check if it is the same as the last one pressed. If it is then we make it 0 to cancel the last command. The new command is then stored in *LC*. Notice Line 16, the *LC* value is converted in place of *k* as in Fig. 4.1. This (and the use of `GetKey` instead of `WaitKey`) is what makes the program continue doing the last command until the same key or a new key is pressed.

To summarize, Lines 07 and 10 to 16 make the program continue to execute the last command until a new one or the same one is issued. This makes the command style a *toggle* action. The rest of the program is similar to the one in Fig. 4.1. The delay of 200 milliseconds in Line 14 is necessary to give the user time to release the button before the program checks again for a button press. Without this delay the user may not have time to release the button before the next check and the program will consider that the user has pushed the button again. Without this delay it would be very difficult for the user to signal the program correctly. Try removing (or commenting out) Line 15 and see what happens.

```
01 rectangle 300,300,500,500,red,red
02 circle 100,100,200,200,blue,blue
03 circle 600,500,700,550,magenta,magenta
04 rectangle 0,0,130,22,blue,blue
05 rlocate 400,200,270
06 //---style 2
07 LC = 0
08 while true
09     XYString 2,2,rGpsX(),",",rGpsY(),",",rCompass(),"    "
10     getkey k
11     if k <> 0
12        if k = LC then k = 0
13        LC = k
14        Delay 200
15     endif
16     C = char(LC)
17     B = rBumper()
18     if (C="f" or C="F") and not(B & 4) then rForward 1
19     if (C="b" or C="B") and not(B & 1) then rForward -1
20     if (C="l" or C="L") then rTurn -1
21     if (C="r" or C="R") then rTurn 1
22 wend
23 End
```

FIGURE 4.2 Second style of remote control.

4.3 Complex Remote Control

In this style of remote control the robot carries out a *series* of actions to accomplish a task specified by the user. For this simulation the mouse will be used as a **laser designator**. If you are familiar with laser targeting devices used by the military you will recognize this style of remote control. The device uses a laser to designate a target for a missile. The missile locks onto the target and moves there. We will emulate this by using the mouse to designate the target we want the robot to go to. The robot will lock onto the mouse position and go there. The robot will also be able to draw on the screen while moving to help you see the actions that took place (this can also make the robot act as a sketcher).

The robot will use its GPS and compass to calculate the difference between its current position and heading and the target's position and direction.

4.3.1 THE MATHEMATICS

Figure 4.3 shows a representation of the calculations that are necessary for this algorithm. The robot's location is represented by the coordinates *Rx, Ry*. *Rx* is the robot's horizontal position on the screen in relation to the top left-hand corner which is position 0, 0. Ry is the vertical position. The target is located at *Tx, Ty*. The difference between the x-coordinates of the robot and target is *dX*. The difference in their *y*-coordinates is *dY*.

As you can see from Fig. 4.3, the two values can be used to calculate the distance *R* between the robot and the target. This is an application of the pythagorean theorem.

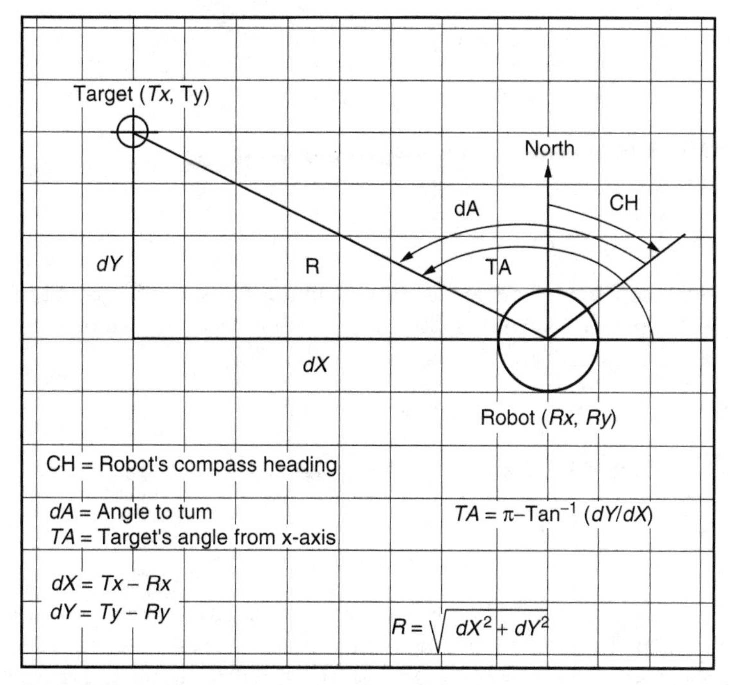

FIGURE 4.3 Laser targeting with the robot.

RobotBASIC has a function that can do this calculation for us, `PolarR` (*dX*, *dY*) which returns the value for *R*.

The function `PolarA` (*dX*, *dY*) returns the angle in relation to the horizontal axis that is formed by the line between the robot's center and the target's center, as shown in Fig. 4.3 (angle *TA*). This angle can be used to calculate a turn direction and amount so the robot can face the target. As you can see in Fig. 4.3 this is the angle *dA*. However, there are two complications.

The first problem is that angle *TA* is measured from the east direction (this is common in computer languages). That is, east is 0°, not 90°, as our robot (and humans) normally think of it. This angle, which is the value returned by `PolarA()` is not a 360° angle like in a compass; it is ±180. The positive angles are measured counter-clockwise from east and negative ones are clockwise from east. So north is 90°, south is −90° and west is +180°. We will have to convert the angle reported by `PolarA()` to a compass heading so that the robot can be turned to that heading. Adding 90° to the angle reported by `PolarA()` will solve this problem, but before doing this another issue has to be resolved.

The angle value returned by `PolarA()` is given in radians, not degrees (again, this is common practice in computer languages). It is simple to convert between degrees and radians in this manner:

$$1° = \pi/180 \text{ radians}$$

So, when you want to convert an angle in degrees to radians you do

Angle_In_Radians = *Angle_In_Degrees* * `pi()/180`

To convert from radians to degrees you do:

Angle_In_Degrees = *Angle_In_Radians* * `180/pi()`

Refer to Sec. C.8 for the function `Pi()`, it essentially returns the value π.

This means we can calculate TA in degrees using:

TA = `PolarA`(*dX*,*dY*)`*180/pi()`

Remember, we also need to convert *TA* relative to north instead of east. Since east is 0° in relationship to the *x*-axis, but it is 90° in relationship to north, we must add 90 to convert *TA* to a compass heading. So, the equation becomes:

TA = `PolarA`(*dX*,*dY*)`*180/pi() + 90`

We can now calculate *dA* (the angle to turn) as *TA* − *CH*, which results in the following formula:

dA = `PolarA`(*dX*,*dY*)`*180/pi() + 90-`*CH*

Since *dA* will be a number from 0 to 360, the robot may have to turn to face the target in the longer direction. To make the turning more efficient we need to check to see if the turn is larger than 180° and if so make the robot turn the other way which would be the shorter angle and thus more efficient (see this later on Lines 40–41 of the code in Fig. 4.4).

```
01 MainProgram:
02   gosub Draw_Obstacles
03   rlocate 400,200
04   rInvisible DarkGray
05   gosub RemoteControl
06 End
//=========================================================
07 RemoteControl:
08   rectangle 0,0,150,23,blue,blue
09   s = " "+rGpsX()+","+rGpsY()+","+rCompass()+" UP"
10   s = s+spaces(16-length(s))
11   xystring 2,2,s  //--display the data
12   p = up
13   repeat
14     readmouse x,y,b
15     if b = 1 then Gosub GotoPoint  //left mouse button
16     if b = 2 //--right mouse button
17        p = not p
18        rpen p
19        delay(300) //--edge detect the mouse button
20     endif
21     if b <> 0  //--any buttons pressed, update display
22        s = " "+rGpsX()+","+rGpsY()+","+rCompass()
23        if p = Up   then s = s+" UP"
24        if p = Down then s = s+" Dn"
25        s = s+spaces(16-length(s))
26        xystring 2,2,s  //--display the data
27     endif
28   until false
29 Return
//=========================================================
30 Draw_Obstacles:
31   rectangle 300,300,500,500,red,red
32   circle 100,100,200,200,blue,blue
33   circle 600,500,700,550,magenta,magenta
34 Return
//=========================================================
35 GotoPoint:
36    dx = x-rGpsX()
37    dy = y-rGpsY()
38    if dx=0 AND dy = 0 then return
39    Theta = PolarA(dx,dy)*180/pi()+90-rCompass()
40    if Theta > 180 then Theta = Theta-360
41    if Theta < -180 Then Theta = Theta+360
42    rTurn Theta
43    Distance = Round(PolarR(dx,dy))
44    for I = 1 to Distance
45      if rBumper() & 4 then break
46      rForward 1
47    next
48 Return
//=========================================================
```

FIGURE 4.4 Complex remote control.

4.3.2 THE PEN

The robot has a pen at its center that can be lowered to leave a trace on the floor. The command to lower and raise the pen is:

`rPen` *Up/Down, Color*

You would type `rPen` *Up* to raise the pen and thus stop drawing and `rPen` *Down,*Cyan to lower the pen and draw with the color cyan.

We will discuss this feature and many uses for it in Chap. 10 and it will be used in Chaps. 8 and 9. For now we will use the pen during our remote control to make the robot draw while it is moving. This effectively converts the robot into a sketcher that can be used to sketch line drawings of any shape. One of the mouse buttons will be dedicated to raising and lowering the pen.

The robot considers any color drawn on the screen to be an obstacle and will report an error if you try to make it move forward into the object. However, there are times when you want certain colors to be considered as nonobstacles. For example, you may have a beacon hanging from the ceiling above the room, or you may have a line drawn on the floor. These colors are not to be considered as objects and we need a way of telling the robot to ignore these colors if it encounters them. Additionally, as you have seen in Chap. 3, sensors like `rBumper()`, `rFeel()`, and `rLook()` will report the presence of obstacles, so if we designate some colors as invisible these sensors will ignore these colors. We do this by using the command:

`rInvisible` *Color{,Color...}*

This command tells the robot to consider the list of colors given as either, lines on the floor, or as beacons up in the air. In effect they become invisible to the robot and its sensors. Some sensors will override this and look for a specified color. We will discuss these sensors later. You can specify a minimum of 1 color or a maximum of 15. You can use the color names as described in Sec. B7.6 or the number corresponding to the color. Using the name is a lot clearer and easier to remember.

When the robot draws with the pen it will leave a trace on the floor in the specified color. If you do not tell the robot to consider this color as invisible it will report a crash error if it encounters the color later. Thus in this simulation we will use the `rInvisible` command to tell the robot to ignore the color drawn by the pen.

Also, as you may notice from reading the description of the `rPen` command in Sec. C.9 you do not need to specify a color when you issue the `rPen` command. If you do not specify a color then the first color on the list given to the `rInvisible` command will be used as the color to draw when the pen is down.

4.3.3 SUBROUTINES

The algorithm in Fig. 4.4 uses a programming construct called a *subroutine*. Think of a subroutine as a tool that completes a task. When you use a *tool* you usually do not care how the tool accomplishes its task. In this case the tool is the subroutine. We will discuss subroutines in detail in Chap. 5.

The program in Fig. 4.4 uses three subroutines, one to do the actual remote control tasks and a subroutine that will be used to calculate the distance and heading as discussed above and also make the robot go there. The third one is to place obstacles in the environment.

In a program you invoke a subroutine by saying: `gosub` *Subroutine_Name*. Once the subroutine finishes its work the program will continue with the next line after the line where the subroutine was invoked. See Secs. B.6 and C.6 for details on `gosub` and other flow-control structures.

4.3.3.1 The Implementation The algorithm is shown in Fig. 4.4 (don't type the line numbers). The result of the algorithm in Fig. 4.4 is shown in Fig. 4.5.

4.3.3.2 The *MainProgram* (Lines 01–06) The main routine calls the subroutine *Draw_Obstacles* then sets up the robot and then calls the *RemoteControl* subroutine. Once there, the subroutine will not end until you halt the program by closing the terminal window. The `rInvisible` command is issued to tell the robot to not consider the color "dark gray" as an obstacle. This color will also be used as a pen color when the pen is lowered since the command on Line 18 does not specify a color and thus the first color in the invisible colors list will be used to draw with the pen by default.

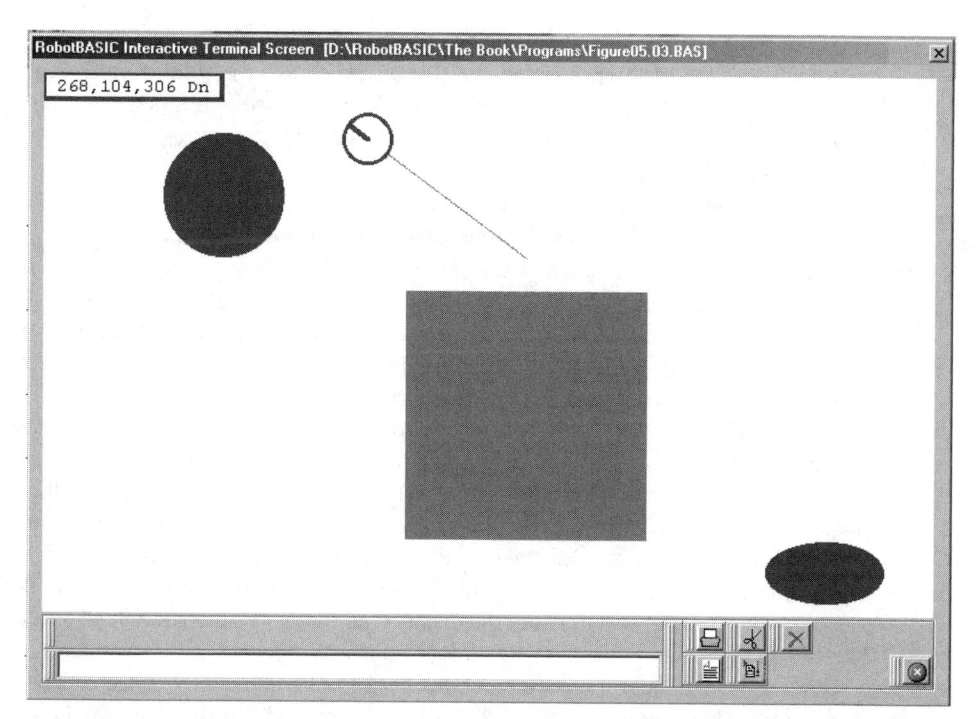

FIGURE 4.5 Result from running the program in Fig. 4.4. Notice the line trailing behind the robot. This is due to the pen being down when the robot moved.

4.3.3.3 The *RemoteControl* Subroutine (Lines 07–29) This subroutine does all the work. It sets up an area at the top of the screen for displaying the current position and heading of the robot and also the state of the pen (Lines 08–11). Then it enters an endless loop (Lines 13–29). The loop is endless because the condition for the `until` is set to *false*, the loop never halts. Of course some condition inside the loop may call a `Break` command and cause a halt, but this does not happen in this program (see Secs. B.6 and C.6).

Line 14 causes the mouse coordinates and button state to be saved in the variables *x*, *y*, and *b*. (See previous section or Sec. C.7 for the `ReadMouse` command).

If you click the left mouse button the statement on Line 15 calls the *GotoPoint* subroutine to cause the robot to move to the point where the mouse was clicked.

If you press the right mouse button the `if-endif` statement on Lines 16 to 20 causes the robot's pen to be *toggled* up or down just like a switch, if it is up it is put down, and if it is down it is put up (Line 17). The statement on Line 19 causes a `Delay` of 300 milliseconds. This is necessary due to the fact that you may press the mouse button for too long and the toggling will occur too fast for you to be able to maintain the desired state. This is the equivalent of making an *edge detector*.

Lines 21 to 27 are executed if any mouse button is pressed. These lines read the robot's position and orientation using `rGpsX()`, `rGpsY()`, and `rCompass()`. Also the pen state (saved in the variable *P*) is already known. These data are put together in a string, which is printed at the top-left corner of the screen (Line 26).

4.3.3.2 The *GotoPoint* Subroutine (Lines 35–48) This subroutine is very important for the action of the program. The subroutine causes the robot to turn in the direction of the point indicated by the user, and then calculates the distance from the robot to that point then makes the robot move to that point. The robot will move as long as no obstacle causes the bumper to be closed.

Lines 36 and 37 calculate *x* and *y* difference between the selected point and the robot's current position. Line 38 exits the subroutine if there is no difference.

In Lines 39 to 42 the angle to turn is calculated and then the robot is turned by that angle. This calculation makes use of the function `PolarA()` discussed previously. This is then used to calculate the difference between the robot's heading and the heading to the point (Lines 39–41). Notice the formula on Line 39. We first convert the angle reported by the `PolarA()` function to degrees using the conversion discussed above. Then we add 90 to it. This is (as discussed) to convert from 0° being east and thus 0 = 90 so we add 90. Then we subtract the robot's heading to get the *difference* between the heading to the point and the robot's heading. The next Lines 40 and 41 convert this to the smallest angle for the robot to turn intelligently to the required heading. Comment out these two lines and observe the effect on the way the robot turns toward the target.

In Line 43 the distance to the point is calculated using the `PolarR()` function. The `Round()` function is used to make the distance an integer instead of a float, so that it can be used as the limit for the `for-next` loop in the next line.

Lines 44 to 47 cause the robot to go forward one pixel at a time while checking to see that the front bumper is not closed. If the bumper ever closes the loop is exited.

If there are any commands or functions that are not clear to you, refer to Secs. C.7 and C.8 for details on how they are used and what parameters and options are available.

Also refer to the program in Fig. 4.4 to see how the function or command is used in light of the discussion above and the details in the appendix.

4.4 Remote Controlled Test Bench

In this section the first style remote control is used to test all the sensors of the robot while moving it around with the remote controller. This will help in understanding how the robot "sees" its environment. We will combine the keyboard and mouse as a remote controller. Study the program code to see how this is done. Essentially, without considering the mouse, the remote controller is similar to what we developed in Fig. 4.1.

You saw in Chap. 3 that the robot has many sensors. The program in Fig. 4.6 will show the status of many of these sensors while the robot moves around. Using this program you can maneuver the robot over lines and in the vicinity of obstacles and observe how all the sensors are affected (see Fig. 4.7). This can help in understanding what the robot sees and can be very valuable while developing algorithms that use sensors to allow the robot to move autonomously.

The program will not be discussed in detail. Many of the techniques used in it will be seen in programs in future chapters and will be discussed then. However, do notice the way the display text is formatted to appear appealing on the screen. Also notice the use of the string manipulation functions `Instring()`, `Length()`, and `sRepeat()` and how the function `Bin()` is used to convert a number to its binary representation.

4.5 Summary

In this chapter you have:

- Seen various methods for remote controlling the robot.
- Explored I/O commands to accept input from the user and display data to the user. `WaitKey`, `GetKey`, `ReadMouse`, `Print`, and `XYString`.
- Examined some mathematical functions and used them in combination with the **GPS** and **compass** instruments of the robot. `PolarA()`, `PolarR()`, and `Pi()`.
- Been introduced to the `rPen` feature on the robot. You will use this feature in more interesting projects in Chaps. 8, 10, and 11.
- Learned about **flow-control** Structures like the `for-next`, `while-wend`, and `repeat-until` loops, and how they can be used to make the program repeat sections of code in a controlled manner.
- Been introduced to the `gosub` command and subroutines and how they can make writing programs easier by dividing the tasks into smaller and easier subtasks. This principle will be discussed in much more details in Chap. 5 and will be used throughout the book.

Now, try to do the exercises in the next section. If you have difficulty read the hints.

```
MainProgram:
  gosub Environment
  rlocate 50,200,90
  rInvisible Cyan
  while true
     getkey k
     readmouse x,y,b
     B = rBumper()
     K = char(k)
     if (k="a" or b = 1) and (not(B & 1)) then rForward -1
     if (k="s" or b = 2) and (not(B & 4)) then rForward 1
     if  k="w" or b = 11 then rTurn -1
     if  k="z" or b = 3 or b = 12 then rTurn 1
     if InString("aswz",k) or b <> 0 then gosub DisplayData
  wend
End
//============================================================
DisplayData:
   xystring 300,0 ,rChargeLevel(),"%      "
   xystring 300,20,rPoints(),"        "
   xystring 300,40,rCompass(),"        "
   xystring 300,60,rGpsX(),",",rGpsY(),"        "
   B = rBumper()
   Bb = sRepeat("0",4-Length(Bin(B)))+Bin(B)+"       "
   xystring 140,100,"rLook()      = ",rLook();"rBumper()     = ",B,":",Bb
   F = rFeel()
   Fb = sRepeat("0",5-Length(Bin(F)))+Bin(F)+"       "
   xystring 140,120,"rRange()     = ",rRange();"rFeel()       = ",F,":",Fb
   S = rSense()
   Sb =sRepeat("0",3-Length(Bin(S)))+Bin(S)+"       "
   xystring 140,140,"rBeacon(red)= ",rBeacon(red);"rSense()      = ",S,":",Sb
return
//============================================================
Environment:
  LineWidth 1
  rectangle 100,80,120,500,red,red
  linewidth 3
  setcolor cyan
  gotoxy 10,200
  lineto 99,200
  gotoxy 50,100
  lineto 50,300
  linewidth 1
  xystring 140,500,"Press 'a' to go backwards     's' to go forwards"
  xystring 140,520,"Press 'w' to turn left        'z' to turn right"
  xystring 140,540,"Red = ",red,"   White = ",white,"  Cyan = ",cyan
  xystring 140,0 ,"rChargeLevel() = "
  xystring 140,20,"rPoints()      = "
  xystring 140,40,"rCompass()     = "
  xystring 140,60,"rGpsX(),rGpsY()= "
Return
//============================================================
```

FIGURE 4.6 Remote controlled test bench.

```
RobotBASIC Interactive Terminal Screen [D:\RobotBASIC\The Book\Programs\Figure05.04.BAS]
rChargeLevel() =  93%
rPoints()      =  12279
rCompass()     =  125
rGpsX(),rGpsY()=  75,217

rLook()       = 4    rBumper()     = 4:0100
rRange()      = 8    rFeel()       = 28:11100
rBeacon(red)= 8      rSense()      = 0:000

Press 'a' to go backwards      's' to go forwards
Press 'w' to turn left         'z' to turn right
Red = 4   White = 15   Cyan = 3
```

FIGURE 4.7 Result of running the program of Fig. 4.6.

4.6 Exercises

1. Rewrite the programs in Figs. 4.1 and 4.2 to use mouse control as well as keyboard control. Also, see if you can change the keyboard commands from using letters to move the robot to using the arrow keys [see the command `GetKeyE` and the function `KeyDown()`].

HINT: See how it is done in Fig. 4.6 and also study the `GetKeyE` command in Sec. C.7. Also see if you can improve on the program using `GetKeyE` and `KeyDown()`.

2. Experiment with the program in Fig. 4.6. Can you predict the sensory data as you are moving the robot around? Why doesn't the robot move forward if you command it to do so when it is next to the red object? Can you point to the lines of code in the program that achieve this?
3. In the program of Fig. 4.4 comment out Line 02 by using //. What will this action achieve? Now run the program and use the robot as a sketching tool to sketch your name for instance. You may need to toggle the pen up or down. Can you think of a way to make the robot draw different colors? What would be needed to achieve this?

HINT: Use the keyboard to specify different colors maybe by using numbers or letters. Also you will need to increase the list of invisible colors to allow for the additional colors so as not to cause a crash.

CHAPTER 5

RANDOM ROAMING

In Chap. 3 we learned about some of the sensors the robot can use to become aware of its environment. You saw some programs that utilized the sensors to make the robot stop before it crashed into walls. However, we often do not want the robot to just stop when it encounters an obstacle. We want it to be able to avoid the obstacle in some manner and continue moving. In Chap. 4 we manually controlled the robot so when the obstacles stopped the robot we were able to decide on how to circumvent them and commanded the robot on what to do to go around the object (by remote control).

The aim of this book is to create an *autonomous* mobile robot. To be autonomous the robot has to be able to decide for itself how to circumvent obstacles. The robot has to be able to avoid or go around obstacles and continue along its route accomplishing the tasks it is supposed to complete all by itself. This chapter is the first one where we will start giving the robot the ability to make decisions. Subsequent chapters will greatly enhance the robot's artificial intelligence (AI) capabilities.

There are various approaches to making the robot avoid obstacles:

- Turn around and travel in a direction away from the obstacle.
- Turn sufficiently to avoid the obstacle but not completely around.
- Negotiate around the obstacle until it clears it and then continue traveling in the same direction as before.
- Wait for the obstacle to move away; assuming the obstacle is a mobile object itself.
- A combination of all or some of the above.

```
rLocate 400,300
while true   // roam forever
   // forward until an object is found
   while rFeel( )=0
      rForward 1
   wend
   // turn 180 degrees plus or minus 30 degrees
   rTurn 150 + random (60)
wend
End
```

FIGURE 5.1 This program causes the robot to roam randomly around the screen.

This chapter will consider the first two options. We will develop algorithms for the other options in later chapters.

5.1 What Is Random Roaming?

Before we can make a robot tackle any challenges, it has to be able to move around its environment without any specified knowledge of the locations of objects. The robot has to be able to avoid obstacles and escape out of corners and tight spots in an intelligent manner. Here we will develop some algorithms that enable our robot to handle moving aimlessly around whatever environment we care to challenge it with. We say aimlessly because in this chapter the robot will have no specific goal to achieve other than meandering around its world without getting stuck in one place for too long or crashing into obstacles. Type the program in Fig. 5.1 and run it.

The inner (second) `while-loop` in the program checks to see if an object in the robot's path has triggered any of the robot's infrared sensors. If no objects are detected, the robot moves forward one pixel. This movement occurs as long as the loop continues to detect no objects in the robot's vicinity. Once an object is encountered, the `while-loop` ends and the program flow continues to the next statement. This statement causes the robot to turn. The number of degrees the robot turns is formed by adding 150 and a random number between 0 and 60. This produces a turn that is between 150° and 210° or 180° ± 30° [see Sec. C.8 for details on `Random()`].

The outer (first) `while-loop` ensures that the two behaviors are repeated endlessly. The condition for the loop is `while` *true*. This means that the loop will always repeat since *true* is always true and will never become false to end the loop.

When you run this program you will see the robot move around the *room*. Each time it encounters a wall, the robot will turn away using a random angle and then move forward again until another wall is encountered. This process will continue until you terminate the program by closing the terminal screen window.

5.2 Some Programming Constructs

In this chapter we use some more programming principles. The following explains these principles to allow for easier comprehension of the programs that will be developed.

5.2.1 LABELS AND SUBROUTINES

The next program we will develop will be divided into sections. Each section will achieve a specific task. The main program will call each section as it becomes needed.

This principle is a powerful strategy; divide a complex task into a set of simpler tasks. Each simpler task can also be divided further. This process makes it easier to complete the project as a series of simple tasks that can be easily accomplished. Some tasks may have been previously accomplished in other projects and can be used again in the current project with only minor modifications. Also you can assign different people to work on each subtask; this way a large project can be finished in less time than if one person was working on it.

The main program acts as a manager program calling the subtasks as they become needed. An example program organized in this manner is shown in Fig. 5.2.

In RobotBASIC you can achieve this kind of structure with **subroutines**. Think of a subroutine as a *tool* that you use to do a certain task. Usually when you use a tool you do not care how it accomplishes its work as long as you know how to use it. To achieve a big project you will use many tools together. You come to a point where you need the tool, so you pick it up and use it. When you finish you put it down and proceed (maybe use another tool). When a project becomes too complex you can summon the aide of specialists and divide the overall project among these specialists who then use tools to do their work. The specialists may utilize additional specialists and so on.

This is exactly how programs should be developed. Programs should have a main routine that calls on subroutines that act as tools or specialists. In RobotBASIC a subroutine is marked as such by surrounding some lines of code with a `Label` and a `Return` statement (see Fig. 5.2). The *label* is the name of the subroutine.

A label has to start with a letter followed by any combination of letters and numbers and has to end with a colon (:) (see Secs. B.5 and C.1). As you can see in Fig. 5.2 we have the labels `Task_1` and `Task_2`, which are markers for subroutines surrounded by the label and the command `Return`.

```
MainProgram:
   //--setup some initial stuff here
   //--do the various tasks
   gosub Task_1
   gosub Task_2
   //Etc. Etc.
   //--do some closing up stuff here
End  //--this is needed to stop the program
//----------
Task_1:
   //do stuff here
Return
//----------
Task_2:
   //do stuff here
Return
//----------
//Etc. etc.
```

FIGURE 5.2 Well structured program.

NOTE: Labels are case sensitive, so *Task*:, *TASK*:, and *task*: are **not** the same. RobotBASIC is generally not case sensitive. Variable names, array names, and labels are the exceptions to this rule.

You invoke a subroutine (use the tool) by using the command `Gosub` followed by the name of the subroutine which is the label that you have given it (without the colon [:]) such as `Gosub` *Task_1*. Once the program issues this statement it will jump to the label and start executing the subroutine from that label until it encounters the command `Return`. The `Return` command ends the subroutine and causes the program flow to go back to the line immediately following the line where the subroutine was called. See Secs. B.5, B.6, C.1, and C.6 for more details on **flow-control structures**.

NOTE: The `End` statement is necessary to stop the program from continuing on to the area of the subroutines.

The more you program the more you will find that you may have already designed a routine in some previous project to achieve what you are trying to do in the current project. If you have designed the routine as a subroutine, then all you have to do in the current project is cut and paste the previously created routine. As your *toolbox* of routines becomes more extensive, you will find that you can develop programs more quickly and more easily.

You will see how all this applies in practice with the program we will develop in the next section for testing various random roaming algorithms.

5.2.2 COMMANDS

RobotBASIC has commands to accomplish many tasks. You have seen in Chap. 4 that there are commands to perform I/O (input and output) with the user. In this chapter you will use commands to do various actions on the screen.

There are commands to clear the screen (`ClearScr`), set the color for drawing on the screen (`SetColor`), position the initial point to start drawing (`GotoXY`), to draw lines on the screen (`LineTo`), and to set the width of the lines being drawn (`LineWidth`).

RobotBASIC has a multitude of commands to help you deal with many different types of programming situations. See Secs. B.4, C.7, and C.10. Remember, commands are not case sensitive.

5.2.3 OPERATORS

Operators are symbols that operate on numbers or strings. There are *math* operators to do things like add (+) and multiply (*). There are *comparison (*or *relational)* operators to see if two things are equal (=) or if one thing is less than another (<). There are *logical* operators that enable you to test if some condition is true AND another is true, or if another condition is true OR another is true. There are *bitwise* operators that will operate on the individual bits of a number (binary e.g., bAnd).

NOTE: Many operators in RobotBASIC have more than one form.

The utility of all these operators will become clear as we proceed with developing programs throughout the book. See Sec. B.7.5 for detailed information on all operators available in RobotBASIC and how they can be used.

5.3 Adding Objects to the Roaming Environment

How will the algorithm in Fig. 5.1 cope when we introduce obstacles into the room? To test this we will develop a subroutine that enables us to place objects of any shape in the room.

Figure 5.4 shows a modification of the program in Fig. 5.1. This new program allows you to draw on the screen with the mouse to simulate placing objects in the robot's environment. Figure 5.3 shows a sample screen with objects that were drawn using the program in Fig. 5.4 (remember, do not type the line numbers).

The sections below explain the details of how the commands, functions, and looping structures work together to achieve the program's action. The details may become challenging,

FIGURE 5.3 The program in Fig. 5.4 allows you to draw objects on the screen.

```
01 MainProgram:
02     gosub DrawObjects  // let the user draw objects on the screen
03     gosub RoamAround
04 End
   //=================================================================
06 RoamAround:
07     while true   // roam forever
08        // move forward until an object is found
09        while rFeel( )=0
10           rForward 1
11        wend
12        // turn 180 degrees plus or minus 30 degrees
13        rTurn 150 + random (60)
14     wend
15 Return
   //=================================================================
17 DrawObjects:  // beginning of subroutine to draw objects
18   rLocate 400,300 // show robot so they know where to draw
19   print "Press the mouse key and hold it while you draw."
20   print "Release when you have completed drawing an object."
21   print "Repeat until you have drawn all the objects you want."
22   Print "Right click anywhere on the screen when finished"
23   Print "The robot will roam randomly while avoiding objects."
24   SetColor GREEN
25   LineWidth 3
26   FirstTime = true
27   while true
28      // wait till the user presses a mouse button
29      repeat
30         ReadMouse x,y,m
31      until m=1 or m=2
32      if FirstTime
33         ClearScr  // clear the screen (remove the text)
34         rLocate 400,300  // put the robot back on the screen
35         FirstTime = false  // only clear screen the first time
36      endif
37      if m = 2 then return
38      gotoxy x,y  // set starting point for drawing
39      while m  // as long as the mouse button is pressed
40         ReadMouse x,y,m  // read a new position
41         LineTo x,y       // and draw a line to it
42      wend
43   wend
44 Return // end of subroutine
```

FIGURE 5.4 This program lets the robot roam around a room, avoiding objects drawn on the screen with the mouse.

but study them carefully because the principles in this program will be utilized many times as you progress through the book. To put things in perspective we will explain the program's action in words and thus give an overall look at the program. Keep this in mind and refer to it as often as you need while reading the discussions in the next sections.

The program is an implementation of the principles discussed in Sec. 5.2. There is a main program that calls subroutines as they become needed. This means that the main program, besides being self-documenting and easy to understand, is a manager for the overall program action.

The first action of the main program is to call the subroutine *DrawObjects,* which allows the user to draw on the screen to simulate objects in the robot's environment. Once the user finishes drawing, the subroutine returns to the main program. The main program then calls the *RoamAround* subroutine, which enters an endless loop that makes the robot move around the room while avoiding obstacles and walls.

The *DrawObjects* subroutine accomplishes the following:

1. Display instructions and the initial location of the robot to the user.
2. The program will then repeatedly do the following things until the user clicks the right mouse button:
   (*a*) Wait for the user to left-click the mouse. If this is the first click, the program clears the screen (to get rid of the instructions) and then replaces the robot at the center of the screen.
   (*b*) As the mouse moves, the routine draws a line from the mouse's previous position to the new one until the user releases the left mouse button.
3. Once the user presses the right mouse button the program will exit the subroutine and return to the line that follows the line where the subroutine was called.

Now with all the above in mind proceed to the next sections to learn how all this is achieved with the functions and commands available in RobotBASIC.

5.3.1 *DrawObjects* SUBROUTINE

This subroutine allows the user to draw on the screen to simulate objects in the robot's environment. It achieves its action by printing instructions to the user of the program then keeps checking the mouse to see what buttons are being clicked. This subroutine also initializes the robot and locates it on the screen (Line 34).

5.3.1.1 Printing on the Screen The first portion of the subroutine (Lines 19–23) consists of `Print` statements that display instructions to the user so that she/he knows what to do. Refer to Sec. C.7 to find out about printing options. Also see the discussion in Chap. 4.

5.3.1.2 Drawing on the Screen Lines 24 and 25 specify the color and width of the lines that will be drawn. On Line 26 a variable *FirstTime* is set to *true*. We will see how this variable is used shortly. The `while`-loop on Line 27 ending on Line 43 surrounds the remainder of the routine causing that code to be repeatedly executed (once for every object that is drawn) until a Return statement is executed.

5.3.1.3 Reading Mouse Data The next section of code (Lines 29 to 31) is a `repeat-until` loop that executes the `ReadMouse` command until the user clicks the left or right mouse button. In this example, `ReadMouse` places the current coordinates of the mouse into the variables *x* and *y* and assigns a number to the variable *m* that specifies if and which buttons were pressed on the mouse. A value of 1 indicates the left button was clicked. A value of 2 means the right button was clicked. Notice the use of the logical OR in the `until`-statement. It causes the loop to wait until *either* of these events occurs and then execution continues with the `if`-statement on Line 32.

The `if`-statement (Line 32) examines the variable *FirstTime* that was mentioned earlier. Remember, it was given a value of *true* to indicate that this is the first time execution has found its way to this point in the program. Since *FirstTime* is *true*, the `if`-statement will execute the lines inside its block (between the `if` and the `endif`, Lines 33–37). These lines clear the screen (to erase the instructions previously printed) and locate the robot again in the middle of the screen so that the user can draw objects in relation to the robot. It also accomplishes another important action. It sets the value of *FirstTime* to *false*. This ensures that the next time through this section of code the program knows it is not the first time and will not clear the screen again.

The next `if`-statement (Line 37) checks the value of *m* to see if the last mouse event was a right-click. If it was, (indicating the user is finished with drawing objects) the program `returns` to the line following the `Gosub`-statement (Line 03). If *m* is not equal to 2, execution continues on to Line 38. Once we are on Line 38 we know two facts. First, the last mouse event was a click on the left mouse button. How do we know this? If the button had been right-clicked we would have returned to the main program. If it had not been clicked at all, the program would still be in the `repeat-until` loop discussed earlier. The second fact we know is that the variables *x* and *y* contain the coordinates for the mouse at the time the button was clicked. The program uses these variables and the `GotoXY` command to establish a starting point for drawing the next object.

The `while`-loop (Lines 39–42) executes as long as the user does not release the left mouse button. Inside this loop, a `LineTo` statement draws a line from the last point used to the current mouse position. This simply means that a line will be drawn wherever the user moves the mouse so long as the left button is pressed. As soon as the button is released the program will return to the beginning of the main `while`-loop (Line 27) and either get a new starting point for a new object (if the user clicks the left mouse button) or terminate the subroutine (because the user clicked the right mouse button).

5.3.2 *RoamAround* SUBROUTINE

This subroutine (Lines 06–15) causes the robot to roam around. It is exactly the same as the code in Fig. 5.1; the only difference is that it is now in a subroutine. The outcome of combining this subroutine with the drawing subroutine is that the robot will now make random turns whenever it encounters randomly placed obstacles or walls.

Most of the time this algorithm for roaming around works properly. Occasionally, especially if you draw objects with sharp points, the robot will cause an error by colliding with an object. This can happen because, as explained in Chap. 3, there are gaps (blind spots) between the infrared sensors.

It might seem strange that the simulator is designed to have blind spots. If you build a real robot it is unlikely that you would purchase enough infrared sensors to completely cover its perimeter and even if you did, the large number of sensors to analyze would make it more difficult to determine what actions should be taken when the sensors are triggered. Programming in RobotBASIC forces you to solve the same problems and face the same challenges you would face while programming a real robot because the robot's sensors simulate *realistic* ones. These thoughts lead to the next section of this chapter where the robot is enabled to avoid objects in a more effective manner.

5.4 More Intelligent Roaming

In previous programs, the robot simply turned away from objects it encountered. In order to give the robot some sort of personality, we added some randomness to the turns, but we can hardly claim that it is intelligent in its decisions. In fact, if you run the program you will observe behaviors that appear unintelligent, especially if there are a lot of obstacles. One such behavior is that the robot will occasionally make its random turn into objects and not away from them. There are several ways the robot can make better decisions.

5.4.1 USING SENSORY INFORMATION MORE EFFECTIVELY

One way of improving the behavior of our robot is to make it decide which is the *best way* to turn, instead of just turning a random amount. It may not be clear what is the best way, but there is a simple idea that produces a very acceptable behavior. If the robot encounters an object and the sensors show it to be on its right side, then the robot should turn left. If the object is on the left side it should turn right. If the object is straight ahead the robot should turn completely around. In all these cases, we will add a little randomness to improve the robot's ability to cope with unforeseen circumstances. However, because the decisions are more intelligent to begin with, we won't need near as much randomness to be effective. In order to implement this improvement we need to be able to examine the status of individual sensors more efficiently. Let's look at some techniques that can help.

5.4.1.1 Making Better Decisions To know if an object the robot encounters is on the left or right we must analyze the value of the individual bits in the sensory data. Figure 5.5 shows some example expressions that can help us analyze the infrared data. All of these expressions can be used as conditions in `if` and `while` statements.

5.4.1.2 Logical Operations The expressions in Fig. 5.5 are valuable, but are limited. RobotBASIC allows you to manipulate expressions using logical conditions. For example, you could test to see if *either* of the two right-hand sensors is triggered individually with a logical OR operation as shown in the following expression.

```
rFeel()=2 OR rFeel()=1
```

Expression	Situation that makes it true
rFeel() = 0	No sensors triggered
rFeel()	Any sensor triggered
rFeel() = 4	Only the front sensor is triggered
rFeel() = 3	Only the two right-hand sensors are triggered (both must be trigger together)

FIGURE 5.5 Example expressions for testing data from the infrared sensors.

NOTE: This expression will not be true if *both* sensors are triggered together because in that case the sensor value will equal 3. In a complex expression like this one it is often important to use parenthesis to make sure certain portions of the expression are evaluated before others. See Sec. B.7.5 for more information on operator precedence.

Notice this is very different from checking to see if `rFeel()` is equal to 3, which means *both* of the right-hand sensors *must* be triggered together and *none* of the others can be triggered. Logical operations are a great help when analyzing sensory data, but there are other ways that can be more efficient or more appropriate in certain situations.

5.4.1.3 Bitwise Operations Below are two expressions that will perform *almost* the same test as the one in the previous section. That statement was true if either of the right infrared sensors were pressed alone. Both statements below will be true if *either or both* bumpers are pressed. Let's see how they work.

```
rFeel() bAND 3
rFeel()   &  3
```

In the above two statements, the RobotBASIC operators & and `bAND` (two options for doing the same operation) cause the infrared values to be bitwise ANDed with the number 3 (binary 0011). Bitwise simply means the values of *each bit position* of the two numbers are ANDed together. This means the answer for each position will be a one, only if that bit position in the first number *and* the same bit position in the second number, are *both* ones. Lets look at some examples in Fig. 5.6 to make this clearer. The number we are bANDing with the sensor value is referred to as a *mask* because it hides some positions (using zeros in the mask) while allowing some to go through unchanged (using ones). As you can see from Fig. 5.6 the expression will be true when either or both of the positions specified by the mask is a one because the only bit positions in the sensor value that are not masked are those where the mask is a one. In this example, the output will be false only if *neither* of the specified sensor positions is triggered.

As we proceed through the text, you will see how the use of bitwise and logical operations can help in analyzing the meaning of all sensory data so that the robot can make decisions on its own. Refer to Sec. B.7.5 for complete information on all the logical and bitwise operations available in RobotBASIC.

Infrared value	0001	0010	0011	1001	1000
The mask (3)	0011	0011	0011	0011	0011
Answer when bANDed	0001	0010	0011	0001	0000
True or False condition	true	true	true	true	false

FIGURE 5.6 Results when various bumper conditions are bitwise ANDed with the number 3 (0011).

```
RoamAround:
  while true
     // forward until an object is found
     while rFeel( )=0
        rForward 1
     wend
     // try to intelligently turn away from the object
     if rFeel()&3 then Ta = -45 // object on right,turn left
     if rFeel()&24 then Ta = 45 // object on left,turn right
     if rFeel()&4 then Ta = 160 // object infront,turn around
     // turn Ta deg. plus a random amount no more than 40 deg.
     rTurn Ta+random(40)*sign(Ta)
  wend
Return
```

FIGURE 5.7 This subroutine shows one method for making our robot more intelligent as it roams the screen.

5.5 Improved Obstacle Avoidance

Armed with more tools for analyzing the infrared data, lets improve the robot's ability to react to objects in its environment. All the improvements will be in the subroutine *RoamAround*. All the algorithms given from now on will be a replacement for this subroutine. In order to test the algorithm, replace the old subroutine in Fig. 5.4 with the new one given and run the program.

5.5.1 A FIRST IMPROVEMENT

Let's see how the robot can use bitwise operations to make better decisions. Look at the subroutine in Fig. 5.7. The first thing you will notice in Fig. 5.7 is the `rTurn` statement near the end. Instead of turning 150° plus a random amount as we did earlier, the program now turns an amount specified by the variable *Ta* plus a random amount. The key to the robot's new intelligence is choosing a proper value for *Ta*.

Inside the main `while`-loop, after an object is encountered, three `if`-statements decide on an appropriate value for *Ta*. If there is an object on the right (if either of the right-side sensors are triggered) a left turn of 45° is specified. Similarly if either of the left-side sensors are triggered a right turn of 45° is used. If the front sensor alone, or in combination with other sensors is triggered, *Ta* is given a value of 160 to make the robot turn almost completely around (180° ± 20°). A random value is still added when the robot turns, but it is much lower than before because the robot is *always* turning in a reasonable direction anyway. Notice the use of the function `Sign`(*Ta*) to ensure that the random number is in the same direction as the turn.

5.5.2 A SECOND IMPROVEMENT

The algorithm in Fig. 5.7 will turn the robot between 45° and 85° when it encounters an object on its left or right. If the turn causes the robot to still be facing an obstacle it will turn again a random amount. This will be repeated until the robot eventually finds a clear

```
RoamAround:
  while true
    while rFeel( )=0 // forward until an object is found
        rForward 1
    wend
    // try to intelligently turn away from the object
    if rFeel()&3 then Ta = -90 //object on right,turn left
    if rFeel()&24 then Ta = 90 //object on left,turn right
    if rFeel()&4 then Ta = 180 //object ahead turnaround
    OldDist=0
    for i=0 to Ta
      rTurn sign(Ta)
      NewDist = rRange()
      if NewDist < OldDist then break
      OldDist = NewDist
    next
  wend
Return
```

FIGURE 5.8 This subroutine turns the robot toward an open space.

path. However, the robot would be a lot more intelligent if, while turning, it had a way of stopping as soon as it senses a possible clear path. This way instead of turning a fixed amount, which may cause it to miss an opening while it is turning, we can make the robot stop turning when it sees an opening.

One way to do this is to have the robot use its range-sensor to measure the distance to objects as it turns. Generally, the distance should get larger as the robot turns away from the object it has just encountered. If we stop the robot turning as soon as the distance starts to decrease, indicating a possible new obstacle, the robot will be able to turn until it avoids the obstacle, but not until it encounters another. This allows the robot to make more intelligent turning decisions. The routine in Fig. 5.8 shows how this can be accomplished.

The algorithm in Fig. 5.8 assigns a value of 90° left or right instead of 45° to the variable *Ta*. We can allow more turn because we are going to stop when the robot sees an opening anyway.

The `for`-loop allows the robot to try and turn the designated number of degrees. The `for`-loop will count up if *Ta* is positive or down if it is negative. The loop keeps track of the last distance read by the range-sensor in the variable *OldDist*. When the new distance read is smaller than the old distance the `Break` statement is used to exit the `for`-loop. Notice the robot is made to turn the value returned by the function `Sign`(*Ta*). This value will be −1 if *Ta* is less than zero, 1 if it is greater than zero and 0 if it is equal to zero.

5.5.3 FURTHER IMPROVEMENTS

The improvements made in this chapter are only suggestions. The robot's behavior should be based on the environment in which it is expected to operate. The programs above can fail, for example, if you draw objects that have sharp points because they can be missed by blind spots in the infrared sensors. It is also possible for the robot to become stuck between two objects that are spaced close enough together to trigger the sensors on both sides of the robot at the same time.

Part of the enjoyment in robotics is finding problems that a robot cannot handle given simple algorithms and trying to design more sophisticated solutions to impart the robot with the intelligence to tackle baffling situations. The exercises below offer ideas for improving the programs in this chapter.

5.6 Summary

In this chapter you have:

- Written programs to roam around while avoiding objects in the environment.
- Learned about subroutines and the `Gosub` statement.
- Learned about the `Print`, `GotoXY`, `LineTo`, `ClearScr`, `ReadMouse`, `LineWidth`, and `SetColor` commands.
- Learned how to develop progressively more complex algorithms that allow the robot to deal with complex situations more intelligently.
- Learned about **bitwise** operators and how to combine them with **logical** operators to interrogate the sensors more efficiently.
- Seen how adding some randomness with the `Random()` function can improve the robot's responses in certain situations.
- Learned that the robot in RobotBASIC has limitations, just like a real robot, and that overcoming these limitations can be challenging, yet rewarding and fun.

5.7 Exercises

1. Modify the program in Fig. 5.4 so that it also uses the bumper sensors. The new program eliminates blind spots that could allow sharp objects to cause errors.
2. The program in Fig. 5.4 currently turns 180° (±30°) if the front sensor is triggered even if other sensors are also triggered. Modify the program so that it will turn 180° (±30°) if only the front sensor is triggered. Consider the differences in the behavior you see between the two programs.
3. Modify the program in Fig. 5.7 so that it will not get stuck between two objects.

HINT: Use bitwise and logical operations to detect such a condition and turn 180°.

4. Experiment with different amounts to turn when the program in Fig. 5.7 encounters an object and note what effect larger and smaller angles have on the robot's behavior. Note the effect of changing or eliminating the random amount.
5. Develop an algorithm to create your own behavior for the robot. Test it in a variety of situations to see how it compares with the behaviors studied in this chapter.

CHAPTER 6

DEBUGGING

Previous chapters introduced RobotBASIC and some of its capabilities through simple programming examples. As the book progresses, programs will become increasingly more complicated making it harder to find errors, not only in the typed code, but also in the logic of the algorithms.

There are three types of errors that can cause problems in a program:

- *Syntax errors.* These are errors in the typed words of the code. For example, you type the command `Prnit` when you actually mean `Print`. RobotBASIC will detect these types of errors and issue a message indicating their nature and location. It will also highlight the error location within the editor.
- *Semantic errors.* These are errors that occur during the running of the program when an illegal operation takes place such as division by zero. For example, you may have a statement like `Speed = Distance/TimeTaken`. If the variable *TimeTaken* becomes 0 some time during the program's execution an error will occur. RobotBASIC will indicate the nature and location of such errors and will highlight the line that caused the error in the editor.
- *Logic errors.* These are errors that cause the program to run in a fashion that you do not expect even though the program is syntactically and semantically correct. This kind of error is usually easy to detect if it affects the program in an obvious manner. Unfortunately, more often, this type of error can be quite subtle and hard to detect or trace to a particular location in the code. For example, you may write

`Speed = Distance * TimeTaken`. This is the incorrect formula for the calculation desired (you should divide not multiply). However, RobotBASIC will not know this and will run anyway since there is no syntactic or semantic error.

Logic errors can only be detected by meticulous testing and analysis of the program. This process is called *debugging*. The term comes from the days when computers were huge machines with electrical and mechanical components as well as a few electronic ones. Real live bugs used to crawl inside some of the electrical and mechanical devices of these machines causing failures. Operators used to go inside these computers to find the bugs and replace the burnt out or jammed component to make the computer run again. Thus the term debugging was coined.

Debugging can be a frustrating process, but RobotBASIC has some unique and powerful debugging features to help ease and facilitate the process. However, before we look at these features, let us explore some of the principles of debugging in general.

6.1 Before You Program

Before you begin writing a program to control a robot, you should consider the problems and situations the robot will face. You must take into account what sensors you want the robot to have and what data you will be able to acquire. Finally, you must decide how the robot will analyze the data it obtains. This means that you have to determine what data patterns are meaningful and what you want the robot to do when it encounters those patterns. This is not always easy.

When a robot navigates through the environment it is likely to produce some unexpected sensory data. Consider the random-roaming programs in Chap. 5. One of the basic behaviors introduced (Fig. 5.7) was that the robot turned right if sensory data showed an object on the left and vice versa. This seems like an algorithm that should work all the time. However, if the environment contains two objects that are just far enough apart to allow the robot to pass between them and, during its random roaming, the robot tries to pass between the two objects in a manner that causes both the left and the right sensors to activate simultaneously, the algorithm will fail.

If you inspect the routine in Fig. 5.7 you may notice that none of the three conditions being tested by the `if`-statements address the situation when the infrared sensors are detecting objects on both the left and right sides of the robot simultaneously. When environmental situations are not anticipated, the robot is likely to react unpredictably at best. The program may sometimes respond with an adequate action, but this only adds to the difficulty of determining the reason for the failure when it occurs.

6.2 Plan Plan Plan

The best way to deal with these situations is to plan ahead so you can anticipate the predicaments the robot may face. One way to do this is to use the remote control program from Fig. 4.6 in Chap. 4. Substitute the environmental situations you want to explore and use the remote control features of the program to move the robot into difficult situations and observe the displayed sensory data. Knowledge of how the robot sees its environment will help you choose the sensors you need and what data patterns to program for.

In a complex or changing environment you might miss some critical situations no matter how much you plan. When this happens you need a way to discover exactly why your robot is getting baffled and what actions it should invoke to deal with these situations when they are encountered.

6.3 Debugging Philosophy

The basic philosophy of debugging a program is composed of a few steps. First, you need to *isolate* the general area of the code where the fault is occurring. Next you need to determine why that portion of the code is not performing as expected (*locating* the specific source of the problem). Finally, you have to *correct* the faulty lines or logic that causes the problem. Let's see how each of these can be accomplished.

6.3.1 ISOLATING THE FAULT

Assume you have a 200-line program that stalls or hangs when it is run. It would be inefficient to look through the entire program hoping to find the problem. Often the reason a program hangs is because it is stuck in a loop, doing the same thing over and over—but appearing to be doing nothing at all to the observer. This means you have already narrowed the problem down to code that lies within a loop. For example, let's assume there are four major loops in the program. Our next goal would be to determine if one of these loops contains the problem and if so, which one. An easy way to do this would be to place some `Print`-statements before and after each loop. These would display something like "Entering loop 1" or "Exiting loop 3" so that when the program is run, you will be able to see how the program is progressing. This procedure should allow you to determine which loop contains the problem. If your program is very large, perhaps containing dozens of loops, you could place similar `print` statements at the beginning and ending of subroutines to initially isolate the problem to a portion of your code. Once you are down to a manageable size, you could then add more `Print`-statements to that area to further isolate the problem.

6.3.2 LOCATING THE FAULT

Once you have the fault isolated to a manageable area, you need to get information that can help determine why the problem is occurring. Without such information you are only guessing at the source of the problem. Typically, the information you need is the value of a variable or sensor. You can use more `Print`-statements to obtain this information.

Ideally, you would like to get this data each time an action occurs in the program. If you can analyze the values of variables and sensors in the isolated area you should be able to determine why the fault is occurring (perhaps an `if`-statement is not showing *true* when you expect it to). The reason for the fault could be that you typed the name of a variable incorrectly or the problem could be with the logic used for dealing with an unanticipated situation in the environment.

6.3.3 CORRECTING THE PROBLEM

How you correct the problem depends on its nature. Sometimes correcting a typing error is all that is necessary. At other times you may have to admit that your initial plan or algorithm was inadequate for the situation. In such cases, the information you obtained during the debugging phase will help you formulate a better algorithm.

6.3.4 PATIENCE PATIENCE PATIENCE

Debugging can often take much longer than you might expect—sometimes much longer than the initial writing of the code; just be patient. The insight gained during debugging is valuable not only for correcting errors in this program, but also in developing this and other programs further and even in improving your problem-solving skills in general. As you gain experience, you will discover that time spent in a careful and systematic design process as well as meticulous coding can save many hours of debugging.

6.4 Debugging with RobotBASIC

You can debug programs with `Print`-statements as discussed in the previous section, but RobotBASIC offers alternatives that are far more efficient. There are several techniques available to you, some of which are similar to those in other languages and some that are unique. Consider again the random roaming subroutine from Chap. 5 (Fig. 5.7). It is shown here in Fig. 6.1 with a `Debug` statement inserted that will be discussed later.

The code shown in Fig. 6.1 works in most situations, but there are environmental conditions that can cause it to fail. The problem is that the infrared sensors that *feel* around the robot have blind spots as discussed in Chap. 3. If the objects drawn to test the subroutine are really small or have very sharp points, the potential for a fault exists. When the robot approaches such an object the point may slip between the sensor detection areas and cause a collision before it can be detected.

```
RoamAround:
  while true
     // forward until an object is found
     while rFeel( )=0
        rForward 1
Debug "An object was detected   ", rFeel()
     wend
     // try to intelligently turn away from the object
     if rFeel()&3 then Ta = -45 // object on right,turn left
     if rFeel()&24 then Ta = 45 // object on left,turn right
     if rFeel()=4 then Ta = 160 // object infront,turn around
     // turn Ta degrees plus or minus 20 degrees
     rTurn Ta+random(40)*sign(Ta)
  wend
Return
```

FIGURE 6.1 This subroutine moves the robot randomly around the screen (see Chap. 5 for the complete program).

If this problem occurs during the process of developing a program, you may be puzzled as to why the robot collides with objects that you believe should have been detected by the code. This could certainly be true if you were unfamiliar with the blind spots associated with infrared sensors. If you could obtain the value of the sensors immediately before, during, and after the collision you would have the data needed to discover the problem.

6.4.1 THE `Debug` COMMAND

In RobotBASIC there is a special form of the `Print`-statement called `Debug`. This command can be used just like a `Print`-statement, but it differs in several important ways. Instead of printing on the terminal screen like the `Print`-statement, the `Debug`-statement prints in a special window that opens the first time it is used. However, it does not just print, it also causes the program to pause execution so that you can see what the robot is doing to cause the data being displayed. Insert the following statement right after the first wend in Fig. 6.1 (comment out the other debug statement already there):

```
Debug "An object was detected    ", rFeel()
```

You also need to place the statement `DebugON` at the beginning of the main program. When the program is run (and you draw some objects) the program will *stop* and the debug window will appear when an object is encountered and show a screen similar to Fig. 6.2. The fact that the program stops is important. It gives you the opportunity to analyze the robot's situation at the instant the `Debug`-statement was executed.

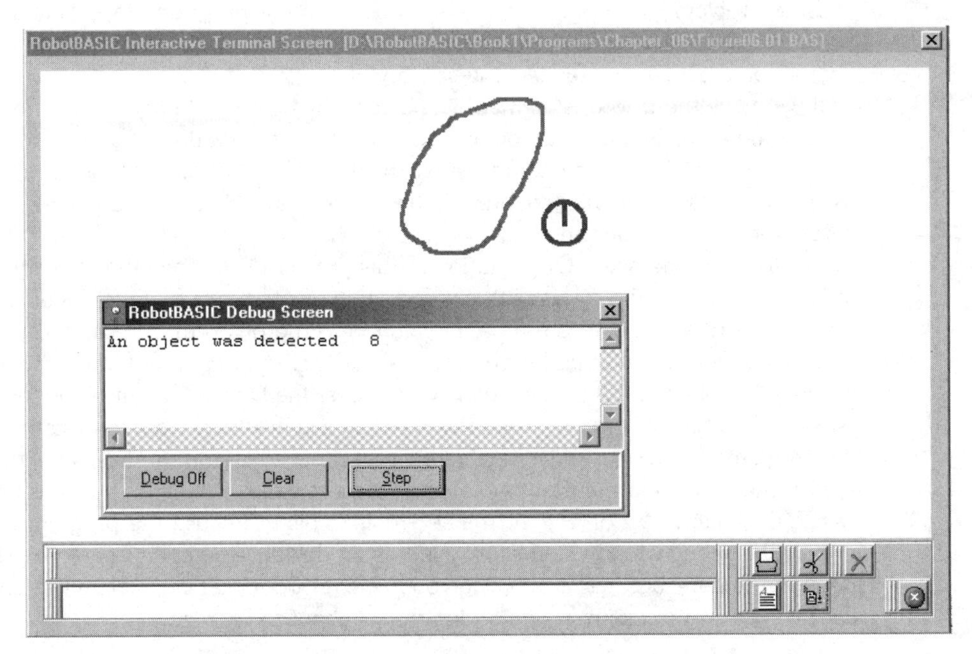

FIGURE 6.2 This Debug screen is typical for the example in the text.

The size of the debug window will probably be larger than what is shown in Fig. 6.2. You may resize it and reposition it as you wish. Notice that the debug window shows that an object has been encountered and that the infrared sensor data is 8. This means that the object activated the sensor at 45° left of the robot's heading. You can also view the terminal screen and see the position of the robot compared to its environment. This allows you to better analyze the sensory data being displayed.

If, while in the process of debugging, you happen to lose the debug window behind other windows, you can bring it back up to the top of all windows by going to the Editor Screen and either pressing *Ctrl+D* or selecting the menu option *Bring Up Debug* from the *Run* menu. These actions have no effect if there is no active debugging session going on.

6.4.2 STEPPING THROUGH A PROGRAM

If the mouse is used to click the Step button on the Debug Screen (or pressed *Enter* on the keyboard), the program will proceed where it left off and continue executing until it encounters the `Debug`-statement again. When it does it will stop and show the new sensory data in the window immediately below the old data. If you continue pressing Step (or *Enter*) you can continue to gather data each time an object is encountered. When the window becomes full you can clear it with the Clear button or just let it scroll upward as new data is added. You can also view a table of all the variables in your program, all in one screen, if you press the *View Variables Table* button. This table can also be viewed even after the program terminates by pressing *Ctrl+B* from within the Editor Screen or by selecting the *View Variables Table* menu option from the *Run* menu.

Although this example is a good introduction to debugging it really won't help us find the problem discussed earlier. This is because the program stops only *after* the infrared sensors have detected an object meaning that a collision will stop the program before you get a chance to view the sensory data.

If we move the `Debug`-statement just before the `wend` statement instead of after it (as shown in Fig. 6.1), we can obtain the necessary information. If you try it, you will see that the program will stop and display the debug window after every move. As long as an object has not been encountered, the sensor data will be 0. It can take some time before the robot reaches an object because you have to press Step each time the robot moves forward one pixel. One solution to this problem is to draw the object you want to test very close to the robot so that it does not have to move far before indicating the sensory information. There is a better solution to the problem of having to press the Step button too many times before the robot arrives at the point of interest where we would like to analyze the data in detail. If you press the Debug Off button on the Debug Screen, the robot will move around as it would if there were no debug statements in the program. This allows you to let the program proceed at normal speed until the robot approaches a situation you want to analyze in more detail. When this happens press the Debug On button on the Terminal Screen. This will cause the program to again display the debug window and allow you to step through the code as before. This is a powerful feature because it lets the robot move around at normal speed until you decide you want to examine something that is about to happen.

NOTE: If you are familiar with the break-point system used by other language debuggers, you will find that this method, while unfamiliar at first, can be a better alternative for debugging a robotic algorithm.

6.4.3 VIEWING THE INFRARED BEAMS

Because of the *blind spots* inherently associated with infrared sensors, RobotBASIC has a special feature to aid with their debugging. If you replace each occurrence of `rFeel()` in a section of code you wish to analyze, with `rDFeel`(red) you will see a very versatile feature when the program is run.

Each time the sensors are read by `rDFeel`(*color*) the robot will display the area being observed by the infrared sensors using beams of a color specified by the value *color*. If you do not pass the function a color by saying `rDFeel()` (notice no color is specified) then the second color on the list of colors given to the `rInvisible` command will be used. This feature allows you to easily see where the blind spots are and helps you make better decisions when you are designing a new algorithm. If you have not specified an invisible colors set then you must use `rDFeel`(*color*) not `rDFeel()`.

6.4.4 VIEWING BUMPER LEDS

If you use `rDBumper`(*color*) in place of `rBumper()` while trying to debug a program, you will see the robot illuminate an LED in the vicinity of where the bumper was touched by an object. This LED will have the color specified by the value *color*. This feature can help in visualizing where crashes are occurring. As in the `rDFeel`(*color*) function you need to specify the color. If you do not specify a color the function will use the second color passed to the `rInvisible` command. If you have not specified an invisible colors set then you must use `rDBumper`(*color*) not `rDBumper()`.

6.5 Summary

In the chapters that follow you will see other ways to use the debug features of RobotBASIC. In this chapter you have:

- ❑ Been introduced to the basic principles of debugging.
- ❑ Learned how to use the `Debug` statement to step through a program while displaying the value of variables and/or sensor data.
- ❑ Learned how to turn the debug feature on and off while the program is executing.
- ❑ Learned how `rDFeel`(*color*) and `rDBumper`(*color*) can help visualize where objects are causing problems while the robot is moving around its environment.

Now, try to do the exercises in the next section.

6.6 Exercises

1. Add the debug features discussed in this chapter to the random roaming programs discussed in Chap. 5. It is not necessary that you find any real faults. The goal is to understand how to use the features. Later chapters will help you develop your debugging skills.
2. In the programs of Exercise 1, try to draw objects that will cause collisions or other problems and use the debug system to find out exactly why the errors occur. Make corrections if you can. Later chapters will offer more opportunities for more intricate debugging.

PART 2

DEVELOPING A TOOLBOX OF BEHAVIORS

In Part 2 we develop a toolbox of utility programs. The programs impart the robot with a collection of behaviors that enable it to handle specific tasks. Each chapter focuses on a single behavior, evolving algorithms that can work in a variety of situations of increasing complexity. In Part 3 we will utilize combinations of these behaviors to create solutions to real-world problems.

We build on the programming skills developed in Part 1 by utilizing new commands and functions from the language as well as show how to use arrays to manipulate data more efficiently. Additional robot commands and functions are introduced along with more sophisticated interrogation and manipulation of the standard sensors on the robot. We also utilize customizable sensors to handle more demanding situations and show how to use advanced features of the standard sensors.

Upon completing Part 2 you will be able to:

- ❑ Create complicated programs and employ advanced programming techniques.
- ❑ Utilize all the sensors on the robot and analyze their data more intricately.
- ❑ Utilize arrays and array commands and functions along with looping constructs to manipulate large amounts of data.
- ❑ Improve on the behaviors introduced in this part as well as create new ones of your own.
- ❑ Appreciate the advantages of using RobotBASIC as a research and development tool so as to minimize abortive efforts in a real-world project.

CHAPTER 7

FOLLOWING A LINE

In Chap. 5 we made the robot move around the screen freely while avoiding objects in the environment. A robot is a device that can be made to do useful work. To be able to achieve its assigned tasks the robot will usually need to move to specific locations where it will perform the required work. There are various ways we can move the robot around:

- Move along a prescribed path defined by a line
- Freely move along a path that the robot determines for itself
- Move to a specific destination while keeping within a specified limited boundary

In subsequent chapters we will explore the second and third options. This chapter will explore the first option. The advantage of having the robot move along a designated path is that we can ensure where the robot will be all the time as it progresses from one location to another. It is also easy to make sure that the robot will have no obstacles along its path or at least avoid having to program it with a sophisticated obstacle avoidance behavior.

An example application for a robot of this kind is an automated waiter that carries food items along a continuous loop starting at the kitchen, winding around and between the tables, and returning to the kitchen. It would not be desirable to have a track that protrudes above the ground due to the risk of customers tripping over the exposed tracks. A robot that can follow a line painted on the ground would be preferable. The line does not have to be visible to humans. Only the robot's sensors need to see it.

Developing a robot that can follow a line on the floor (perhaps black tape on a white floor) is a common activity at many robotics clubs. The project is straightforward enough that it usually can be understood and accomplished by novice robot enthusiasts, yet it is complex enough to introduce them to many aspects of robotics.

7.1 The Base Program

In this chapter we will develop a few algorithms to perform line following, but before we can do this we need to develop a base program in which we will place the code that implements the various algorithms. The base program sets up the robot and the environment and then starts the robot on its way to follow the line using the algorithm that we want to test.

The first thing we need to do is to draw a line on the screen for the robot to follow. Next we need to create and place the robot on the screen. Finally we want the robot to start executing the line-following algorithm we are trying to test.

The code in Fig. 7.1a contains three subroutines called *InitializeRobot, DrawLine,* and *FollowLine*. All the *MainProgram* does is call each of these in turn. The third line after the *MainProgram* label makes the robot move forward 10 pixels. The purpose of this will be discussed below.

Notice how the use of subroutines makes it easier to understand what the program accomplishes. The subroutine names indicate what the subroutines do. The main program becomes an overall manager. Of course the actual details of each subroutine's actions may need further explanation, but if you keep this policy of *modularization* throughout your programs, whenever possible, the programs become self-documenting.

The subroutine *DrawLine* creates a line for the robot to follow. The first statement sets the width of the lines (in pixels). The next sets their color and the one that follows, positions the cursor on the screen. A series of `LineTo` commands draw the line one segment at a time. Refer to Sec. C.7 for details of these commands. Also see Sec. 7.4 for a better way to implement this routine.

The *IntitializeRobot* subroutine positions the robot. The command `rLocate` *x,y,heading* creates the robot and places it on the screen at the specified location and heading. Since the robot's default radius is 20 pixels and this routine places the center of the robot 30 pixels to the right of the start of the line, the front edge of the robot will be 10 pixels away from the line. This is why we need to forward the robot 10 pixels before we start the line following routine. This action brings the front of the robot to the beginning of the line in preparation for following it. In a later improvement (Sec. 7.5) this action will not be necessary. RobotBASIC normally issues an error if the robot bumps into a color on the screen (collision with some obstacle). Since the robot must be able to move over the line, we must tell the system the color of the line so that it can differentiate it from an obstacle. We do this with the `rInvisible` *`Green`* command. *Green* is used here because the line color was set to green in the *DrawLine* subroutine. Refer to Sec. C.9 for a detailed discussion on the `rInvisible` command.

The final action of the *MainProgram* is to call the *FollowLine* subroutine. This subroutine is the code that actually performs the task of following the line. All the routines we will develop in this chapter will be replacements for this subroutine. Figure 7.1b shows

```
MainProgram:
  gosub DrawLine
  goSub InitializeRobot
  rForward 10    // move the robot over to the line
  goSub FollowLine
End
//==========================================================
InitializeRobot:
  //-- place the robot at the beginning of the line
  //-- and face it left 90 degrees
  rLocate  200, 71, -90
  rInvisible Green    //-- Green is a line not an object
Return
//==========================================================
DrawLine:
  linewidth 4
  setcolor Green
  gotoxy 170,71
  lineto 160,72
  lineto 145,80
  lineto 140,90
  lineto 130,100
  lineto 125,110
  lineto 120,140
  lineto 130,200
  lineto 140,250
  lineto 130,270
  lineto 145,300
  lineto 200,350
  lineto 300,325
  lineto 450,375
  lineto 450,450
  lineto 600,450
  lineto 600,400
  lineto 650,200
  lineto 500,350
return
//==========================================================
FollowLine:
  //-----Line Following algorithm
  //--we will put code here to make the robot follow a line
Return
```

FIGURE 7.1a This code draws a line on the screen and places the robot at its start.

the output screen when the program in Fig. 7.1a is run. For now *FollowLine* is left empty so no line following will occur.

7.2 An Initial Algorithm

RobotBASIC's robot has three line sensors. One is mounted directly in front of the robot, and the other two are spaced 10° left and right of the front sensor. Figure 7.2 shows this setup. The scale has been enlarged to make the setup obvious. Refer to this diagram to visualize the action of the algorithms that will be developed.

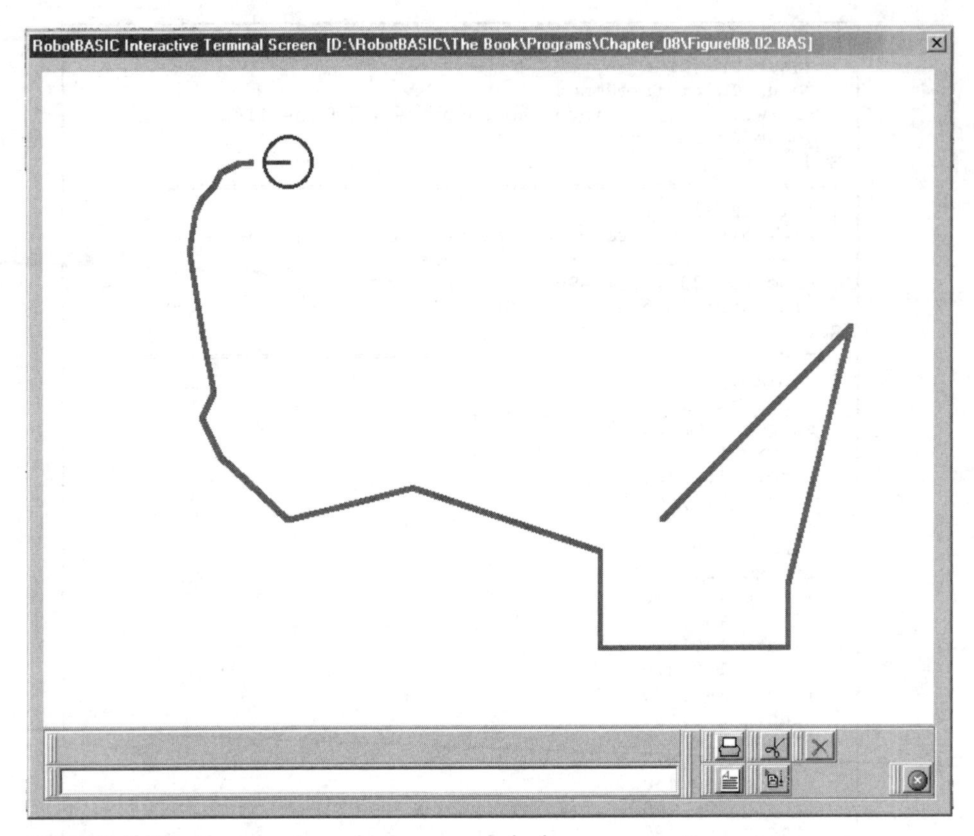

FIGURE 7.1b The robot is ready to approach the line.

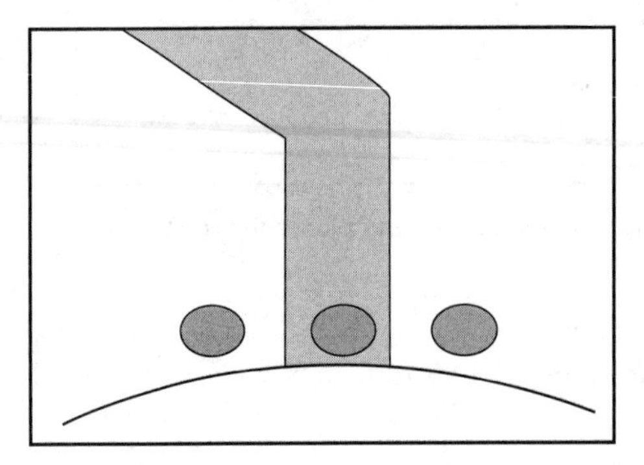

FIGURE 7.2 The line-sensors setup in RobotBASIC.

With three sensors there are many choices for how to develop a line-following robot. You could, for example, use only the front sensor and have the robot constantly swing left and right as it attempts to keep the sensor on the line. Such an algorithm can work, but the robot follows the line with an oscillating snaking sort of motion that is far from efficient.

On deeper analysis, you might decide that a better implementation would be to use the two outside sensors. In this case, the robot should try to keep the line between the two

Example	Action
if rSense( ) & 1	true if right sensor sees the line
If rSense( ) = 4	true if only the left sensor sees the line
if rSense( ) & 6	true if left OR middle OR both sensors see the line
if rSense( )	true if any sensor sees the line
a = rSense( )	
if (a = 2)	true if only the middle sensor sees the line
if a & 7	true if any sensor sees the line
if a = 7	true only if ALL the sensors see the line

FIGURE 7.3 The `rSense()` function reads the three line sensors.

outside sensors. It can do this by turning a little to the right every time the sensor on the right detects the line and turning left when the left sensor is triggered.

7.2.1 READING THE LINE SENSORS

All three line sensors are accessed simultaneously with the single function `rSense`(*Color*). If you do not specify a color by using `rSense()`, the first color in the list of invisible colors passed to the command `rInvisible` will be considered as the line color to be sensed. You *must* specify a list of invisible colors before you do any line sensing, and the color of the line must be in the list (preferably the first one on the list). If you do not do this, the robot will not be able to sense the line since it will consider it an obstacle and will report an error if you make the robot move over the line.

The `rSense()` function returns a number from 0 to 7. This number represents the sensory condition (on/off) with the least significant bit being the right-most sensor. Each sensor is *on* if it senses a line underneath it and is *off* otherwise. In the situation of Fig. 7.2, `rSense()` would return a value of 2 (010 in binary) because only the center sensor is seeing the line. A value of 6 (110 in binary) means that the left and front sensors are sensing the line while the right sensor is off the line. This could happen if the line made a sharp turn to the left as shown in Fig. 7.2.

We can determine the status of a specific sensor by using a *binary* AND (&) operator as shown in the examples of Fig. 7.3.

7.2.2 A FIRST ATTEMPT

Figure 7.4 shows a subroutine that can follow a line using the above logic. The routine simply turns right or left depending on where it sees the line. The `while`-loop creates a loop that, in this case, continues forever (or until the user stops the program). To test

```
FollowLine:
  while true
    if rSense() & 1 then rTurn 1
    if rSense() & 4 then rTurn -1
    rForward 1
  wend
Return
```

FIGURE 7.4 This subroutine will follow a relatively straight line.

```
FollowLine:
  c=0
  while c<50    //exit loop if line is not seen for 50 tries
    if rSense() & 1 then rTurn 1
    if rSense() & 4 then rTurn -1
    rForward 1
    if rSense()      // if any sensor sees the line
       c = 0         // start the counter over
    else
       c = c + 1     // increment counter if no line is seen
    endif
  wend
Return
```

FIGURE 7.5 This subroutine knows when the end of the line has been reached.

any of the routines given from now onwards, replace the *FollowLine* subroutine in the base program of Fig. 7.1, with the new figures given (Fig. 7.4 in this case). If you test this subroutine you will see that it fails if the line turns too sharply. We will address this problem shortly.

7.2.3 AN IMPROVEMENT

One problem with the routine in Fig. 7.4 is that the robot does not know when it loses the line and continues moving until it crashes into a wall. Figure 7.5 shows how the robot can determine when it is no longer on the line. The robot constantly checks to see if any of the sensors are seeing the line. Since it is possible that a thin line could be between two of the sensors (and thus make the robot incorrectly assume it has lost the line), the algorithm will continue trying to follow the line until the sensors have not seen the line 50 times in a row. If you substitute this code into the base program, you will see that the robot stops shortly after losing the line. This is an improvement, but we still need to find a way to keep the robot on the line.

7.3 Sharp Turns Cause a Problem

As mentioned earlier, the algorithms in Figs. 7.4 and 7.5 fail if the line turns sharply. The robot will do just fine if the line is relatively straight, but it will lose the line when there is a sharp turn in it.

7.3.1 POSSIBLE SOLUTIONS

In order to solve this problem, we need to understand exactly why it is happening. The robot fails to follow the line when the line turns faster than the robot is turning—in this case more than about a 45° change because the algorithm makes the robot turn about 1 pixel left or right for each pixel that it moves forward. When this happens the robot moves past the turn and will not see the line on any of the sensors. It continues moving forward and loses the line.

There are relatively straightforward approaches to solving this problem. We can, for example, make sure the robot stays on the line by ensuring that it does not move forward past a turn in the line. This can be done by continuing to turn until it is safe to move forward. Another solution would be to let the robot move past the turn in the line, but give it a means of finding its way back to the line.

Either of the above strategies can provide a possible solution for the robot, but they do it in a very different manner. The differences will be reflected in the behavior you see as the robot attempts to follow a line using the above methodologies. In the first case, the robot will appear to slow down when it sees a sharp turn because it executes more turning than forwarding. In the second algorithm, the robot will constantly move forward, but try to make its way back to the line after it has lost it due to a sharp curve.

You might think that the second strategy is better. After all, it should allow the robot to reach the end of the line more quickly if we can implement it properly. However, consider for a moment that the robot in question could be a car driving down a road and not just following a line on the screen. The second algorithm would indeed let the car take a shorter path to the end of the road, but it does so by letting the car take short-cuts by driving off the road when the road makes a sharp turn and then getting back on the road a little further on.

It is important to realize that neither of these strategies is necessarily better than the other. Each one has advantages and disadvantages depending on the situation and the environment.

One of the advantages of using a simulator is that you can test and improve a variety of algorithms very quickly and test them in various environments just as easily. We will develop two algorithms to implement both strategies discussed above.

7.3.2 A FIRST STRATEGY

Figure 7.6 shows a subroutine that implements the first strategy discussed above. If you run the program, you will see the robot behaving exactly as predicted. For simplicity the code does not check if the end of the line has been reached.

Compare Figs. 7.4 and 7.6, the subroutine in Fig. 7.4 lost the line in a fast turn because the robot only checks once (with an `if`-statement) to see if it needs to turn, and only turns once (if needed) before proceeding forward. This means that the robot can lose the line if the line turns faster than the robot can turn.

```
FollowLine:
  while true
    rForward 1
    while rSense() & 1
      rTurn 1
    wend
    while rSense() & 4
      rturn -1
    wend
  wend
Return
```

FIGURE 7.6 This routine keeps the robot on the line even at sharp turns.

```
FollowLine:
  while true
    if rSense() & 1
      rTurn 1
      LastTurn = 1 //remember which direction we WERE turning
    endif
    if rSense() & 4
      rTurn -1
      LastTurn = -1 // remember which direction we WERE turning
    endif
    rForward 2  // since we don't care if we lose the line,
                // move forward twice
    if rSense()=0
      rTurn 3*LastTurn // if we lose the line make a BIG
    endif              // turn back towards it
  wend
Return
```

FIGURE 7.7 This routine lets the robot find the line after it has lost it in a turn.

In Fig. 7.6, the robot uses a `while-wend` loop to continue to turn as much as is necessary to stay on the turning line before moving forward. This means that the robot will not move forward until it has turned sufficiently to remain on the line. Extremely sharp turns, such as the last one in Fig. 7.1b, still present a problem, but the robot performs well in most situations.

7.3.3 A SECOND STRATEGY

The second strategy is implemented in Fig. 7.7. The routine allows the robot to leave the line when it turns sharply and then reacquire it a short distance later.

Once the robot loses the line it cannot use the line sensors to determine which way to turn. To solve this problem, we need a way for the robot to *remember* which way it was turning the *last* time it saw the line. This will normally be the direction the robot should turn if it has lost the line. Each time the robot makes a normal turn (the `if`-decisions in Fig. 7.7) the subroutine stores the turn direction in the variable *LastTurn* to remember which direction the robot was turning. Later in the routine, if none of the sensors are on (indicating we probably have lost the line), the robot will be able to head back toward the line. Extremely sharp turns are still a problem even for this algorithm.

In this algorithm, since it is acceptable to lose the line, we can speed up the robot's progress by moving it forward 2 pixels at a time. While the robot is still over the line it will turn only 1° at a time to stay on it, but if the line is lost, the robot will take a 3° turns to help it get back on course. Notice that these choices for how much to move or turn are somewhat arbitrary. With a little experimentation, you can determine the optimum values for your situation. This reminds us again of the advantage of using a simulation. With RobotBASIC you can change the values and see how the robot responds to your changes very quickly.

7.3.4 VERY SHARP TURNS

The routine in Fig. 7.6 stays on the line nicely and even handles 90° turns. However, if the line turns much more than 90°, the robot can still lose the line. There are many ways to solve this problem. Figure 7.8 shows a potential solution.

```
FollowLine:
  while true
    rForward 1
    if rSense() = 3
      rForward 20    //move the centre over the corner
      while rSense() = 0
        rTurn 1     //turn back to the line
      wend
    endif
    if rSense() = 6
      rForward 20    //move the centre over the corner
      while rSense() = 0
        rTurn -1    //turn back to the line
      wend
    endif
    //-- reposition over the line
    while rSense() & 1
      rTurn 1
    wend
    while rSense() & 4
      rturn -1
    wend
  wend
Return
```

FIGURE 7.8 This routine deals with sharp turns in a unique way, allowing it to not only handle very sharp turns, but also acquire the line if it finds it while roaming randomly.

The principle is that when one of the outside sensors *and* the middle sensor are *on* simultaneously, the robot assumes that the line must be making a sharp turn. When this situation is detected, the robot moves forward so that its center is near the point where the line turns. The robot then turns until its outside sensor finds the line again. Having done all this, the routine proceeds as before with the `while`-loops keeping the robot on the line in the same manner as in Fig. 7.6.

When you run the program you will see that the new algorithm does in fact handle very sharp turns. You will also see that this new turning behavior happens even on moderately sharp turns, making the robot correctly follow more complex lines. However, the robot's movement is now somewhat erratic which may not be acceptable in some situations.

There are many factors that can cause an algorithm to fail. The above algorithm, for example, will not work properly if the line width is reduced from 4 to 3 pixels. Any algorithm is only a potential solution until it has been thoroughly tested in a variety of expected environments. A robot's ability to perform properly depends on the programmer's ability to predict the situations the robot is likely to face. Subsequent chapters will explore this idea further.

7.4 Random Roaming with Line-Following (Racetrack)

In the previous sections we assumed that the robot is already over the line before starting the line-following procedure. What if the robot is not over the line? You can test for such a situation by commenting out the `rForward 10` statement in the *MainProgram* in Fig. 7.1a.

A better algorithm would allow the robot to roam around until it finds the line and then commence the line-following routine. In this section we are going to combine random roaming as you have seen in Chap. 5 with the line following algorithm developed in Fig. 7.7.

The requirement is to have the robot roam around until it encounters the line. Once it encounters the line it will acquire the line and follow it. The random-roaming algorithm developed in Chap. 5 (Fig. 5.3) will be modified slightly to allow for checking if a line is being sensed and stop roaming when it is. This way the robot can be placed anywhere and it will roam around until it encounters the line after which it starts following the line.

The new program in Fig. 7.9 is a complete program except that it does not have the line-following subroutine (add one of your choice from those previously discussed). It is a modification of the base program in Fig. 7.1a. The changes are as follows:

```
MainProgram:
  gosub DrawLine
  goSub InitializeRobot
  goSub RoamAround
  goSub FollowLine
End
//==============================================================
InitializeRobot:
  repeat
     readmouse X,Y,b
  until b = 1
  rLocate  X,Y //locate the robot where the user clicked the mouse
  while b = 1
     readmouse X,Y,b
     rTurn 1 //turn the robot while the mouse is being clicked
     delay 50
  wend
  rInvisible Green   //-- Green is a line not an object
Return
//==============================================================
RoamAround:
 while True
   // move forward until an object is found
   while (rFeel( )=0) and (rBumper( )=0) and !rSense()
      rForward 1
   wend
   if rSense() then Break
   // turn 180 degrees plus or minus 30 degrees
   rTurn 150 + random (60)
 wend
Return
//==============================================================
DrawLine:
  linewidth 4
  setcolor Green
  Data Coord;-257, 158,  492, 166,  591, 249,  616, 401
  Data Coord; 565, 506,  345, 515,  257, 364,  118, 413
  Data Coord;  62, 243,  185, 217,  256, 157
  MPolygon Coord
return
//==============================================================
```

FIGURE 7.9 Roaming around to acquire the line then follow it.

- In the *MainProgram* a line is added just before calling the *FollowLine,* to call the subroutine *RoamAround,* and the line that caused the robot to move forward 10 pixels is removed.
- The *InitializeRobot* subroutine has been changed to allow the user of the program to place the robot anywhere and give it an initial heading by using the mouse. This allows for good testing of the roaming and line acquisition.
- A new subroutine *RoamAround* has been added to do the work of roaming.
- The *LineFollow* subroutine is exactly the same as the one in Fig. 7.8 and is not repeated here. You will need to incorporate Fig. 7.8 (or another from the ones developed earlier) with Fig. 7.9 to make the program work.
- The *DrawLine* subroutine is modified to draw using the `mPolygon` command. This is a much more compact way of drawing lines and shapes on the screen. We will discuss this in detail later.

7.4.1 THE ROAMAROUND SUBROUTINE

The *RoamAround* subroutine is very similar to the one given in Fig. 5.3 of Chap. 5. The only change is an added check to see if any line sensors are on. If there are any active line sensors, the roaming is abandoned by using the `Break` statement to break out of the `while`-loop. This of course causes the `Return` statement to be executed, which causes the subroutine to return back to the main program, which then calls the *FollowLine* subroutine.

7.4.2 THE *InitializeRobot* SUBROUTINE

The *InitializeRobot* subroutine allows the user to place the robot at any position on the screen and to give it an initial heading.

The `Repeat-Until` loop keeps tracking the mouse until the user presses the left button (b=1). The robot is then placed at the position indicated with the default heading of North. If the user keeps the left mouse button clicked the `while`-loop will keep rotating the robot as long as the button is pressed. The `Delay 50` statement is necessary to avoid turning the robot too rapidly for accurate control. Finally the `rInvisible` command is used to set the color of the line as a nonobstacle.

7.4.3 THE `Data` STATEMENT AND `mPolygon` COMMAND

The original *DrawLine* subroutine of Fig. 7.1a used a series of `LineTo` statements following a `GotoXY` statement to draw the desired line. This works, but the program can become quite long if we have a substantial shape with many points.

The RobotBASIC command, `mPolygon` can draw any number of consecutive lines in one command. This makes specifying the points compact and the execution is much faster than the `LineTo`-method. We are going to use this construct to draw a racetrack for our robot to follow. (Refer to Sec. C.7 for all the commands used below. Also refer to Sec. B.7.4 for more information about arrays.)

To use `mPolygon` you have to create a one-dimensional array that contains the coordinates of the points that will form the lines to be drawn. You do this in RobotBASIC with a series of `Data` statements. The `Data` command creates an array and puts the data that follows it into successive elements of the array. An *array* is a collection of data elements. Let's say we have five numbers that we want to use in a program. We could use five different variable names in this manner:

```
Num_1 = 9
Num_2 = 8
Etc., etc.
```

This can become quite tedious if we have a lot of numbers to specify. Also later on in the program it can be quite hard to refer to the numbers since each has a distinct name. A better way to do all this is:

```
Data Nums; 9,8, etc., etc.
```

Nums is the name of the array (of course, you can use any name you want here). The semicolon (;) is necessary after the name of the array. The comma (,) is necessary to separate the various data elements. You can have as many elements as you desire. If the data is too long to fit on one line you can have as many `Data` statements as you need on as many lines as you need. So long as they all have the same array name the elements will be put in the same array.

Later on in the program you can refer to the *n*th element by saying *Nums[n-1]*. The reason you use *n-1* and not *n* is because the counting of element numbers starts at 0 not at 1. So the first element is element number 0, the second element is element number 1 and so on. So the *n*th element would be element number *n-1*.

For an example, let's convert the data in the *DrawLine* subroutine of Fig. 7.1 to an array. We would use the statements you see in Fig. 7.10. This is much more efficient than before.

The `for`-loop draws the line, but instead of using many `LineTo` statements we now use only one inside the loop to draw the line. The function `MaxDim()` is used to find out the number of elements in the array. Of course we could have counted these by hand and

```
DrawLine:
  Data Points; 170, 71, 160, 72, 145, 80, 140, 90, 130,100
  Data Points; 125,110, 120,140, 130,200, 140,250, 130,270
  Data Points; 145,300, 200,350, 300,325, 450,375, 450,450
  Data Points; 600,450, 600,400, 650,200, 500,350
  LineWidth 4
  SetColor Green
  GotoXY Points[0],Points[1]
  for I = 2 to MaxDim(Points,1)-1 Step 2
    LineTo Points[I],Points[I+1]
  next
Return
```

FIGURE 7.10 A better line drawing method.

```
DrawLine:
  Data Points;-170, 71, 160, 72, 145, 80, 140, 90, 130,100
  Data Points; 125,110, 120,140, 130,200, 140,250, 130,270
  Data Points; 145,300, 200,350, 300,325, 450,375, 450,450
  Data Points; 600,450, 600,400, 650,200, 500,350
  LineWidth 4
  SetColor Green
  mPolygon Points
Return
```

FIGURE 7.11 Using `mPolygon` to draw polygons.

put the number (38) instead of using the function. The `MaxDim()` function eliminates the inconvenience of having to recount every time we add or remove points.

You might be wondering why the loop count started at 2 not 1. This is because we used the first pair of elements to `GotoXY` to the start of the line. Also notice the use of `Step 2` in the `for`-loop. This is because the elements are in pairs of point coordinates (*x*, *y*). Notice the count goes up to one less than the number of elements. This is (as discussed above) because the element number counting starts at 0 not 1. Put the subroutine in Fig. 7.10 in place of the *DrawLine* subroutine in the base program of Fig. 7.1a and verify that it performs the same action. The routine is now only 12 lines, much smaller than the original 23-line program.

We can do even more saving. In Fig. 7.11 there is no `for`-loop to draw the lines, no `GotoXY` or LineTo commands. There is only one command after defining the data, `mPolygon`. This one command does the same action as the code in Fig. 7.10. Actually, this command does a lot more; see Sec. C.7 for details. Try this new subroutine by placing it into the base program of Fig. 7.1a in place of the old *DrawLine* subroutine.

Notice that the first data element in the routine of Fig. 7.11. is now a negative number. This is to designate it as a `GotoXY` point not a `LineTo` point for the purposes of the `mPolygon` command. Read all about this in Sec. C.7.

Now that you are familiar with the `mPolygon` command, refer back to Fig. 7.9 to see how it is utilized there.

7.5 Summary

In this chapter you have:

- Been introduced to several algorithms for following a line.
- Seen how proper interpretation of the sensory data can improve the robot's performance while carrying out complex tasks.
- Learned how arrays, `Data`, and `mPolygon` can be used to draw more efficiently.
- Seen how an array is a more efficient way to store and manipulate data.
- Learned that some algorithms may work properly under certain environmental conditions but fail if these conditions are modified.

Now, try to do the exercises in the next section.

7.6 Exercises

1. Run the programs in this chapter to see how they perform. Add debugging statements to help analyze the robot's behavior, and find out why some of the algorithms fail on sharp turns.
2. Try to determine the optimum values for the `rForward` and `rTurn` commands (as discussed in Sec. 7.3) for the routine in Fig. 7.7.
3. Choose your favorite algorithm from this chapter (or develop one of your own) and combine it with the *DrawObjects* subroutine in Fig. 5.2 of Chap. 5 so that your robot can follow *any* line that the *user* draws.
4. The routine in Fig. 7.8 is particularly sensitive to the width of the line being followed. It works great if the line has a width of 4 pixels. Try other line widths and explain the behavior that occurs. Check to see if the line width affects any of the other algorithms in this chapter.
5. Modify the line-following algorithm of your choice so that the robot will check for objects that might block its path. When one is found, the robot should turn 180° and follow the line in the reverse direction. Try out the algorithm you develop with obstacles on the line.
6. The line following algorithm given in Fig. 7.9 works most of the time but it does not guarantee that the robot will keep going around the racetrack in the same direction. The way the algorithm works may cause the robot to back track. Can you write a new algorithm to prevent this?

HINT: Some memory of the direction of travel may be necessary.

7. The new main program in Fig. 7.9 calls *RoamAround* then calls *FollowLine*. If *FollowLine* ever finishes as in Fig. 7.5, then the main program will go to the next line, which is End. Convert the main program so that it will not end, but keep roaming then following a line then roaming and so on endlessly.

HINT: Use a `while`-loop.

CHAPTER 8

FOLLOWING A WALL

There are occasions when it may become necessary for the robot to follow the contour of an object:

- If a robot encounters an object while moving along an intended path, it might go around the object by navigating around the perimeter of the object.
- A robot that delivers mail in an office environment, for example, might follow a wall down a hallway, visiting each office in turn to deposit its cargo and collect new mail.
- A strategy for solving a maze of corridors is to keep following the walls around in one direction (left or right).

In this chapter you will learn various strategies for enabling the robot to follow the perimeter of an object while staying close to it but not crashing into it or moving too far away.

8.1 Constructing a Wall

Before we can examine the algorithms we will need a relatively complex contour with which to test our strategies. Also we will need a base program, which we will use throughout with only a few changes to accommodate the various algorithms.

The robot will start by moving forward until it encounters an obstacle. When it encounters an obstacle the robot will abandon the forward-moving behavior and start the wall-following behavior.

Figure 8.1 shows a template with a main program and three subroutines. The main program calls the subroutines in order as they become needed and locates the robot on the screen. The line that sets the variable *TurnDir* will be needed in later sections and will be discussed there. We set the list of invisible colors and put the pen down. We put the pen down in order to leave a trail behind the robot while it is following the wall. This helps in observing the robot's behavior and gauging the algorithms' effectiveness (or lack thereof), as you will notice later. You have seen the pen feature in Chap. 4 and will learn more about it in Chaps. 10 and 11 (see Sec. C.9 for more details).

The first color on the list of invisible colors will be used by the pen to draw when it is lowered since no color was specified when the `rPen` command was issued. Similarly the second color will be the default color used by the `rDFeel()` function. We will use `rDFeel()` later in the chapter.

The subroutine *DrawWall* does exactly that using `mPolygon` and the array *Wall* created by the series of `Data` statements, as in Chap. 7.

The subroutine *RoamAround* makes the robot move forward until it encounters an obstacle. When the robot encounters an obstacle the routine is terminated, which causes the program flow to go back to the main program, which then starts *FollowWall*. This subroutine is left empty for now. We will develop various wall following algorithms that will be substitutions for this routine.

```
MainProgram:
   gosub DrawWall
   rLocate 100,300,50
   rInvisible Cyan,Red
   rPen Down
   gosub RoamAround
   TurnDir = -1
   gosub FollowWall
End
//============================================================
DrawWall:
    ClearScr
    LineWidth 4
    Data Wall;-161, 177,  220, 124,  375, 155,  485, 275
    Data Wall; 624, 300,  668, 370,  517, 412,  499, 320
    Data Wall; 499, 321,  389, 387,  361, 311,  369, 283
    Data Wall; 348, 235,  334, 275,  318, 223,  251, 319
    Data Wall; 161, 177,  247,-193
    MPolygon Wall,Blue
Return
//============================================================
RoamAround:
  while not(rBumper()&4)
    rForward 1
  wend
Return
//============================================================
FollowWall:
   //we will develop this later
Return
//============================================================
```

FIGURE 8.1 This code draws a wall for the robot to follow and starts moving it forward.

```
FollowWall:
  while true
     // anything on right makes you turn left
     while rFeel() & 3
       rTurn -1
     wend
     rForward 1
     rTurn 1
  wend
Return
```

FIGURE 8.2 This code is a basic algorithm for following a wall.

8.2 A Basic Algorithm

In order to understand how the robot can follow a wall, imagine that you are blindfolded and are asked to stay close to a wall as you follow it to a desired destination. You would probably put out one hand (your right hand if the wall was on your right) to help you know that the wall is still there. As the distance between you and the wall becomes larger your hand will eventually stop touching the wall. You would then need to turn to your right and move forward to get closer to the wall again. If you find yourself getting closer to the wall you would have to bend your arm. To maintain your arm stretched out you will need to turn away from the wall to avoid running into it.

Figure 8.2 shows one method for telling the robot how to achieve the above logic. Replace the *FollowWall* subroutine of Fig. 8.1 with the one in Fig. 8.2.

The outer `while-loop` makes the robot follow the wall forever. The inner `while-loop` turns the robot away from the wall as long as either of the infrared sensors on the right side of the robot can detect the wall. The robot then moves forward and turns back toward the wall.

8.2.1 PROBLEMS WITH THE BASIC ALGORITHM

There are two shortcomings with this algorithm. If you look at Fig. 8.3 you will see both of them. The first problem is that the robot tends to move in arcs around the wall rather than following it in a parallel line. The second problem is that the robot crashes into the first hard turn it encounters.

Let us analyze why the logic failed. However, before we can do this let's observe the infrared sensors while the algorithm is running. If you replace the `rFeel()` function with the `rDFeel()` function you will be able to observe the infrared beams while the robot is moving. Since no color is specified the second color on the invisible colors list (red) will be used to display the infrared beams [read about `rDFeel()` in Sec. C.9].

Observing the infrared beams is a great help in analyzing what the robot sees, and can give real insight into why it fails in situations you think should work. Combining `rDFeel()` with `Debug` statements can help you figure out many complicated and puzzling situations.

You will notice that the robot is moving in arcs due to the way infrared beams are tested. We are testing for either or both of the right beams [`rFeel() & 3`, `3` = 00011 in binary]. This means that the robot will turn away from the wall until the right-hand beam is not

FIGURE 8.3 Simple algorithm fails.

sensing the wall, which is almost 90°. The robot then moves forward and turns. This forwarding and turning is the reason we get the arcs. The robot will move in an arc until it encounters the wall again and turn away 90° and so on.

The reason for the crash is that the 90° and 45° right sensors did not sense the wall at the angle in the wall you can see in Fig. 8.3. This means that the robot will continue forwarding. Unfortunately there is no way for the robot to know that there is still part of the wall ahead and thus will crash into it.

8.2.2 IMPROVING THE ALGORITHM

To prevent the robot from turning too far away from the wall we will ignore the 90° sensor. Also to give the robot the ability to see ahead of it we will test the front sensor. So instead of testing for `rFeel() & 3` we will test for `rFeel() & 6`. Replace the value 3 with 6 (binary 00110) in Fig. 8.2 and run the program again.

As you can see from Fig. 8.4, the robot does indeed follow the wall in a straight line. However, if you look closely, you will notice that the robot still tends to loop around sharp corners. This happens because the robot cannot turn fast enough to follow the sharp turn because it only turns 1° for every 1 pixel forward move. We will solve this problem shortly, but first let's examine a more critical problem.

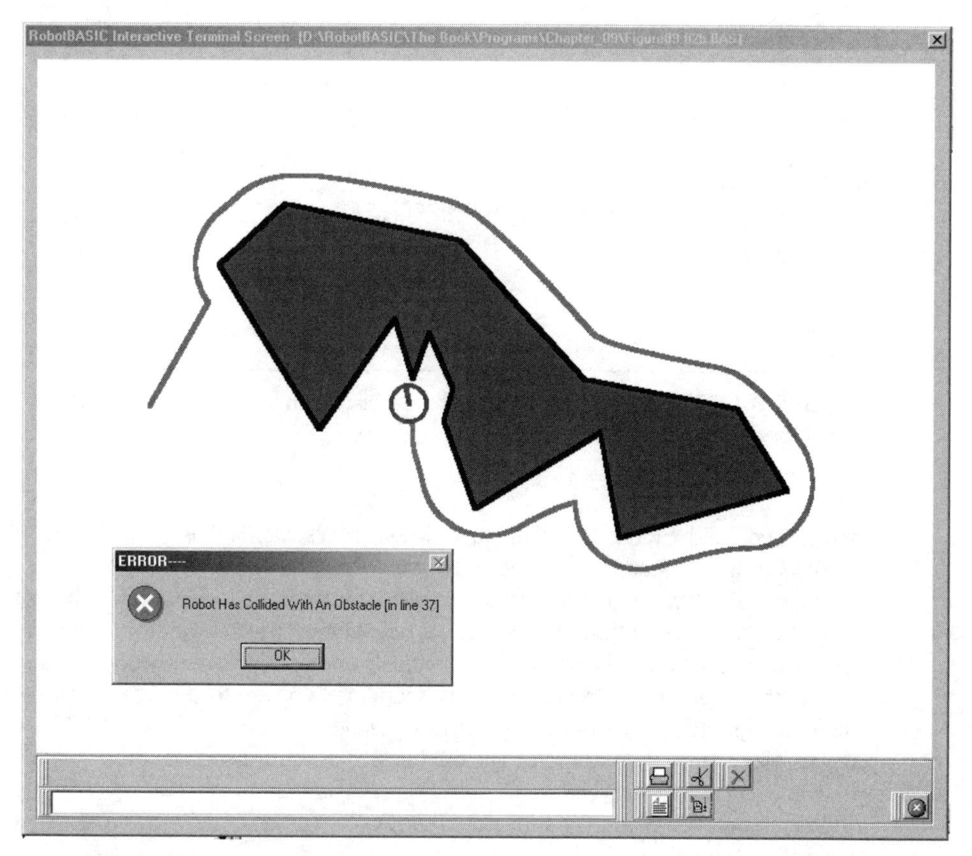

FIGURE 8.4 An improved simple algorithm.

The robot does not crash at the first sharp bend in the wall but it does crash later on. The front infrared sensor failed to detect the sharp protrusion in the wall. If you use `rDFeel()` you will see that it just misses detecting the sharp v in the wall. So, even though testing the front sensor helped, it still fails to catch all situations.

You already know how to solve this problem from previous chapters. If you test for the bumper sensor along with the front sensor you should be able to catch most of the awkward situations.

8.2.3 USING THE BUMPERS

We will show how to use the bumpers with the infrared sensors to follow a wall, but before we give the new code, let us consider what modifications are needed to change the behavior from following a wall on the right to following it on the left.

First we will need to change the sensors used. Second, we need to turn in the opposite direction. If the wall is too close to the left we turn right and we turn left if the robot is too far from the wall.

However, we want to be able to change between turning left or right easily without changing more than a variable in the program. This is why we have the line `TurnDir = -1` in

```
FollowWall:
  if TurnDir > 0
     FN = 6
  else
     FN = 12
  endif
  while true
     while (rFeel()&FN) or (rBumper()&4)
       rTurn -TurnDir
     wend
     rForward 1
     rTurn TurnDir
  wend
Return
```

FIGURE 8.5 This program turns away if the wall is seen by either the infrared sensor *or* the bumper sensor.

the main program in Fig. 8.1. This variable acts as a switch. If it is -1 the robot will follow the wall to the left and if it is 1 the robot will follow the wall to the right.

We will also have to add some code to allow for this switch. Figure 8.5 shows the algorithm that achieves all this. Notice that we now check to see if the front bumper is closed as well as checking for the infrared sensors. Also notice how we set the variable *FN* to be used in the `rFeel() & FN` statement. Remember if you are following the wall to the right then the right 45° and front infrared sensors need to be considered (i.e., 00110 = 6), and if you follow the wall to the left then the left 45° and front sensors are to be tested (i.e., 01100 = 12).

Run the program and try changing *TurnDir* to 1 and see how the robot now follows the wall to the right instead of to the left.

The program in Fig. 8.5 performs reasonably well. As mentioned earlier though, the robot still arcs far away from the wall when it rounds a sharp corner. The algorithm can still be improved further.

8.3 Staying Close on Sharp Corners

The reason the robot arcs too far away from the wall on sharp corners is due to the wall angling away from the robot at a sharp angle, and since the robot only turns 1° as it moves forward 1 pixel, it will not be able to turn fast enough to catch up with the sharply turning wall.

8.3.1 INITIAL ALGORITHM

The algorithm in Fig. 8.6 is a first attempt at solving this problem. The code makes the robot turn more degrees while it is forwarding. This needs to be done only if the wall is turning sharply away. If we do it all the time the robot will go back to moving in arcs. We need a way for determining if the wall is turning away from the robot too sharply.

This is exactly what Lines 13 to 21 in Fig. 8.6 do. We test to see if none of the infrared sensors are seeing the wall or if only the 90° sensor in the direction of the turn is sensing

```
00 FollowWall:
01    TurnAmount = 5
02    if TurnDir > 0
03       FN = 6
04       SFN = 1
05    else
06      FN  = 12
07      SFN = 16
08    endif
09    while true
10      while (rFeel() & FN) or (rBumper() &4)
11        rTurn -TurnDir
12      wend
13      if rFeel()=SFN or rFeel()=0
14        // too far from wall or no wall
15        rForward 1  // forward always to prevent stall
16        while not rFeel()
17            // turn back quickly to find wall again
18            rTurn TurnAmount*TurnDir
19            rForward 1
20        wend
21      endif
22    wend
23 Return
```

FIGURE 8.6 Staying close on sharp corners.

the wall. This is the purpose of the variable *SFN*, it is set to indicate which 90° sensor needs to be tested depending on the direction of the turn (Lines 04 and 07). Also notice in Line 18 the robot turns 5° at a time as set by the variable *TurnAmount* on Line 01 (you can experiment with this value).

If you test this new algorithm turning to the left (*TurnDir* = −1) the robot will work and turn around the corners a little tighter than before. However, if you try the program turning to the right (*TurnDir* = 1) the robot will stall.

8.3.2 FINDING THE PROBLEM

To determine why the robot stalls we will use the debugging feature in RobotBASIC. Add the following statement to Fig. 8.6 immediately before Line 13.

```
Debug rFeel()
```

Run the program and let the robot go to the point where it stalls. Press the button on the terminal screen when the robot stalls. Step the program a step at a time and note the values of the sensors at the stall point. Notice that the value returned by `rFeel()` is 5 (00101) meaning that the front and right sensors are triggered, but not the 45° sensor.

Since the code in Fig. 8.6 uses the 45° sensor to decide when to turn away from the wall it is easy to see why it fails to turn. The question now becomes "Why doesn't the program in Fig. 8.5 also fail?" Closer examination of the program in Fig. 8.5 provides the answer. In Fig. 8.5 the robot always moves forward 1 pixel each time it goes through the outer `while`-loop. This forward movement causes the 45° sensor to trigger and the robot turns as desired.

8.3.3 SOLVING THE PROBLEM

In the program of Fig. 8.6 the robot does not move forward because neither of the `rForward` commands will execute when `rFeel()` returns a value of 5. To solve the problem we need to remove Line 15 and place it before Line 13. This ensures that forwarding always occurs, just as before.

This is exactly the kind of error that creeps into a program as you develop an algorithm. It is important that you keep this fact in mind when debugging your programs. In fact, let's examine the program a little further.

Once we move the `rForward` statement, it becomes obvious that the `if`-statement is deciding when the `while`-loop will execute. The `while`-loop, however, already makes its own decision as to whether to loop or not and the decision used in our example is more restrictive than the one in the `if`-statement. This implies that the `if`-statement could have only checked for `rFeel()=0`. Since `rFeel()=0` is exactly the same as `NOT rFeel()`, it becomes apparent that the `if`-statement and its associated endif (Lines 13 and 21) are not needed and should be eliminated. Additionally, since the `if`-statement is removed there will no longer be a need for the variable *SFN*, so Lines 4 and 7 can also be removed. You should analyze the above discussion carefully because understanding it can help you become a better programmer.

As the tasks you want a robot to achieve become progressively more complex, situations will arise that are hard to predict. Without the debugging capabilities of a simulator many of these situations are nearly impossible to find. Often many programmers *randomly* change lines of code trying to resolve problems that have no apparent logical reasons for why they occur. This is surely not a good strategy for problem solving. The development cycle using a real robot is extremely unavailing for solving intricate problems. Without seeing the environment from the point of view of the robot it becomes very difficult to understand its behavior and program against problematic situations.

With a real-world robot it is often hard to test algorithms comprehensively due to the difficulty of constructing various environments that may cause problems. A robot that seems to work well in the limited testing environments will often fail when it is faced with an unanticipated situation. With a simulator you can construct random environments. You can construct numerous varieties of environments. You can test the robot inside these environments for many hours or days. This helps ensure that your algorithms have been well exercised and the possibilities of failure are reduced, but as we have seen above, the debugger cannot substitute for thorough reasoning.

8.4 A Different Approach

All of the algorithms so far have used the infrared sensors as the primary means of *feeling* for the wall. While these sensors are very effective and may be the only type of sensors available in a real-robot situation, they are lacking in two ways. First they leave gaps that can cause problems. You can use more sensors, placed closer together, but this can get expensive. The second problem is that the range of the sensors is not easily adjustable and you cannot control how far the robot stays away from the wall. You can adjust the power (or even the frequency in some cases) applied to the infrared emitters to make the detectors sense shorter or longer distances, but this requires more complex electronics.

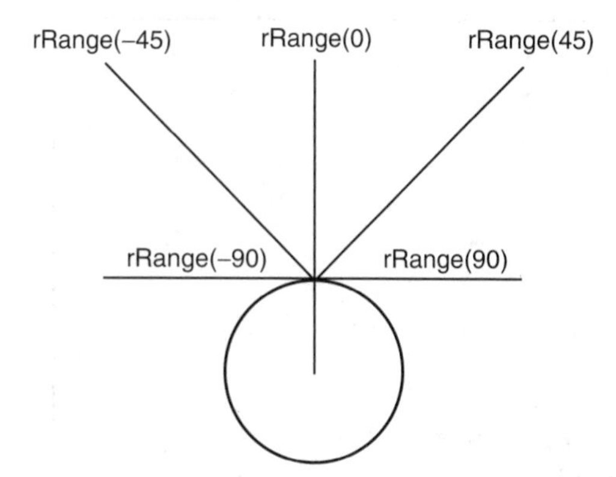

FIGURE 8.7 `rRange`(*Angle*).

Using a ranging sensor instead of infrared sensors can eliminate both of the above shortcomings. The `rRange()` function in RobotBASIC can be used in two ways. If you use `rRange()` without an argument, it assumes you want to measure the distance to objects directly in front of the robot. If you give `rRange()` an angle, however, it will measure the distance to objects at the angle you specify.

Figure 8.7 shows how the measurement of the distance is in reference to the front of the robot and how the angle pivots around the front of the robot. You can give the function any angle from +90° to −90°. The distance to an object is relative to the front of the robot along the direction given.

In real life such a sensor can be easily built by mounting a distance-measuring device on a stepper-motor (or servo-motor) turret so that the software can turn the ranging hardware without turning the entire robot. However, if you are going to measure only one or two angles then instead of a motor to turn the single sensor, it might be cheaper and more convenient to just mount one or two sensors at the angles required. If you need to adjust the sensors to tweak the robot's behavior just turn them by hand to the proper angles. The advantage of using the range sensor is that you can make the robot stay at a specified distance from the wall. Also you can use only one sensor and you do not need to worry about gaps since the ranger can be turned to any angle needed. In the algorithm developed below only three angles are necessary. It is a lot easier to maintain the robot parallel to the wall using a ranger than using infrared sensors.

The first action of the algorithm in Fig. 8.8 is to set some parameters that define the behavior and response of the robot. These parameters are crucial and you should experiment with different values to see how the robot may stay closer or further away from the wall. If the robot gets too close to the wall, however, it may still run the risk of crashing into real sharp corners despite the use of the `rBumper()` sensors. Also, when you get too close to the wall the robot seems to rely more on its bumpers to avoid the wall. This means that the robot would be scraping along the wall.

Lines 6 to 14 ensure that the robot is angled parallel to the wall in the direction as defined by the variable *TurnDir*. The robot will have the wall to its left (*TurnDir* = −1) or to its right (*TurnDir* = 1). Also, Lines 12 to 14 ensure that the ranger will be sensing the wall at the required value (*RangeLimit*).

We will follow a different strategy around corners than we have done before. We will conclude that the robot is at a corner if the ranger returns a value greater than a certain

```
01 FollowWall:
02    RangeLimit   = 30
03    NoWallLimit = 35
04    RangerAngle = TurnDir*90
05    while True
06       while rBumper()&4
07          rTurn -TurnDir
08       wend
09       rTurn -2*TurnDir
10       rForward 1
12       while rRange(RangerAngle) > RangeLimit
13          rTurn TurnDir
14       wend
15       while true
16         if rRange(RangerAngle) >= NoWallLimit
17            for WF_I = 1 to 45
18               if rBumper()&4 then break
19               rForward 1
20            next
21            rTurn TurnDir*90
22            for WF_I = 1 to 20
23               if rBumper()&4 then break
24               rForward 1
25            next
26         endif
27         while rRange() <=5
28           rTurn TurnDir
29         wend
30         while rRange(RangerAngle) < RangeLimit
31           rTurn -TurnDir
32         wend
33         while rRange(RangerAngle) > RangeLimit
34           rTurn TurnDir
35         wend
36         if not (rBumper()&4)
37           rForward 1
38         else
39           Break
40         endif
41       wend
42    wend
43 Return
```

FIGURE 8.8 This program follows a wall using only the `rRange()` sensor.

limit set by *NoWallLimit* (Line 16). If it is on a corner the robot will forward until it is clear of the corner (Lines 17–20) then turn 90° toward the wall (Line 21) then forward half a robot diameter (Lines 22–25). This ensures that the ranger is sensing the wall again. All the forwarding will only be done so long as there is no obstacle in the way. This allows for awkwardly turning walls (we shall see this later).

Lines 27 to 29 check for a wall ahead of the robot. Lines 30 to 32 turn the robot away from the wall if it is getting too close. Lines 33 to 35 turn the robot towards the wall if it is getting too far. Lines 36 to 40 forward the robot if there is no obstruction. If there is an obstruction we exit out of the inner `while-loop` so as to renegotiate the wall as if the robot was approaching it anew.

This algorithm will work with ease around the example wall considered so far. However, we shall test this algorithm by giving it a horrendously difficult environment.

If you look at Fig. 8.9, you will see various types of corners that we would like to consider for rigorous testing of the algorithm. Also there will be various combinations of tight valleys and protrusions to task the sensing and logic of the robot.

The replacement *DrawWall* subroutine given in Fig. 8.10 draws a very complex environment that should be used to test any wall following algorithm to see if it can cope with all the baffling situations that the robot may encounter.

Use this new wall with all the algorithms we have developed and see which ones perform best. Do not forget to try both right and left following. Some algorithms may be successful when following left but not when following right or vice versa.

Figure 8.11 shows the result of running the algorithm in Fig. 8.8 with the new wall. Notice that the robot stays close to the walls and visits all tight spots that can accommodate the size of the robot. Also note how well it stays close to sharp corners.

Try the algorithm in Fig. 8.8 with simple objects like circles, rectangles, and triangles. Vary the parameters and notice how easy it can be for the robot to follow these simple contours very closely when there are no hard corners to negotiate.

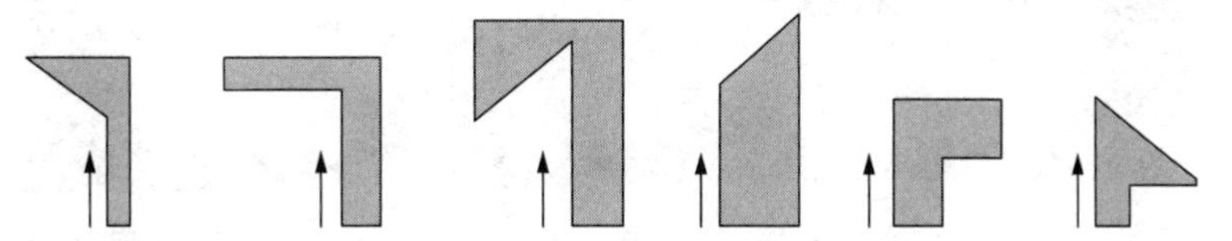

FIGURE 8.9 Various corner types. The arrows indicate the robot's travel direction.

```
DrawWall:
    ClearScr
    LineWidth 2
    Data Wall;-666, 555,  664, 493,  705, 490,  702, 461
    Data Wall; 667, 462,  668, 431,  701, 428,  699, 400
    Data Wall; 631, 402,  630, 358,  686, 359,  683, 289
    Data Wall; 651, 291,  644, 231,  693, 228,  693, 169
    Data Wall; 583, 166,  665, 101,  666,  35,  547,  30
    Data Wall; 461,  85,  466, 166,  404, 163,  383, 125
    Data Wall; 368, 162,  332, 159,  316, 205,  296, 160
    Data Wall; 266, 159,  222,  65,  167, 147,  101, 146
    Data Wall; 146, 254,   56, 211,   60, 316,   56, 366
    Data Wall; 202, 369,  202, 370,  203, 387,  202, 388
    Data Wall; 121, 385,  122, 402,  169, 402,  170, 411
    Data Wall; 125, 414,  125, 440,  255, 442,  317, 337
    Data Wall; 308, 434,  395, 401,  401, 310,  464, 315
    Data Wall; 457, 408,  456, 408,  456, 409,  286, 482
    Data Wall; 127, 475,  135, 551,  270, 558,  359, 520
    Data Wall; 391, 558,  479, 485,  524, 489,  540, 559
    Data Wall; 666, 556, -734, 580,   87, 570,   33, 371
    Data Wall;  32, 178,  185,  40,  517,  10,  691,  17
    Data Wall; 718, 162,  720, 392,  736, 579,  313, -95
    MPolygon Wall,Blue
Return
```

FIGURE 8.10 A hard wall to follow.

FIGURE 8.11 Hard wall-following with ranging sensor.

8.5 Summary

In this chapter you have:

- ❑ Learned how to make the robot follow a wall using a variety of algorithms and sensors.
- ❑ Seen the process of developing and debugging an algorithm.
- ❑ Discovered how unpredictable situations can arise and be difficult to analyze.
- ❑ Learned that programming a robot requires not only an understanding of the sensors but also the ability to deal with them using both logical and bitwise operators.
- ❑ Learned about the extended `rRange()` function.
- ❑ Seen further uses for arrays, `mPolygon`, and `Data` commands.

Now, try to do the exercises in the next section.

8.6 Exercises

1. Run the subroutines in this chapter to see how they perform. Try adding debugging statements to help you analyze the robot's behavior.
2. Try all the algorithms in this chapter with the wall given in Fig. 8.10. Can you determine why some algorithms fail? If an algorithm does not fail can you see why it may be considered better or worse than another that also did not fail?
3. In the algorithms of Figs. 8.8 and 8.6 (modified as discussed) try changing the parameters to see how they affect the robot's behavior. In Fig. 8.8 how would you make the robot stay closer to the wall? What would be the minimum number to give *RangeLimit*? Why?

HINT: The `rRange()` function returns a value relative to the front. The front is in line with the robot's center. So an `rRange(90)` value of 20 will mean that the object is almost touching the robot at its right side.

4. Draw circles, rectangles, and triangles and then test the algorithms with these types of simple objects. Which algorithms follow the contour faithfully (i.e., the robot draws a circle around a circle, a rectangle around a rectangle, etc.)? With the algorithm of Fig. 8.8 you can specify exactly how far the robot will be. Try following the objects at various distances (closer and further).

CHAPTER 9

AVOIDING DROP OFFS AND RESTRICTED AREAS

In previous chapters, our robot was programmed to roam anywhere it could. In this chapter we will discuss ways of confining the robot to a certain area (or out of an area) but without erecting walls, fences, or other barriers that protrude out of the ground. The robot should be able to move anywhere while avoiding obstacles but should not be allowed to move out of (or into) a specified area. There are various ways to delimit such a confinement for a robot:

- Draw lines around the boundary or use infrared beams, or bury an electric wire in the ground around the perimeter, or any other means where the robot can use *sensors* to know the limit has been reached and take action to stay away from the boundary.
- If the area is surrounded by a drop off, such as a table or the passageway of the second floor of a house, we give the robot the ability to detect a drop off to enable it to stay away from cliff edges.
- Give the robot a set of coordinates defining the prohibited/allowed areas and then have the robot use a GPS (global positioning systems) to calculate if it is allowed to enter/exit an area.

In this chapter we will develop some algorithms that simulate a means of confining the robot using the standard sensors that our robot has, but we will also explore the possibilities that more advanced sensors can provide.

9.1 Good Robot

Most people are familiar with the electronic fence systems that confine dogs to a yard. A perimeter is defined by some means (usually a radio signal) and a device on the dog's collar triggers some kind of pain to the dog (how cruel) that makes the dog shy away from the perimeter effectively confining the dog to the yard and house. We are not going to be so mean to our robot.

We will surround the desired perimeter with a red line and tell the robot not to move beyond the line. This line simulates a buried wire carrying a very low power radio frequency (RF) signal. A real robot would need a small radio receiver whose antenna is placed so that it can detect the wire. The simulated robot senses the line with the `rSense()` function we have explored in Chap. 7, simulating the ability to sense the RF signal. The details of the physical mechanism are not important to developing the algorithm for the behavior. Once the algorithm for teaching the robot the required behavior has been established, the actual mechanism for marking the boundary can be whatever is suitable. The aim is to explore how to make the robot recognize a boundary and stay within it.

9.1.1 AN INITIAL ALGORITHM

The program in Fig. 9.1 is an attempt at achieving the requirement discussed above. The idea is to use `rSense()` to detect when the robot is approaching the boundary and then turn away. The turning direction depends on which sensor combination is triggered:

- If the right sensor is triggered or the right and front together then turn left.
- If the left sensor is triggered or the left and front together then turn right.
- If all the sensors or just the front one are triggered, the robot turns around. To prevent the possibility of repeating a behavior endlessly, we will give the robot some randomness by turning 180° ± 30°.

Let the program run for a while. Does the robot behave?

```
MainProgram:
  linewidth 7
  rectangle 50,50,650,550,red
  rlocate 100,100,90
  rInvisible red
  goSub Confine
End
//=========================================================
Confine:
   while true
      while not rSense()
         rforward 1
      wend
      S = rSense()
      if S = 4 OR S = 6 then rTurn 1
      if S = 1 OR S = 3 then rTurn -1
      if S = 2 OR S = 7 then rTurn 150 + random(60)
   wend
Return
```

FIGURE 9.1 The robot does not behave quite right.

Notice the use of the line `S = rSense()`. This saves the sensors' condition in the variable *S*. We then use this variable to test for various situations. We could have written each `if`-statement like this:

```
if rSense() = 4 OR rSense() = 6 then rTurn 1
```

There are three `if`-statements where the sensory data is read twice. This means that if the style above is used, the function `rSense()` would be called six times. This is very wasteful and slow. The sensory data **is not changing** between executing each of the `if`-statements. If we save the value returned by `rSense()` in a variable there is no need to call the function more than once.

In real life the `rSense()` function would be implemented by the use of electronics and some way of communicating between the electronics and the microprocessor. The action of calling and interrogating the sensor is time consuming and battery utilization would be higher than needed. In this manner of calling the sensor only once (since its data is not changing), we save time and battery life.

Notice that `rSense()` is used without specifying a color. This means that the first color on the invisible colors list (red) will be redeemed as the line color for the sensors.

Notice the condition of the `if`-statements. We use `S = 4 OR S = 6`, this means that we are checking `if` the left sensor is on by itself *or* both the left and front sensors are on *simultaneously*. We cannot use `S & 6` because this condition would be true if *either* the left *or* the front sensor is on. This is not what we want. We want to turn if the left sensor is triggered by itself or if it is triggered along with the front sensor. So the condition `S = 4 OR S=6` is not the same as `S & 6`.

The algorithm in this program is not completely effective. The robot sometimes steps over the line and given enough time it will eventually escape completely from the confined area. This is due to a combination of hardware and software deficiencies.

The hardware is not sufficient for the situation, the three sensors of `rSense()` do not sense the line early enough to check if the robot is approaching the line at too shallow an angle. By the time the sensors trigger, part of the robot's body has already passed over the confining line.

The second problem is a software problem in combination with the hardware limitation. The random turn of $180° \pm 30°$ can cause the robot to turn in a direction that puts the sensors away from the line but also outside the area. This is due to the body of the robot being on top of the boundary line. When the robot turns it is possible for the front of the robot to be outside the boundary and so the sensors would be outside the boundary and unable to sense the boundary line. The problem is the robot will not know that it is facing outside the area and since the sensors are not triggered, it will happily go forward until it escapes the confinement area altogether.

If we are willing to accept that some of the body of the robot can go over the line, so long as the robot never escapes totally, can we adjust the algorithm to make the robot behave better?

9.1.2 IMPROVING THE ALGORITHM

We are willing to accept the robot going over the line a little but we do not want it to escape, ever. Replace the third `if`-statement in the program of Fig. 9.1 with this line and run the program to see what the new behavior is.

```
if S = 2 OR S = 7 then rTurn 170+random(20)
```

The difference is the amount of randomness in the turn around. The turn is now $180° \pm 10°$ instead of 30°. The reasoning is that since the sensors are 10° on either side of the front, if we turn the robot around and keep it within this limit (10°) of where it came from (if it is over the line) it will still sense the line and turn back in.

Another way the same effect can be achieved is by widening the line. If you change the number in the statement `LineWidth` 7 to 20 the boundary would become too wide for the robot's front to be outside the line when it turns the $180° \pm 30°$.

There is still a problem. The robot seems to spend most of its time around the perimeter straddling the line. This is predictable. If the turning around is almost 180° whenever it senses the line with all or just the front sensor, then it will almost always go back to where it came from, and if that happens to be from a long run while straddling the line until it comes to the corner, then it will just turn around and go back to the other corner.

This is not very effective behavior. The robot should cover more of the inside area, especially if it is supposed to be doing some useful work in that area. Therefore, even though we have solved the problem of escaping, we have not solved it in a satisfactory manner. The algorithm needs further refinement.

There is also another problem. If the robot approaches a corner at exactly 45° to it, then the left and right sensors would trigger together, but not the front one. `rSense()` would return a value of 5. Have we handled the value 5 in our algorithm? There is no code in the algorithm that handles this situation. What is the result of this? Since the sensors are triggered, no forwarding can occur, and since the situation is not handled, no turning occurs either. This means that the robot will stay where it is without turning or forwarding; it is stalled. How should this situation be handled? Can it be included in the third `if`-statement? The answer to these questions depends on what action we would like the robot to take in the situation.

9.1.3 A BETTER ALGORITHM

The program in Fig. 9.2 solves most of the problems mentioned above, except for the tendency of the robot to protrude outside the line limit. In the situation being simulated this is not a problem. It is tolerable that some of the robot's body exits the area as long as most of it stays confined.

The algorithm in Fig. 9.2 is very acceptable in that the robot does not get stuck in a corner, nor does it escape. The robot also covers the inside area more efficiently because it tends to spend more time inside the area than around the periphery.

The reason this algorithm works better than the previous one is the use of random numbers. The robot does not just forward, it also turns on occasion (10% of the time as set by the variable *F_RandomTurn_Percent*). This causes the robot to cover more of the inside area. Additionally when it is turning to avoid the line it turns more of an angle with an additional random value. This makes the robot turn away from the line at a greater angle, minimizing the possibility of approaching the line at too shallow an angle, which causes the problem of some of the body going outside the line due to hardware limitations.

The set of constants defined at the top of the program are very important. The value for *F_RandomTurn_Percent* defines what percent of time the robot will turn in addition to forwarding. The amount of turn is defined in *F_RandomTurn*. The constant *LineAvoidA* defines the amount the robot turns to avoid the line. In addition, a random amount is added

```
00 //-----Constants
01    F_RandomTurn            = 3
02    F_RandomTurn_Percent    = 10
03    LineAvoidA              = 3
04    LA_RandomTurn           = 3
05    TurnAroundA             = 60
06 //=========================================================
07 MainProgram:
08    linewidth 7
09    rectangle 50,50,650,550,red
10    rlocate 100,100,90
11    rInvisible red
12    GoSub Confine
13 End
14 //=========================================================
15 Confine:
16     while true
17        while not rSense()
18           rforward 1
19           if random(10000) < 10000%F_RandomTurn_Percent
20               rTurn random(F_RandomTurn)
21           endif
22        wend
23        S = rSense()
24        if S = 4 OR S = 6
25           C_Ta =  (LineAvoidA+random(LA_RandomTurn))
26           SN = 1
27        elseif S = 1 OR S = 3
28           C_Ta = -(LineAvoidA+random(LA_RandomTurn))
29           SN = 4
30        elseif S = 2 or S = 7
31           C_Ta = 180-TurnAroundA/2+random(TurnAroundA)
32           SN = 1
33        elseif S = 5
34           rTurn 180-TurnAroundA/2+random(TurnAroundA)
35           continue
36        endif
37        for i=0 to C_Ta
38           rTurn sign(C_Ta)
39           if rSense()& SN
40              rTurn 180
41              break
42           endif
43        next
44     wend
45 Return
46 //=========================================================
```

FIGURE 9.2 A better behaving robot.

to this, which is defined in *LA_RandomTurn*. The turn amount to avoid the line when it is approached head on is 180° ± an amount set by the value in *TurnAroundA*. Experiment with these values to see the effect on the effectiveness of the algorithm. These numbers can make the robot behave quite differently.

Lines 19 to 21 determine if the robot should turn in addition to forwarding. Turning occurs only *F_RandomTurn_Percent* of the time. The amount of turn is random up to *F_RandomTurn* degrees.

In Lines 24 to 32 the amount of turn is set in the variable *C_Ta* but no actual turning occurs until later. The amount of turn is a fixed amount plus an additional random amount as set by the variables.

On Lines 26, 29, and 32 the variable *SN* is set to indicate which sensor will be used to test in Line 39 (see later for explanation). If turning to the left is to be done then the left sensor will be used (*SN* = 4) and if turning to the right then the right sensor (*SN* = 1) will be used.

The actual turning is carried out by the `for`-loop (Lines 37–43). This loop turns the robot as indicated by the variable *C_Ta* a degree at a time. Notice the use of the function `Sign()` to make sure the turn direction is the correct one. The loop tests to see if the sensor in the direction of the turn ever touches the line. If it does then there is a possibility of the robot's front ending up outside the line. In this case the robot is made to turn 180° (turn around) and the loop is abandoned.

In Lines 33 to 36 the stalling situation discussed at the end of the previous section is handled. This situation does not cause the robot's front to turn outside the line and the actual turning is carried out on Line 34. On Line 35 a `Continue` statement is used to make the program flow go back to the top of the `while`-loop and avoid going into Lines 37 to 43, which are only needed in the other situations.

In the next section we will see what we can do to avoid the robot's body protruding outside the line as much as it does in this section.

9.2 Cliff Hanger

In the previous section we developed an acceptable algorithm that was adequate for the situation it was designed to tackle. However, the algorithm did allow a portion of the robot's body to briefly move outside the confinement area. This would be unacceptable if the area was a table for instance, or a landing on the second floor. The robot would just topple over the moment its center of gravity went over the table. Even if its center of gravity remained on the table, if one of its wheels goes beyond the edge of the table it would cease creating traction while the other wheel would continue doing so, which would turn the robot toward the ledge eventually causing it to fall over the edge.

In the table top situation it would not be acceptable for the robot to have more than a very small fraction of its body protrude over the edge of the table (never any of its wheels). In this section we will explore how to achieve this. Obstacle avoidance will also be added to the behavior so the robot can avoid obstacles at the same time that it is avoiding the edge of the table or landing.

To detect an edge in a real life situation there are many sensors that can detect a change in light intensity or color variation, which indicates that there is a change in the surface that is being sensed. This can tell us that the floor is dropping. Such sensors require a constant color and reflectivity surface. Another method is to have a sonar or infrared sensor that can measure the distance to the floor. An increase in the distance indicates that the floor is dropping away. The problem with this method is that many of the available sensors do not measure small distances accurately, which is necessary for this application.

This simulation will make use of specialized ground sensors. RobotBASIC has downward-facing color sensors accessed with the function `rGroundA()` that returns the color the sensors are sensing. With these sensors you can look at the ground around the robot's perimeter and see what color it is. The function enables you to create almost any combination of sensors you wish around the perimeter of the robot. We are going to create 5. Read about this function in Sec. C.9 before proceeding.

We are going to set up a sensor at 90° to the left, another at 45° to the left, one on the front and one on the right at 45° and a final one at 90° to the right. We will write code to make the sensors return a value of 0 if the sensor is seeing the table and a value of 1 if it is not (i.e., over the edge).

It is necessary to be familiar with binary numbers and bitwise operators to appreciate the actions of the program being developed here. Review Chaps. 3, 4, and 5 if you need a refresher in these concepts.

As you have seen in Chap. 3, the `rFeel()` function returns a number where the first five bits from the right represent the state of each of the five infrared sensors on the robot. We are going to set up the special ground sensors to return a number in exactly the same manner. To summarize:

- Each sensor looks at the ground just at the perimeter of the robot and is redeemed to be on if there is no table underneath it (returns a color not equal to the table color).
- There are five sensors 90° and 45° to the left and right and one at the front.
- The sensors are combined in one number that ranges from 0 to 31, where the most significant bit (MSB) is 90° degrees to the left. The least significant bit (LSB) is 90° to the right.
- The number 14 (01110) means that the sensors at the front and at 45° to the left and right **do not** sense a table underneath them, while the others are still sensing the table.

The subroutine *TestSensors* in the listing in Fig. 9.3 (Lines 52–59) shows how this is achieved. Notice the use of the | (bOR) operator to create the final number. Also notice how the `for`-loop creates the five sensors at the correct angle. The subroutine will set up the variable *Sensors* to have the required value.

So in place of using `rSense()` we call the subroutine *TestSensors*. Then we use the same kind of checking on the variable *Sensors* as described in the previous section.

To allow the sensors to work correctly we will make the table white and the floor gray. But we will need to issue an `rInvisible Gray` command as you will see on Line 12. This is due to the fact that the robot will think that the gray color is an obstacle (see Sec. C.9 for a detailed discussion on the Robot color arrangement). If the gray color is set as an invisible color the robot will not consider it an obstacle and will be able to move into that area (dropping off the table). This way we would not be cheating in the simulation.

Review the program in Fig. 9.3 and run it before proceeding with the following discussion. The program's main routine sets up the environment (*SetTable*), locates the robot, sets the required color parameters, and then calls the *TableRoam* subroutine (Lines 17–45) which will be executed continuously due to the `while`-loop.

The *TableRoam* subroutine is not very different from the *Confine* subroutine in Fig. 9.2. The action is very similar except for using *Sensors* [not `rSense()`] and the added obstacle avoidance. The `elseif`-statements (Lines 32 and 36) reflect the fact that there

```
01 //-----Constants
02   TableColor            = White
03   F_RandomTurn          = 3
04   F_RandomTurn_Percent  = 10
05   LineAvoidA            = 20
06   LA_RandomTurn         = 30
07   TurnAroundA           = 60
08 //=====================================================
09 MainProgram:
10   goSub SetTable
11   rlocate 100,100,90
12   rInvisible Gray  //---no cheating
13   goSub TableRoam
14 End
15 //=====================================================
16 //--  Roam On Table for ever
17 TableRoam:
18   while true
19     goSub TestSensors
20     S = Sensors
21     if not S
22       if not(rBumper() & 4)
23         //---only forward if no obstacles
24         rForward 1
25         if random(10000) < 10000%F_RandomTurn_Percent
26           //--add a bit of turning for better coverage
27           rTurn random(F_RandomTurn)
28         endif
29       else
30         gosub Reverse  //--avoid obstacle
31       endif
32     elseif ((S & 24) AND not(S & 3))
33       //-- only left sensors and not right sensors
34       //-- don't care about front sensor
35       rTurn LineAvoidA+random(LA_RandomTurn)
36     elseif ((S & 3) AND not(S & 24))
37       //-- only right sensors and not left sensors
38       //-- don't care about front sensor
39       rTurn -(LineAvoidA+random(LA_RandomTurn))
40     else
41        //---catch all other conditions
42        rTurn 180-TurnAroundA/2+random(TurnAroundA)
43     endif
44   wend
45 Return
50 //=====================================================
51 //-- Creates 5 ground sensors
52 TestSensors:
53   Sensors = 0
54   for i = 0 to 4
55      if rGroundA(90-i*45) <> TableColor
56         Sensors = Sensors | (2^i)
57      endif
58   next
59 Return
60 //=====================================================
61 //-- Makes a random reverse and turn
62 Reverse:
```

FIGURE 9.3 Effective table roaming.

```
63    for i = 1 to (random(10)+10)
64      //--reverse so long no object on the
65      //--back and only max 20 pixels
66      if rBumper() & 1 then break
68      rForward -1
69    next
70    //--turn min 45 degrees max 90
71    //-- in a random direction
72    if random(1000) >= 500
73      rTurn 90-random(45)
74    else
75      rTurn -90+random(45)
76    endif
77 Return
78 //=====================================================
79 //--Sets the table environment
80 SetTable:
81    rectangle 0,0,800,600,gray,gray
82    rectangle 50,50,650,550,TableColor,TableColor
83    circle 200,200,250,250,black,black
84    circle 400,400,450,450,blue,blue
85    rectangle 250,400,300,480,magenta,magenta
86    rectangle 450,300,480,410,cyan,cyan
87 Return
```

FIGURE 9.3 *(Continued)*

are now five sensors and thus the numbers being compared are a little different. Here is the logic:

- Turn right to avoid the edge if either of the left sensors reports a drop off but not any of the right sensors. We do not care about the front sensor, it can be on or off, no matter if we still want to turn right.
- Turn left to avoid the edge if either of the right sensors reports a drop off but not any of the left sensors. We do not care about the front sensor, it can be on or off, no matter if we still want to turn left.
- If neither of the above conditions is true, but there are sensors that are detecting an edge in a combination that has not been covered above, we will execute an about face. The amount of turning will be 180° ± an amount that is set by the parameters at the top of the program.

If you are familiar with other programming languages, the construct used in this subroutine is similar to a `Case` statement, which we do not have in RobotBASIC but can be easily emulated with the `if-elseif-else-endif` construct (see Sec. C.6 for more details on emulating the `Case` construct).

The algorithm implements obstacle avoidance with `rBumper()`, if there are no obstacles it moves forward and turns in the same manner as described in the previous section. If there is an obstacle it calls the *Reverse* subroutine (Lines 62–77). This subroutine causes the robot to backup a distance specified by a random number between 20 and 10 pixels but only if there is no object behind it. After backing up, the robot turns a random amount in a random direction between 45° and 90°.

Run the program and notice how the robot never protrudes over the table edge. It covers the table quite effectively, roaming around almost everywhere while avoiding obstacles in a well behaved manner. Remember not to type the line numbers.

9.3 GPS Confinement

In this section we will explore a method that uses the GPS feature on the robot to confine it to a particular area. Some algebra will be necessary to develop this algorithm. The principles are not complex and all the calculations will be explained in detail.

The GPS system on the simulated robot is accurate to a pixel. Assuming that the robot is 40 pixels in diameter, simulating a real robot of about 1 ft in diameter, then one pixel would be 0.6 in. This means that the simulated GPS system is far more accurate than is possible with a real world GPS. The resolution of a GPS affects the accuracy with which an algorithm can control a robot. The algorithm developed below is not affected greatly by this fact, however, there are solutions for the problem.

The U.S. GPS system relies on satellites in geosynchronous orbits around the earth that send microwave signals to any terrestrial receiver that knows how to read the encoded data. The receivers use the data from three or more satellites to triangulate a position on the surface of earth. This process is accurate to hundreds of feet (more or less depending on the receiver). This means that if you use this system you would be limited by this resolution.

One way to increase accuracy is to use an augmentation system that supplements the satellite data with local position references that enable some receivers to be accurate down to feet. Such systems are used by some airports to enable aircrafts to land precisely on the centerline of a runway without the pilot ever seeing the runway. These systems can be expensive and may still not be as precise as might be needed in some situations.

Another way to have a very accurate PS (positioning system) is to create one that implements an LPS (local PS). This system uses the same principles as the GPS but uses *local* radio transmitters and receivers to calculate positions in reference to a *local grid*. These systems can be accurate down to inches or centimeters and are not overly costly and can be used effectively over a wide area.

The following algorithm will concentrate on developing the principles of using a PS and how to utilize the information to control the robot. Later chapters will explore solutions around accuracy and even how to eliminate the need for a PS altogether.

9.3.1 THE SPECIFICATIONS

We will develop an algorithm that allows the robot to roam around an area as in Fig. 9.3, but we will use the GPS to obtain the robot's position and use the information in combination with an array of coordinates that delimit the border of the area in which the robot must stay. The algorithm will determine if the robot is about to cross any of the imaginary borders and turn away.

To achieve this goal, we have to specify a few things:

- We will assume the array of coordinates defines a fully enclosed area if all of the points are connected together with lines. The coordinate points are consecutive points. There is a line from the first point to the second, then from the second to the third and so on.

- The GPS reports the robot's center, so we will allow for the robot's radius in calculating the condition of crossing the boundary line (with a buffer zone).
- The boundary lines are imaginary, but we would like to have some indication of where they are for display purposes. Therefore, we will draw the lines on the screen to show where they are. We will designate these lines as invisible using the `rInvisible` command. This prevents the robot from seeing the imaginary drawn boundary as an obstacle. The algorithm will not use these lines in any other way to ascertain the boundary position; only the array of coordinates will designate the boundaries.
- We will have obstacles in the area and therefore we will incorporate obstacle avoidance in the algorithm.
- To simplify the math, we will not consider in what way the robot is about to cross the boundary. We will design an algorithm to make the robot think that the boundary lines are walls. The robot will avoid these virtual walls as if they were obstacles.
- The above means that the algorithm for obstacle avoidance and boundary avoidance are the same (subroutine *Reverse* of Fig. 9.3). However, the way of determining the boundary crossing (as opposed to object detection) is different and will rely on the GPS to determine if the robot is about to violate the boundary limits. The secret for the new algorithm will be the subroutine *TestViolation* and the array of coordinates (we will call it *Boundary*).

Figure 9.4 is the first part of the program to be discussed below. Figure 9.6 holds the second part, which will be discussed later. To run the program you need to combine Figs. 9.4 and 9.6.

9.3.2 MAIN PROGRAM

The main program (Fig. 9.4) calls the subroutine *DrawBoundary* (Line 10) to allow us to see the boundary, but, as you will see later, this subroutine also sets up the boundary array. We designate the *BoundaryColor* constant (Line 13) as an invisible color so as to prevent the robot from seeing the border as an obstacle. If you do not do so then the robot will avoid the boundary as an obstacle and we would not be testing our GPS avoidance algorithm (cheating).

The *SetEnvironment* subroutine is exactly the same as the one you have seen in the previous section.

9.3.3 *RoamAround* SUBROUTINE

This subroutine in Fig. 9.4 is very similar to the one you have seen in Fig. 9.3. The differences are in:

- Lines 21 to 22, where we call the *TestViolation* subroutine. We will discuss this routine later but for now just accept that it sets the variable *Violation* as *true* if the robot is about to cross the virtual boundary.
- Line 36 where instead of using `if`-statements we just call the *Reverse* routine as if the boundary was an obstacle as we have done in previous examples. This is due to a limitation of the *TestViolation* subroutine that will be discussed later. The *Reverse* subroutine is exactly the same as before.

```
01 //-----Constants
02   F_RandomTurn            = 3
03   F_RandomTurn_Percent    = 5
04   LA_RandomTurn           = 30
05   LineAvoidA              = 20
06   TurnAround              = 60
07   BoundaryColor           = Gray
08 //===========================================================
09 MainProgram:
10   goSub DrawBoundary
11   goSub SetEnvironment
12   rlocate 330,200
13   rInvisible BoundaryColor  //---no cheating
14   goSub RoamAround
15 End
18 //===========================================================
19 RoamAround:
20  while true
21   goSub TestViolation
22   S = Violation
23   if not S
24     if not(rBumper() & 4)
25       //---only forward if no obstacles
26       rForward 1
27       if random(10000) < 10000%F_RandomTurn_Percent
28         //--add a bit of turning for better coverage
29         rTurn random(F_RandomTurn)
30       endif
31     else
32       gosub Reverse  //--avoid obstacle
33     endif
34     continue  //--go back to top of loop
35   endif
36   gosub Reverse
37  wend
38 Return
39 //===========================================================
40 //-- Makes a random reverse and turn
41 //===========================================================
42 Reverse:
43   for i = 1 to (random(10)+10)
44     //--reverse so long no object on the
45     //--back and only max 20 pixels
46     if rBumper() & 1 then break
47     rForward -1
48   next
49   //--turn min 45 degrees max 90
50   //-- in a random direction
51   if random(1000) >= 500
52     rTurn 90-random(45)
53   else
54     rTurn -90+random(45)
55   endif
56 Return
57 //===========================================================
58 //--Sets the environment
59 //===========================================================
```

FIGURE 9.4 GPS confinement part I.

```
60 SetEnvironment:
61   circle 200,200,250,250,black,black
62   circle 400,400,450,450,blue,blue
63   rectangle 250,400,300,480,magenta,magenta
64   rectangle 450,300,480,410,cyan,cyan
65 return
66 //=======================================================
67 //-- define & draw the boundary
68 //=======================================================
69 DrawBoundary:
70       LineWidth 1
71       setcolor BoundaryColor
72       Data Boundary; -89,  77,  255,  34,  685,  64
73       Data Boundary; 766, 210,  665, 333,  721, 451
74       Data Boundary; 740, 565,  341, 566,   28, 495
75       Data Boundary; 125, 313,   39, 242,  105, 194
76       Data Boundary;  46, 111,   89,  77
77       MPolygon Boundary
78       setcolor black
79       Boundary [0] = -Boundary[0]
80 Return
81 //=======================================================
```

FIGURE 9.4 *(Continued)*

9.3.4 *DrawBoundary* SUBROUTINE

This subroutine (Fig. 9.4) carries out two actions. It draws the boundary lines so that they can be seen for the purpose of observing the robot as it avoids them. The boundary lines are there only to help *us* see what is happening. The robot cannot see them, since they are designated as an invisible color (Line 13).

This routine also creates the boundary array called *Boundary* using the `Data` command (Sec. C.7). The array will contain a set of number pairs. Each pair is a coordinate (x, y) of a boundary corner. These points, when connected with lines, define the boundary to be avoided. The points are considered to be a sequence, where there is a line between the first and second point, and between the second and third and so on. The shape can be any shape you desire.

The first point in the array must have a negative x value because the `mPolygon` command is used to draw the connecting lines. See Sec. C.7 for details on this command. This negative value must be changed back to positive so it can be used correctly in the calculations performed later in the program. This is the purpose of Line 79.

To have a closed boundary, the last point in the array has to be the same coordinates as the first point. This ensures that a closed boundary is created. The algorithm could have eliminated the need for respecifying the first point and assume a closed polygon, but this way is more versatile since you may need to specify an open boundary. This is an example of a design consideration. Either the code assumes things and thus it will be less versatile but the user has less work, or the user has more work and the code assumes less things but is more general and can be applied to more situations. Here we opt for the latter policy.

9.3.5 *TestViolation* SUBROUTINE

This subroutine (Fig. 9.6) is the secret of the algorithm. This subroutine checks to see if the robot is about to cross a boundary line and sets the variable *Violation* to *true* if that is the case or *false* if it is not.

To simplify the math we will not calculate *where* on the robot's body the collision with the boundary line occurs. This limitation means that we will not have the ability (as in Figs. 9.3 or 9.2) of knowing whether to turn left or right for more intelligent boundary avoidance. But we can consider the boundary line as if it is a wall and avoid it like we would avoid an obstacle.

The algorithm is not perfect but it is a good balance between complexity and effectiveness. The algorithm is effective in that the robot never exits the boundary and is reasonably fast.

The principle is easy to understand even though the implementation is a little complex. If you are not familiar with the mathematical concepts, do not worry. You can read through to get an idea about the actions that the subroutine accomplishes. You can always use a tool if you know *what* it does without necessarily knowing *how* it does it. Just think of this subroutine as a *tool* that you can use. You don't need to know the details of how it works as long as you are able to utilize it when you need to.

The basis for this routine is the fact that the robot's perimeter is a circle and we test each boundary line for intersection with this circle. A line can intersect a circle either at two points or one point (tangent to the circle). If no boundary lines intersect the extended circle around the robot then there is no violation and *Violation* is set to *false*. If any of the lines intersect (at either one or two points) *Violation* is set to *true*. The intersection is not a physical one since the lines do not actually exist. The intersections are mathematically calculated as we show below.

In order to implement this idea we need to know about quadratic equations, slopes, intercepts, and circle formulas. The equation of a line is

$$Y = mX + b$$

where m is the slope of the line and b is the y-axis intercept. The equation of a circle is

$$(X - C_x)^2 + (Y - C_y)^2 = R^2$$

where C_x and C_y are the coordinates of the circle's center and R is its radius.

When we calculate algebraically the points where a line intersects a circle we get a *quadratic equation*. This equation either has no solution, which means that the line cannot intersect the circle (Line A in Fig. 9.5a), one solution, which means the line just

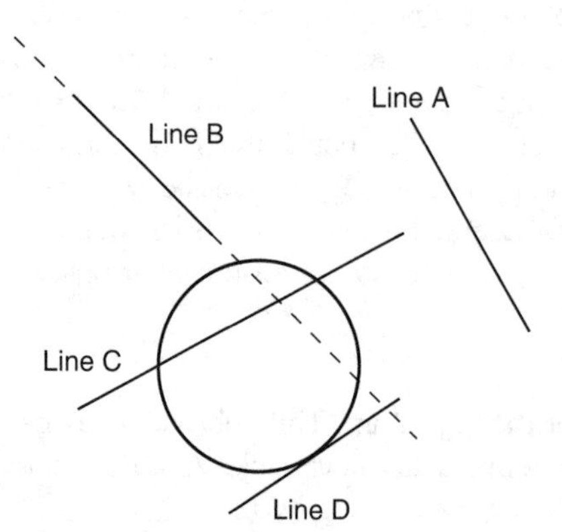

FIGURE 9.5a Circle-line intersection.

touches the circle (Line D), or two solutions, which means the line intersects the circle fully (Line B or C).

The equation is derived as follows: a point that is an intercept between a circle and a line must satisfy both the line and the circle equation. We know from the line equation that $Y = mX + b$. So we substitute the value of Y in the circle equation by $mX + b$. So we now have

$$(X - C_x)^2 + (mX + b - C_y)^2 = R^2$$

This equation can be manipulated to give the following:

$$(1 + m^2)X^2 + [2m(b - C_y) - 2C_x]X + [C_x^2 + (b - C_y)^2 - R^2] = 0$$

This is a quadratic equation like $AX^2 + BX + C = 0$ where the solution is

$$X = \frac{-B \pm \sqrt{B^2 - 4AC}}{2A}$$

In the above equation we substitute

A with $1 + m^2$
B with $2m(b - C_y) - 2C_x$
C with $C_x^2 + (b - C_y)^2 - R^2$

If $B^2 - 4AC$ is greater than or equal to zero then there is a solution. The line touches the circle if it is zero (Line D) or fully intersects the circle at two points (Line B or C) if it is a positive number.

In the above calculations the circle's radius and center coordinates will be the robot's radius (plus a buffer) and GPS coordinates. The line's equation will be calculated from the coordinates that define each line in the border as follows:

The two points that define each line-segment at the border are X_1, Y_1 and X_2, Y_2. We calculate the slope of the line $m = (Y_2 - Y_1)/(X_2 - X_1)$ and the intercept $b = Y_1 - mX_1$. So now we have the equation for the line and can do the calculations to determine if any boundary line-segment intersects the extended circle around the robot. However there is a problem.

If you observe Fig. 9.5b you will see that it is possible to have some boundary lines *appear* to be crossing the robot's path.

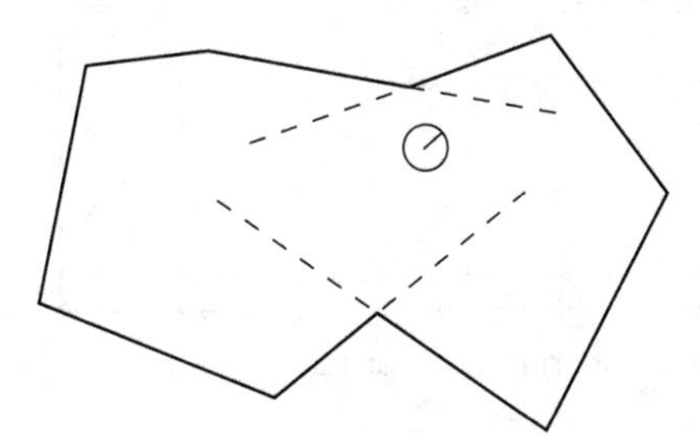

FIGURE 9.5b The robot is about to approach the extended borderline.

A line defined by the previous equation has infinite length. When we derived the equation for the boundary lines we did so to enable the above calculation, however, even though the infinite line may intersect the circle (Line B in Fig. 9.5a and also as in Fig. 9.5b), the segment that we are interested in (the actual boundary line) may not do so.

It is necessary to check further to see if the intercept between the line and the circle is actually within the segment being considered. This is accomplished by checking to see if the calculated intercept x-coordinates are within the x-coordinates of the two points that define the line-segment. If both calculated intercept x-coordinates are outside the boundary line-segment then the boundary line does not intersect the circle even though its infinite extension does, as is about to occur in Fig. 9.5b (see the Line 47 in Fig. 9.6).

Figure 9.6 shows how all of the above can be implemented in code. The math can be a little complex, but once the final equations have been developed, the implementation is fairly straightforward.

```
15 //=========================================================
18 TestViolation:
19    Violation = false
20    rGPS Rx,Ry
22    for j = 1 to MaxDim(Boundary,1)/2 -1 //--for each line
24      mm = (j-1)*2
25      X1 = Boundary[mm]
26      Y1 = Boundary[mm+1]
27      X2 = Boundary[mm+2]
28      Y2 = Boundary[mm+3]
29      If j=1 then X1 = -X1 //due to MPolygon specs
30      //--Line formula Y = mX+b
31      m = 1.0*(Y2-Y1)/(X2-X1)
32      b = Y1-m*X1
33      //--quadratic X=(-B+Sqrt(B*B-4*A*c))/2/A
34      //--            X=(-B-Sqrt(B*B-4*A*c))/2/A
35      //--if there is a solution then the
36      //--circle and line intersect (possibly)
37      A = 1+m^2
38      B = 2*m*(b-Ry)-2*Rx
39      C = Rx^2+(b-Ry)^2-625 //625=(20+5)^2
40      BB = B^2-4*A*C
41      if (BB) >= 0
42        BB = Sqrt(BB)
43        XX1 = (-B+BB)/2/A  //--first intercept
44        XX2 = (-B-BB)/2/A  //--second intercept
45        //--check if intercept is actual
46        //--not on extended line
47        if Within(XX1,X1,X2) OR Within(XX2,X1,X2)
48          Violation  = true
49          break //no need to check any more boundary lines
50        endif
51      endif
52    next
53 Return
54 //=========================================================
```

FIGURE 9.6 GPS confinement part II (line numbering starts at 15 there is no missing code).

- Lines 25 to 28 are where we obtain the *x*, *y* coordinates of both ends of the line (this will be done for each line that defines the boundary).
- Lines 31 to 32 are where we calculate the values *m* and *b*.
- Lines 37 to 39 are where we set the quadratic equation coefficients that will be used to solve the formula. Notice (Line 39) that the value 625 = 25^2. 25 is the robot's radius (20) plus an added buffer of 5. We write 625 directly so as to save time. This algorithm is very calculations intensive and any time saving we can do (such as this one) would help make the algorithm faster. This is an example of compromise between efficiency and good programming practice. In good practice we ought to have set a constant for the robot's radius (e.g., *R_Radius* = 20) and a constant for the buffer (e.g., *Buffer* = 5) then we would write Line 39 as:

  C = Rx^2 + (b −Ry)^2−(R_Radius + Buffer)^2

  Unfortunately this requires a lot more time to execute than just writing the number as a literal. Also since this line of code will be executed many times (for each boundary-line and every time the robot moves) the time penalty would be high.
- Line 41 determines if the formula has a solution. If it does, the program enters the inside of the `if`-block (Lines 42–55). If it does not, the next boundary is considered (using the `for-next` loop).
- Lines 42 to 44 are where the equation is solved to obtain the *x*-coordinate of the two possible intercept points (they could be equal which means the same point i.e., one point).
- Lines 47 to 50 are where we check to see if the intercept points are within the boundary line-segment and not just part of the extended infinite line. If either point is, then there is a violation and we abort any further checking by exiting out of the `for-next` loop using the `Break` statement. Otherwise we proceed with checking the other lines [see Sec. C.8 for details on the `Within()` function].

Combine the code from Figs. 9.4 and 9.6 (without the line numbers) then run the program. Remember, the robot does not see the boundary lines except through the mathematics described in this chapter. It may appear that the robot is avoiding the boundary as if it was an object but this is only because we made boundary avoidance the same as obstacle avoidance. Convince yourself of this by commenting out Line 77 in Fig. 9.4. This prevents the boundary plotting and you will see the robot avoiding an invisible wall.

9.4 Summary

In this chapter we have explored:

- ❑ Various methods for keeping the robot within or outside of an area.
- ❑ The advanced ground sensor capability of the robot.
- ❑ How to use the GPS to confine the robot within an area.
- ❑ Mathematical concepts for calculations involving circles and lines.

Now, try to do the exercises in the next section. If you have difficulty read the hints.

9.5 Exercises

1. Add obstacle avoidance to the program in Fig. 9.2, include obstacles and see the results.

HINT: Study the routines in Fig. 9.3; you can use the same logic.

2. Write an algorithm that enables the robot to roam around a table while avoiding obstacles (as in Fig. 9.3), but also not entering a zone on the table designated by a border as in the program of Fig. 9.2.

HINT: Try combining the two algorithms of Figs. 9.2 and 9.3.

3. Change the data in the *DrawBoundary* subroutine to create different boundaries. Does the robot always stay inside? Does it ever exit the area? What if you try a set of data where there is a break in the virtual wall?
4. Line 31 in Fig. 9.6 has a formula that says: $m = 1.0*(Y_2-Y_1)/(X_2-X_1)$, why not just say: $m = (Y_2-Y_1)/(X_2-X_1)$? What is the reason for using 1.0 in the formula?

HINT: See Secs. B.7.1 and B.7.5 for details of on integer and floating-point numbers and operations on these numbers.

5. In Fig. 9.6, comment out Line 47 and Lines 49 to 50. Now run the program and see what happens. Can you explain the reason for the robot's behavior?

HINT: It seems to be avoiding boundaries that are not there. Why?

6. The subroutine *TestViolation* does not consider where the boundary line touches or crosses the robot's perimeter. If it had done so we could have used a better avoidance mechanism such as the one in Fig. 9.3. Can you modify the subroutine to give the *Violation* a value that indicates the place where the robot's body is touching the boundary line (instead of just *true* or *false*)?

HINT: Calculate the angle an intercept point makes in relation to the robot's center point and its center line (left [−] and right [+]) and set *Violation* to that value [you can use `PolarA()` with additional calculations]. However, this will slow the program appreciably.

CHAPTER 10

VECTOR GRAPHICS ROBOT

The robot in RobotBASIC has a feature that facilitates many possibilities for innovation. This feature is a pen at its center that can be lowered to allow the robot to leave a trace on the floor as it moves (you have seen some uses of this feature in previous chapters). With the proper program we can convert the robot into a device for drawing vector graphics (meaning we specify a line to be drawn with a starting point, an angle, and a length).

This feature can be used for a variety of projects:

- Draw and write on the floor.
- Show area covered by the robot for a sweeper simulation.
- Display the effectiveness of an algorithm by showing the path taken.
- Solve mazes by leaving a *breadcrumbs* trail.

In this chapter we will develop a few applications utilizing this feature to draw and write on the floor. In Chap. 8 we used the pen feature to observe the effectiveness of the algorithms and in Chap. 4 we used it to observe the trajectory of the robot and could have used the robot as a remote controlled sketching tool. The other options will be explored in subsequent chapters.

There are many industrial applications for a robot that can draw on a surface. A robot could cut intricate and complex designs out of metal sheets if a laser cutter is substituted for the pen. Imagine a robot that can draw the yard lines and other messages and designs on a football field once the desired data has been given to it.

10.1 DrawBot

The command to lower and raise the pen is:

`rPen` *ExprN1 {,ExprN2}*

If *ExprN1* is 0, the pen will be raised and, if it is any number other than 0 the pen will be lowered. You can also use the constants `Up` and `Down` (see Sec. B.7.6). *ExprN1* has to be a number. The pen is up when you first initialize and `rLocate` the robot. *ExprN2* is optional. If it is specified then the color of the pen will be set to *ExprN2*. If it is not specified then the color of the pen will be set to the first color on the invisible colors list. If you have not specified an invisible colors list then the pen color will be black. *ExprN2* should be a valid color number (see Sec. B.7.6).

If you don't place the pen's color in the list of invisible colors the robot will crash if it ever encounters its own trace. You must specify any colors you are likely to draw on the floor with the pen as invisible colors so the robot will be able to drive over them. You can reissue the `rPen` command with a different color to draw as many colors as you desire.

The command `LineWidth` will set the pen's width. This command was discussed in Chaps. 5 and 7. If you are unfamiliar with these commands refer to Secs. C.7 and C.9 for details.

The pen is exactly at the center of the robot and, when lowered, will draw a line trailing behind the robot whenever an `rForward` command causes the robot to move. Obviously, since the pen is at the center, an `rTurn` will not create any trail, but as you turn and forward the trail will display the path the robot has taken.

Let us see how these commands can be used to make the robot draw a square on the floor. Type the program in Fig. 10.2 and run it. You will observe the result shown in Fig. 10.1. As you can see, it is very simple to make the robot draw. The robot can be made to behave like a vector plotter. You may not have seen these devices before, but they are used in many engineering offices. They can draw using a pen and instructions to move to specified *X, Y* coordinates on a flat surface and to lower or raise pens of various colors. These devices are aptly called xy-plotters. Our robot can behave as an xy-plotter with programs like the one in Fig. 10.2.

10.1.1 DRAWING CIRCLES

The program in Fig. 10.3 makes the robot draw a circle. Type it and see the result.

The subroutine *DrawCircle* is what makes the circle. Notice the line `rForward` *fStep*. Rather than forwarding a fixed distance, the subroutine forwards a distance defined by the variable *fStep*. This variable must be assigned a value prior to calling the subroutine. The same thing is done in the line `rTurn` *tStep* and in the `for-next` loop. Here the rate of turn is specified by *tStep* and again has to be assigned prior to calling the routine. The limit of the loop is 360/*tStep* rather than 360 so that if the rate of turn is changed the routine still continues to turn 360° and no more. What will happen if *tStep* is set to −1?

Experiment with changing these numbers and try to predict the results. First try changing only *fStep* and see the outcome. Then only change *tStep* and observe the action.

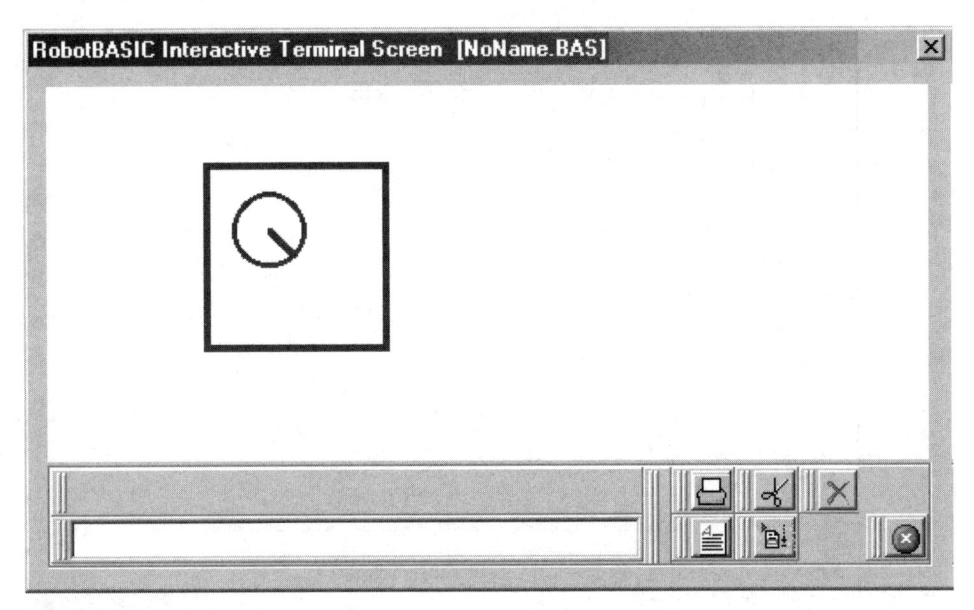

FIGURE 10.1 The robot can draw.

```
//---Pen test
rLocate 300,300,90   //--Initialize the robot facing East
rInvisible Magenta   //--magenta color for the pen
LineWidth 4          //--line width of 4 is good here
rPen Down            //--start drawing
for i = 1 to 4       //--draw a square (4 sides)
  rForward 100
  rturn 90
next
rPen Up              //--stop drawing
rTurn 45             //--move out of the way
rForward 50
End
```

FIGURE 10.2 Program using the pen to draw a square as shown above.

Finally try changing both. You may have to change the initial position of the robot. Also experiment with making the numbers negative.

10.1.2 DRAWING RECTANGLES

The program in Fig. 10.4 causes the robot to draw rectangles. Type it and observe the results. As in the previous section, there are two variables *RectWidth* and *RectHeight* that define the shape of the rectangle. Additionally, the initial position and heading of the robot determine the orientation of the rectangle. Experiment with changing these variables and observe the results.

```
//---Draw Circles
MainProgram:
  //---Change these values to change the position of the circle
  R_Init_X = 100
  R_Init_Y = 300
  R_Init_Heading = 0

  rLocate R_Init_X,R_Init_Y,R_Init_Heading
  rInvisible Magenta
  LineWidth 4

  //---Change these values to change the size of the circle
  fStep = 1
  tStep = 1

  gosub DrawCircle

  rTurn 90    //--move out of the way
  rForward 40
End
//=======================================================

DrawCircle:
  rPen Down //--start drawing
  for i = 1 to 360/tStep
     rForward fStep
     rturn tStep
  next
  rPen Up //--stop drawing
Return
//=======================================================
```

FIGURE 10.3 Drawing circles.

10.1.3 DRAWING TRIANGLES

So far we have not had to use much math to draw rectangles or circles. However, drawing triangles will require a little math. If you find some of the math here to be too complex, do not worry. Do try to understand it, but more importantly you should understand the overall algorithm. The math details are not as important as the final outcome.

In Chap. 4 we introduced some of the math functions in RobotBASIC, and also introduced the concept of converting from angles in radians to angles in degrees. Review that information if necessary (Sec. 4.2, Chap. 4).

To draw a triangle we need to know the angles and sides of the triangles. You can define a triangle by two sides and their included angle, or by all the angles, or by all the sides, or by two angles and a side, and so on. You can see where this is going. You need three parameters that can be any combination of sides and angles. This implies that there are eight possible combinations. The ultimate outcome though, is to be able to calculate the lengths of the three sides and the corresponding three angles.

Once the lengths of the sides and angles are known the robot can draw a triangle by moving the length of the first side, turning 180° minus the angle, then moving the length of the next side, turning again 180° minus the angle, and then finally moving the last length. The subroutine (Lines 21–28) in Fig. 10.5 shows this.

Remember, line numbers are only for the purpose of reference during the coming discussion.

```
//---Draw Rectangles
MainProgram:
  //----change these to make the rectangle have
  //----different orientation and position
  R_Init_X = 100
  R_Init_Y = 300
  R_Init_Heading = -45
  rLocate R_Init_X,R_Init_Y,R_Init_Heading
  rInvisible Magenta
  LineWidth 4
  //---change these variables to change size of the rectangle
  RectWidth = 100
  RectHeight = 50
  gosub DrawRectangle
  rTurn -45     //--move out of the way
  rForward 40
End
//=======================================================
DrawRectangle:
  rPen Down //--start drawing
  for I = 1 to 2
     rturn 90
     rForward RectWidth
     rTurn 90
     rForward RectHeight
  next
  rPen Up //--stop drawing
Return
//=======================================================
```

FIGURE 10.4 Drawing rectangles.

```
02 MainProgram:
03   //----change these to make the triangle have
04   //----different orientation and position
05   R_Init_X = 400
06   R_Init_Y = 300
07   R_Init_Heading = -90
08   rLocate R_Init_X,R_InitY,R_Init_Heading
09   rInvisible Magenta
10   LineWidth 4
11   //---change these values
12   Side = 150
13   Angle = 50
14   Data Sides;2*Side*cos(Angle*pi(1)/180),Side,Side
15   Data Angles;Angle,180-2*Angle,Angle
16   gosub Draw Triangle
17   rTurn -45     //--move out of the way
18   rForward 40
19 End
20 //=======================================================
21 Draw Triangle:
22   rPen Down //--start drawing
23   For i = 1 to 3
24      rForward Sides[i-1]
25      rTurn 180-Angles[i-1]
26   next
27   rPen Up //--stop drawing
28 Return
29 //=======================================================
```

FIGURE 10.5 Drawing an isosceles triangle.

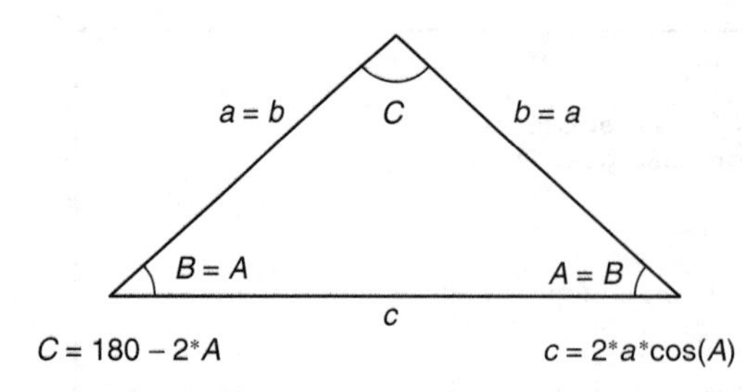

FIGURE 10.6 Isosceles triangle angle $A = B$ and side $a = b$.

The `Data` statements on Lines 14 and 15 create two arrays that hold the sides and angles. See Secs. B.7.4 and C.7 for details on arrays and on the `Data` command.

Arrays are an efficient method of storing and referring to a collection of data. Instead of referring to the three sides of the triangle as *Side_1*, *Side_2*, and *Side_3* we can say *Sides*[*n*], where *n* is the side number. Except, *n* has to start with 0 not 1 so *Side_1* is the same as *Sides*[0] and *Side_2* is the same as *Sides*[1] and so on. For a small collection like this one it might not be too tedious to refer to each element of the collection by its own name. For a large collection it is better to use an array because you can iterate through the data using `for-next` or other looping constructs.

If you look at Lines 24 and 25, this is exactly what is happening. The `for-next` loop refers to each side and angle as *Sides*[*i*-1] and *Angles*[*i*-1]. The reason for the subtraction is because the `for`-loop counts from 1 to 3 and we need 0 to 2, as discussed above.

Now let's go back to Lines 14 and 15. The `Data` statement creates the arrays *Sides* and *Angles* with three sets of data each as specified. How is this data determined?

Observe Fig. 10.6, all the details of how to calculate the lengths and angles are given. In Lines 12 and 13 we specify the *Angle* (*A* or *B*) and *Side* (*a* or *b*). The angle *C* is calculated as shown and so is the side *c*. However, you still have to convert to radians to be able to use the `Cos()` function. All the trigonometric functions in RobotBASIC use and return angles in radians. See Sec. 4.3 (Chap. 4) for an explanation on how to convert between degrees and radians.

This program allows you to draw any isosceles triangle if you know the angles and sides that are equal. Also by changing the parameter for the robot location and heading you can draw the triangle at any orientation and position. Try experimenting with the various variables and note the results.

What is needed to draw a different type of triangle? The program in Fig. 10.5 can be modified to calculate the three sides and angles so that the same *DrawTriangle* subroutine can still be used. All you need is to put the correct data in the *Sides* and *Angles* array. How should the lengths and angles be calculated? Refer to your math textbooks, all the formulas are there, they are similar to the formulas given in Fig. 10.6.

10.1.4 DRAWING ANY SHAPE

As you may have noticed from the previous sections, all that is needed to draw any shape is the correct set of instructions to make the robot turn and move. Combining this with the ability to put the pen up and down you can make the robot draw any shape.

One way to achieve this easily is to use the `Data` command to create an array of data pairs where each pair represents a command for the robot. The first element of the pair

```
01 //---Draw Any Shape
02 MainProgram:
03    //----change these to make the shape have
04    //----different orientation and position
05    R_Init_X = 400
06    R_Init_Y = 300
07    R_Init_Heading = 90
08    rLocate R_Init_X,R_Init_Y,R_Init_Heading
09    rInvisible Magenta
10    LineWidth 4
11    Data SomeShape; "p",Down, "f",165, "t",-120, "f",58
12    Data SomeShape; "t",-60,    "f",50,  "t",-30,  "f",100
13    Data SomeShape; "p",Up,     "f",40
14    for I = 1 to MaxDim(SomeShape,1)/2
15       J = (I-1)*2
16       if SomeShape[J] = "f" then rForward SomeShape[J+1]
17       if SomeShape[J] = "t" then rTurn    SomeShape[J+1]
18       if SomeShape[J] = "p" then rPen     SomeShape[J+1]
19    next
20 End
```

FIGURE 10.7 Drawing any shape.

specifies an action to be taken. The second element provides a parameter associated with the action. Let's look at some examples.

We can use the pair "f", 20 to tell the robot to move forward 20 pixels. The pair "t", 40 means turn 40° and "p", Up tells the robot to put the pen `Up`. Based on these commands, what shape would this set of data draw?

```
Data SomeShape; "p",Down, "f",165, "t",-120, "f",58
Data SomeShape; "t",-60,   "f",50, "t",-30,  "f",100
Data SomeShape; "p",Up,    "f",40
```

Type the program in Fig. 10.7 and see what happens (don't type the line numbers).

In Line 15 we calculate an index into the array *SomeShape* to retrieve the right pair of data. The outcome is that we can now say *SomeShape[J]* to obtain the command element of the pair and *SomeShape[J+1]* to retrieve the value element of the pair. Why this formula for *J*? Remember that the first element of the array is 0. So the elements 0 and 1 are the first pair, 2 and 3 the second, 4 and 5 the third, and so on. Can you see a pattern? We have 0, 2, 4, 6,... for the first element of the pairs and 0 + 1, 2 + 1, 4 + 1, 6 + 1,... for the second. So if we are counting using a `for-next` loop using 1, 2, 3, 4,... then to get 0, 2, 4, 6,... we use the formula in Line 15. The `if-then` statements (Lines 16–18) are to make sure that the right command is executed.

The limit for the `for-next` loop in Line 14 uses the function `MaxDim()` allowing the program to determine how many elements there are in the *SomeShape* array (see Sec. C.8). Since the data is in pairs, we divide the number of elements by two to calculate the number of pairs.

If you replace the data in the `Data` statements on Lines 11 to 13 you can create any shape. You can have as many `Data` statements as you need to draw the shape, you are not limited to just the three in the program.

10.2 ABC Robot

In the program of Fig. 10.7 we developed a program that draws any shape given an array of data pairs. If we create shapes that are letters, we can give the robot the ability to write words. Remove Lines 11 to 13 from the program in Fig. 10.7 and in their place put the lines in Fig. 10.8 then run the program.

The principles used in this program will be extended to develop a program that is able to accept a string and write it on the screen. The size of the font will be scalable and the robot will write the string at any angle. But before we can do any of this, we need to develop a font array and a way of making the robot write any letter individually. See Fig. 10.9 for a sample output of such a program.

10.2.1 THE SPECIFICATIONS

First we will need an array of fonts. This array should be two-dimensional. For example we can have `Dim` *Letters*[27,100] (see Sec. C.7 for the `Dim` command). Why 27? We have 26 letters and we will allow for the space character (you can expand this to include other symbols like numbers if you wish). Why 100? This number is arbitrary for now. The number of instructions that will be needed to create each letter might be smaller than 50 or more. This number is set to the maximum likely to be needed. Remember the instructions are a pair of data (the command and its value) so we will need double the amount of instructions for the limit of the number of elements in the row. Once we have designed our fonts array we would have a better idea what limit will be necessary, and we will change the 100 to whatever is appropriate.

Once the array *Letters*[ ] is created we can access the commands to draw the letter by indexing into the array in this manner: *Letters*[*Letter_Number,n*]. So how do we obtain the *Letter_Number* value? In computers each letter has a code number called the ASCII code. We can obtain any character's ASCII code in RobotBASIC using the function

```
t = "t"
f = "f"
p = "p"
d = 10*sqrt(2)
//--A
data SomeShape; p,down, t,-90, f,50, t,45, f,d, t,45
data SomeShape; f,40, t,45, f,d, t,45, f,20, t,90, f,60
data SomeShape; f,-60, t,-90, f,30, t,-90, p, up, f,15
//--B
data SomeShape; p,down, t,-90, f,60,t,90, f,50, t,45, f,d/2
data SomeShape; t,45, f,15, t,45, f,d/2, t,45, f,50, f,-50
data SomeShape; t,-135, f,d, t,45, f,20, t,45, f,d/2, t,45
data SomeShape; f,55, p,up, f,-60, t,180, f,15
//--C
data SomeShape; t,-90, f,10, p,down, f,40, t,45, f,d, t,45, f,40
data SomeShape; t,45, f,d, t,45, p,up, f,40, t,45, p,down, f,d
data SomeShape; t,45, f,40, t,45, f,d, p,up, f,-d, t,-45
data SomeShape; t,180, f,80
```

FIGURE 10.8 The robot can write (well, only three letters so far).

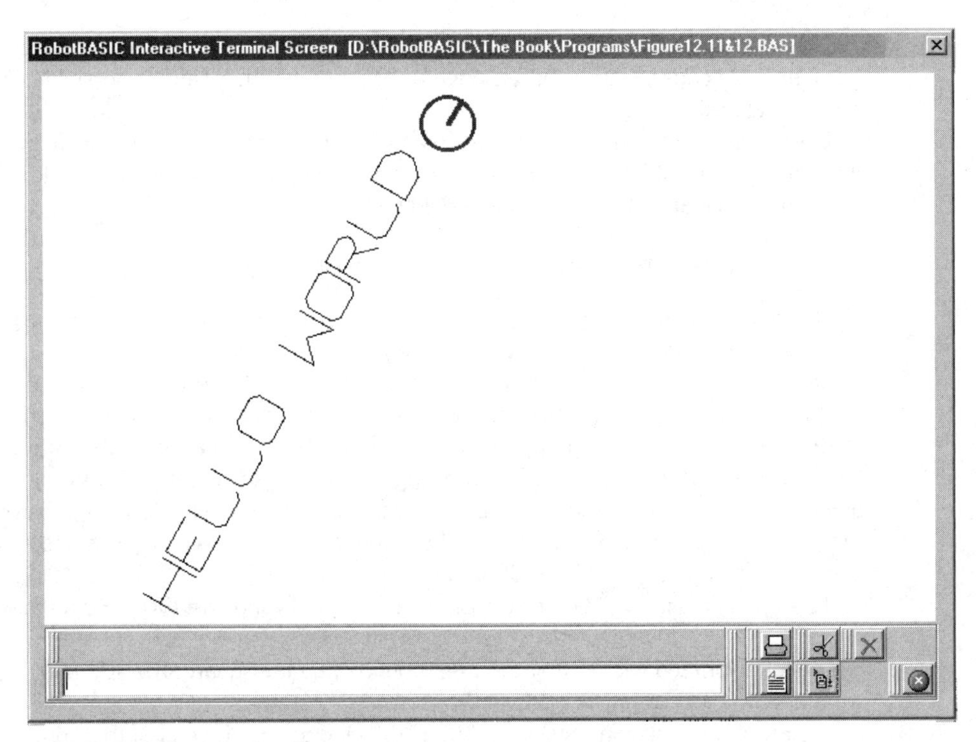

FIGURE 10.9 Writing robot.

`Ascii()`. The ASCII code for "A" is 65, for "B" is 66, and so on, but we need to index into the array of letters at 0 for "A", 1 for "B", and so on. Can you see a pattern?

Letter_Number of any letter = Ascii(of the letter) `-Ascii("A")`

The index position of any letter into the array of letter fonts can now be calculated. This array will have only capital letters. What if the message to be printed has lower case letters? We will use the function `Upper()` to convert all the letters in a string to upper case letters. What if the message contains unsupported characters? We will test to see if the character about to be printed is within the limits `Ascii("A")` to `Ascii("Z")` and if it is not, a space will be printed in its place.

The number of instructions to draw each letter will vary from one letter to another. This means that we will need a way of knowing how many instructions there are in each row in the letters array. We could do this by having the number inside the array, or create another array that has this number for each letter. Here is a better method. The data pairs will have a number telling us the instruction whether to move, turn, or raise/lower the pen. We can add one more instruction to indicate the end of instructions. So, for example, we can have Forward = 1, Turn = 2, Pen = 3, and End = −1000. Now a set of instructions can look like this:

```
Data Temp_Letter; "A",3,up, 1,20, 2,45,..........,-1000
```

Notice the −1000 at the end of the data. It is to indicate the end of instructions for this letter. We chose −1000 because we want a number that cannot possibly be used in the normal instructions.

Unfortunately the above is not very readable, you might confuse the instruction with its quantity and will have to constantly keep referring to the codes table to know what instruction is what. A better solution would be to do this:

```
Data Temp_Letter; "A",p,up, f,20, t,45,..........,e
```

Notice the use of variables *p, f, t,* and *e*. This is a lot easier to understand. All we need is to add a few statements that set a variable $f = 1$, $t = 2$, $p = 3$, and $e = -1000$. This is the same as using the constant red to stand for the number 4 (see Sec. B.7.6).

We need to establish a convention for the fonts. We will assume that the letter prints starting at where the robot happens to be and at −90° orientation to the current robot heading, also that the pen is up. The drawing instructions will create the font for the letter and position the robot at the end of the letter facing the same direction it was before, and the pen in the up position. The font will occupy a 6 × 6 pixels area in its base scale, so a scale of five will make it 30 × 30 and so on.

To summarize; if the robot is at position 100, 100 facing east (90°) and a scale of 10:

- The letter printed will occupy the area bounded by the square 40, 40, 160, 100 and be oriented vertically (90 − 90 = 0).
- The robot will end up at position 160, 100 facing east at the end of the printing of the letter and the pen will be up.

Figure 10.10 shows a program that makes use of all the principles discussed above. The subroutine *Create_Font* is not shown because it will be discussed later.

The main program sets up the robot and sets the colors, scale, and message to be printed (the orientation is defined by the robot's initial heading). Then it calls the subroutine *Print_Message*. This subroutine makes use of all the conventions discussed so far.

10.2.1.1 *Print_Message* Subroutine Look at Fig. 10.10 carefully. You should be able to recognize most of what we have discussed in this routine. You cannot run this program without combining it with the code given in Fig. 10.11.

First the message is converted to upper case. Then we iterate for each letter in the message with the `for`-loop. Notice how we use the function `Substring()` to obtain each letter from the message string.

We find the index into the array of letters by the same formula we discussed above. Any character that is not from A to Z will be made into the last letter in the array, which is the space character. This means that the routine will write a space for any character in the message that is not a valid letter. The `while`-loop will execute each instruction for printing the letter until it finds the instruction code of −1000, which is, by the above convention, the indicator of the end of the drawing instructions.

Notice that no `for-next` loop is used to obtain the instructions since it is not known ahead of time how many there are. Also, notice the formula for indexing to obtain the instruction:

*Letters[L,Inst_No*2 + 1]*

```
MainProgram:
   goSub Create_Font
   Scale = 5
   rLocate 100,500,30
   rInvisible DarkGray
   Message = "Hello World"
   gosub Print_Message
   rForward 30
End
//=======================================================
Print_Message:
  Message = upper(Message)
  for i = 1 to Length(Message)
     L = Ascii(SubString(Message,i,1))-Ascii("A")
     if L < 0  or L > 26 then L = 26
     Inst_No = 0
     while true
       Inst = Letters[L,Inst_No*2+1]
       if Inst = f then rForward Scale*Letters[L,Inst_No*2+2]
       if Inst = t then rTurn    Letters[L,Inst_No*2+2]
       if Inst = p then rPen     Letters[L,Inst_No*2+2]
       if Inst = e then break
       Inst_No = Inst_No+1
     wend
     rForward Scale
  next
Return
```

FIGURE 10.10 Letter-writing program.

We multiply by two because we have pairs of elements that are retrieved two at a time. The one is added because we need to allow for the fact that the first element in the font array is not an instruction (see Fig. 10.11), rather it is the letter itself. This will be explained below.

Finally after drawing the letter the robot is forwarded to make a slight gap (*Scale*) between the letters. When you combine this program with the code for the *Create_Font* subroutine (Fig. 10.11) and run it you will see the result shown in Fig. 10.9.

10.2.1.2 *Create_Font* Subroutine

This subroutine populates the array *Letters*[ ] with the data for creating each letter as per the specifications in the preceding discussion.

There are a lot of `Data` statements to create one-dimensional array *letters*[ ] that contains all the letters in one long row. This is because it is a lot easier to enter the data in this manner than having a separate statement for each element of *Letters*[ ]. Remember, that array and variable names are case sensitive, so *letters*[ ] and *Letters*[ ] are not the same. After the data is loaded into *letters*[ ] we reformat it as a two-dimensional array by copying it into *Letters*[ ] using the code that follows the `Data` statements.

```
Create_Font:
  f = 1
  t = 2
  p = 3
  e = -1000
  d = sqrt(2)
  //-----------Fonts
  data letters; Char(Ascii("A")+26), f,6,  e
  data letters; "A", p,down, t,-90, f,5, t,45, f,d, t,45
  data letters; f,4, t,45, f,d, t,45, f,2, t,90, f,6
  data letters; f,-6, t,-90, f,3, t,-90, p,up, e
  data letters; "B", p,down, t,-90, f,6, t,90, f,4, t,45, f,d
  data letters; t,45, f,1, t,90, f,5, f,-5
  data letters; t,-135, f,d, t,45, f,2, t,45, f,d, t,45
  data letters; f,5, p,up, f,-6, t,180, e
  data letters; "C", t,-90, f,1, p,down, f,4, t,45, f,d, t,45
  data letters; f,4, t,45, f,d, t,45, p,up, f,4, t,45, p,down
  data letters; f,d, t,45, f,4, t,45, f,d, p,up, f,-d, t,135
  data letters; f,5, e
  data letters; "D", p,down, t,-90, f,6, t,90, f,4, t,45, f,d*2
  data letters; t,45, f,2, t,45, f,d*2, t,45, f,4, t,180, p,up
  data letters; f,6, e
  data letters; "E", p,down, f,6, f,-6, t,-90, f,3, t,90, f,4
  data letters; f,-4, t,-90, f,3, t,90, f,6, t,90, p,up, f,6
  data letters; t,-90,  e
  data letters; "F", p,down, t,-90, f,3, t,90, f,4, f,-4
  data letters; t,-90, f,3, t,90, f,6, t,90, p,up, f,6, t,-90, e
  data letters; "G", t,-90, f,1, p,down, f,4, t,45, f,d, t,45
  data letters; f,3, t,45, f,d, p,up, f,d, t,45, f,2, t,90
  data letters; p,down, f,2, f,-1,  t,-90, f,1, t,45, f,d, t,45
  data letters; f,3, t,45, f,d, p,up, f,-d, t,135, f,5, e
  data letters; "H", p,down, t,-90, f,6, f,-3, t,90, f,6, t,-90
  data letters; f,3, f,-6, t,90, p,up, e
  data letters; "I", p,down, f,3, t,-90, f,6, t,-90, f,3, f,-6
  data letters; t,-90, p,up, f,6, t,90, p,down, f,3, f,-3, p,up
  data letters; t,180, e
  data letters; "J", t,-90, f,1, t,135, p,down, f,d, t,-45
  data letters; f,2, t,-45, f,d, t,-45, f,5, t,-90, f,2, f,-4
  data letters; p,up, t,-90, f,6, t,-90, e
  data letters; "K", p,down, t,-90, f,6, f,-3, t,63, f,6.6
  data letters; f,-4.6, t,90, f,4.5, f,-4.5, p,up, t,-90
  data letters; f,-2, t,-63, f,-3, t,90, f,6, e
  data letters; "L", t,-90, f,1, p,down, f,5, f,-5, t,135
  data letters; f,d, t,-45, f,4, t,-45, f,d, t,135, p,up
  data letters; f,1, t,-90,  e
  data letters; "M", t,-90, p,down, f,6, t,135, f,3*d, t,-90
  data letters; f,3*d, t,135, f,6, t,-90, p,up, e
  data letters; "N", t,-90, p,down, f,6, t,135, f,6*d, t,-135
  data letters; f,6, p,up, f,-6, t,90,  e
  data letters; "O", t,-90, f,1, p,down, f,4, t,45, f,d, t,45
  data letters; f,4, t,45, f,d, t,45, f,4, t,45,  f,d,  t,45
  data letters; f,4, t,45, f,d, f,-d, p,up, t,135, f,5,  e
  data letters; "P", t,-90, p,down, f,6, t,90, f,4, t,45, f,d
  data letters; t,45, f,1, t,45, f,d, t,45, f,4, p,up, f,-6
  data letters; t,-90, f,3, t,-90,  e
  data letters; "Q", t,-90, f,1, p,down, f,4, t,45, f,d, t,45
  data letters; f,4, t,45, f,d, t,45, f,4, t,45,  f,d,  t,45
  data letters; f,4, t,45, f,d, f,-d, p,up, t,135, f,5, t,-135
  data letters; p,down, f,d, p,up, f,-d, t,135, e
  data letters; "R", t,-90, p,down, f,6, t,90, f,4, t,45, f,d
```

FIGURE 10.11 *Create_Font* subroutine.

```
data letters; t,45, f,1, t,45, f,d, t,45, f,4, f,-3, t,-135
data letters; f,3*d, t,-45, p,up, e
data letters; "S", t,-90, f,1, t,135, p,down, f,d, t,-45, f,4
data letters; t,-45, f,d, t,-45, f,1, t,-45, f,d, t,-45, f,4
data letters; t,45, f,d, t,45, f,1, t,45, f,d, t,45, f,4, t,45
data letters; f,d, p,up, t,45, f,5, t,-90, e
data letters; "T", f,3, t,-90, p,down, f,6, t,-90, f,3, f,-6
data letters; t,-90, p,up, f,6, t,-90, e
data letters; "U", t,-90, f,1, p,down, f,5, p,up, t,90, f,6
data letters; t,90, p,down, f,5, t,45, f,d, t,45, f,4, t,45
data letters; f,d, p,up, f,-d, t,-45, f,-5, t,45, p,down
data letters; f,d, p,up, f,-d, t,135, e
data letters;  "V", t,90, f,-6, p,down, f,3, t,-45, f,3*d
data letters; t,-90, f,3*d, t,-45, f,3, p,up, f,-6, t,90, e
data letters;  "W", t,90, f,-6, p,down, f,6, t,-135, f,3*d
data letters; t,90, f,3*d, t,-135, f,6, p,up, f,-6, t,90, e
data letters;  "X", t,90, f,-6, p,down, t,-45, f,6*d, t,45
data letters; p,up, f,-6, t,45, p,down, f,6*d, p,up, t,-135
data letters; f,6, e
data letters;  "Y", f,3, p,down, t,-90, f,3, t,-45, f,3*d
data letters; t,-45, p,up, f,-6, t,-45, p,down, f,3*d, p,up
data letters; t,-45, f,3, t,-90, f,3, e
data letters;  "Z", t,-90, f,6, t,-90, p,down, f,-6, t,-45
data letters; f,6*d, t,45, f,-6, t,180, p,up, e
//-----------Fonts
Letter_No = 0
Dim Letters[27,100]
Inst_Count = 0
for i = 0 to MaxDim(letters,1)-1
    if Inst_Count=0
      Letter_No = Ascii(upper(letters[i]))-Ascii("A")
      Letters[Letter_No,Inst_Count] = letters[i]
      Inst_Count = Inst_Count+1
      continue
    endif
    Letters[Letter_No,InstCount] = letters[i]
    Inst_Count = Inst_Count+1
    if letters[i] = e
      Inst_Count = 0
    endif
next
Return
```

FIGURE 10.11 *(Continued)*

In the array of fonts the first element in each row is the letter itself. This is useful for two reasons. First it is a good self-documenting practice. Second we can use the ASCII code of the letter to be able to put the data in the correct row in the array *Letters*[ ].

The code iterates into the *letters*[ ] array until it finds an end of instruction command (e), and then looks for the next letter to calculate the row into which to put the data that follows.

10.3 Summary

In this chapter you have learned:

- How to draw on the screen with the robot using `rPen`.
- How to draw shapes such as circles, rectangles, triangles, or any other shape.
- About the trigonometric and other mathematic functions in RobotBASIC.
- About string manipulation and formatting functions like `Substring()`, `Ascii()`, `Upper()`, and `Length()`.
- How to utilize arrays, the `Data` and `Dim` commands, and the `MaxDim()` function.
- How to make the robot write messages.

Now, try to do the exercises in the next section. If you have difficulty read the hints.

10.4 Exercises

1. Most of the programs in this chapter do not avoid obstacles while doing the drawing action. Can you combine obstacle avoidance with the programs in Sec. 10.1?

HINT: See Chaps 4 and 5.

2. The program in Fig. 10.3 draws a circle starting at the robots position rather than a circle whose center is the robot position. Can you write a program that makes the robot draw a circle (or an arc of a circle) given the radius and centre coordinates of the circle?

HINT: The formula for a circle is:

$$X = R \text{ Sin } \theta + C_y$$
$$Y = R \text{ Cos } \theta + C_x$$

Where R is the circle's radius, C_x and C_y are the circle's center coordinates, θ goes from 0 to 360 (or any part of 0 to 360 for a partial circle).

3. Modify the main program in Fig. 10.10 to accept a message from the user then allow the user to place the robot on the screen at any heading.

HINT: Use the `Input` command. For placing the robot you can either use the `Input` command or you can make use of the mouse as in Fig. 7.9 (Chap. 7).

4. Modify the program of Fig. 10.10 to write a message in a circle. See Fig. 10.12 for a sample output.

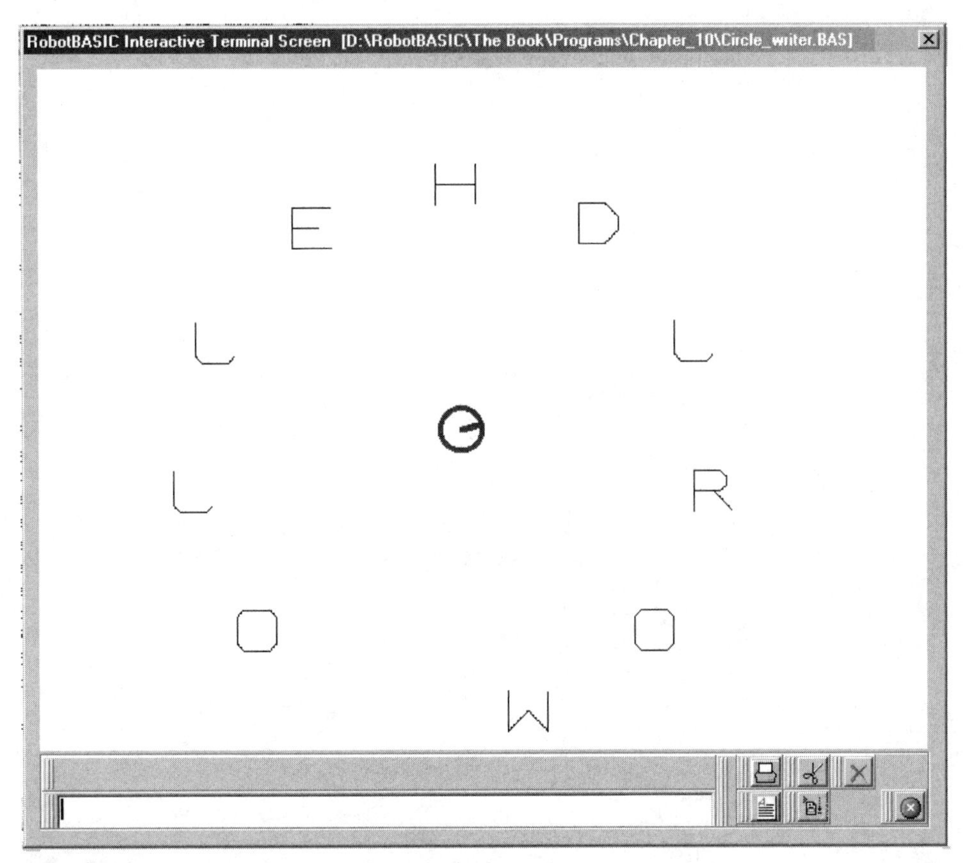

FIGURE 10.12 Sample output desired in Exercise 3.

HINT: You will need to reposition the robot for each letter at the right position and at the right heading (see Exercise 2).

5. The array of fonts (Fig. 10.11) does not define the numerals 0–9. Add the required code to achieve this. You will need to change some of the code in the *Print_Message* subroutine to allow for the new fonts. Notice that the ASCII code for 0–9 is not contiguous with the ASCII code for the letters.

PART 3

COMPLEX COMPOUND BEHAVIORS

Part 3 capitalizes on all the skills and experiences gained in Parts 1 and 2 to build complete, useful, and interesting projects. The reader, at this point in the text, is expected to be fully competent with the language and simulator. Single behaviors are combined to create compound complex behaviors. Many of the programs developed in previous chapters are modified to function in combination, allowing for smooth transitions from one behavior to another, creating an overall behavior that achieves a multifaceted and realistic job.

We acquaint the reader with more advanced ideas in computer science by introducing topics such as graphs, lists, stacks, queues, and simple databases. We also explore the software-engineering skills required to proficiently handle the process of conceiving, specifying, designing, and building a complete practical project.

Each chapter evolves the project by considering initial designs and then improving on the designs up to a certain level. Possible additional improvements are discussed and suggested for the readers to accomplish on their own.

Upon completing Part 3 you will

- Have full ability in utilizing all the features of the robot simulator.
- Have advanced knowledge of the RobotBASIC language.
- Have the skills to program complex and realistic projects.
- Be familiar with some software engineering skills required during the life cycle of a robotics project.
- Appreciate the concepts of some computer science topics related to the field of artificial intelligence (AI).

CHAPTER 11

MOWING AND SWEEPING ROBOT

In previous chapters the robot roamed around an area in a random manner while intelligently avoiding obstacles and staying within the confines of a specified area. If the robot is equipped with a vacuum cleaner or a mowing blade it could vacuum the floor or mow the lawn while it is doing the roaming. This makes our robot a very useful device.

In this chapter you will learn how to make the robot:

- Vacuum an office area with partitions, desks, cabinets, and chairs.
- Mow a lawn with trees and flowerbeds. The lawn will not necessarily be surrounded by a physical barrier.

11.1 Sweeper Robot

In this section we want to use the robot to vacuum an office floor. An office area can be cluttered with many obstacles, but the problem of confining the robot is not an issue since the area is walled. Confinement becomes a matter of obstacle avoidance rather than actively checking for confining devices.

11.1.1 THE BASE PROGRAM

To test our algorithms we will use the same office throughout. Figure 11.1 shows code for a base program where the *MoveRobot* subroutine is left blank for now. This subroutine will be developed in the subsequent sections.

The subroutine *DrawOffice* uses the `Data` command to create an array of locations and types of furniture (cabinets and desks). We then use a `for-next` loop to draw the furniture at the required location. The command `DrawShape` *VarS,X,Y,Scale* is used to draw a shape defined in a variable *VarS* at a location *X, Y* and with a scaling factor *Scale*. The variable *VarS* contains a string, which is a set of instructions for moving up, down, left, right, and so on. (See Sec. C.7 for details.) We also use more `Data` statements along with the `mPolygon` command to shade the furniture and to draw the office partitions. Notice that all objects are drawn as they would be seen by the robot. This is as if we have taken a slice through the room at a few inches above the floor and parallel to the floor. Thus chairs, for example, appear as four little circles, which are the legs (see Fig.11.3).

The *InitRobot* subroutine locates the robot and puts the pen down. To indicate the effectiveness of the vacuuming, the pen feature on the robot will be used to leave a trail showing where the robot has cleaned. In this manner we will have a visual indication of the amount of coverage and a way of gauging the effectiveness of the algorithm.

The robot's size is set to 11 pixels so that we can establish a scale for the office area. If the robot's radius is 1 ft then the office would have a 4628 ft^2 area. This is a realistic size for the robot and office.

11.1.2 A FIRST ATTEMPT

The first algorithm we will try is a simple one where we will make the robot roam randomly around the area (as in Chap. 5) while avoiding obstacles. Figure 11.2 shows the code for the subroutine *MoveRobot*. This subroutine replaces the one in the base program of Fig. 11.1.

The code of this subroutine makes the robot move around while avoiding obstacles, using a combination of the infrared sensors and the bumper sensors. This combination will work in most cases and will enable the robot to avoid small objects such as chair legs. Also the use of random numbers allows for avoiding a situation where the robot becomes stuck repeating the same sequence of moves.

Notice that instead of using `rForward 1` directly inside *MoveRobot* we call a subroutine *ForwardRobot*. For now this code forwards the robot one pixel, but this routine will accomplish more complicated actions with later improvements of the algorithm.

When you run the program you will notice that this algorithm does not prevent the robot from going over an area that has already been vacuumed. You will also notice that the robot may eventually get in a situation where it is repeatedly vacuuming the same area.

11.1.3 AN IMPROVEMENT

A possible improvement for the algorithm of Fig. 11.2 is to minimize the possibility of going over an area that has already been vacuumed. With a real robot this might be difficult to achieve with simple sensors. One possible way is to have a method of monitoring the level of dirt being sucked by the vacuum pump. Another, less desirable way, is to spray

```
MainProgram:
  gosub DrawOffice
  gosub InitRobot
  gosub MoveRobot
End
//=================================================================
InitRobot:
  RR = 11
  LnClr = Cyan  //Line Color
  rlocate 400,300,random(360),RR
  rInvisible LnClr
  rpen down
  linewidth (RR-3)*2
Return
//=================================================================
DrawOffice:
  LineWidth 15
  Data Walls;-165,140,165,0,-357,245,0,245,-590,513,590,600
  Data Walls;-165,140,255,140,-360,140,517,140
  Data Walls;-644,140,797,140,-517,140,517,0,-474,245,699,245
  Data Walls;-474,246,474,419,797,419,-357,247,357,470,113,470
  MPolygon Walls
  Cabinet_H = "rrrddddllluuuu"
  Cabinet_V = "dddllllluuurrrr"
  Desk_H    = "rrrrrrrrrrddddllllluuuullllldddduuuuu"
  Desk_V    = "ddddddddddlllllluuuurrrruuuuullllurrrrr"
  LineWidth 1
  //Desks & Cabinets Locations
  Data Furniture; "CH",478,0,"CH",597,559,"CH",769,370
  Data Furniture; "CH",0,252,"CH",551,559,"CH",0,0
  Data Furniture; "CV",564,0,"CV",40,569,"CV",214,569
  Data Furniture; "CV",548,569,"CV",169,569,"CV",156,0
  Data Furniture; "DV",348,300,"DV",800,496,"DV",800,0
  Data Furniture; "DH",481,252,"DH",259,0
  //Draw them
  for I = 0 to MaxDim(Furniture,1)-1 step 3
    if Furniture[I] = "CH" then ss = Cabinet_H
    if Furniture[I] = "DH" then ss = Desk_H
    if Furniture[I] = "CV" then ss = Cabinet_V
    if Furniture[I] = "DV" then ss = Desk_V
    DrawShape ss,Furniture[I+1],Furniture[I+2],10
  next
  //Shade them
  Data FF_Furniture;10,-17,10,-271,488,-21,573,-579,782,-395
  Data FF_Furniture; 612,-585,140,-19,544,-24,22,-584,151
  Data FF_Furniture; -586,198,-580,530,-591,337,-34,776,-80
  Data FF_Furniture; 323,-388, 565,-274, 772,-580
  MPolygon FF_Furniture,gray
  //Tables
  Circle 59,69,109,119,darkgray,darkgray
  Circle 118,329,168,379,darkgray,darkgray
  //Chairs
  Data Chairs;275,67,699,16,500,319,245,316,693,512,75,279
  for I = 0 to MaxDim(Chairs,1)-1 step 2
     X = Chairs[I]
     Y = Chairs[I+1]
    Sp = 35  //leg spacing
    LD = 4   //leg diameter
```

FIGURE 11.1 Base program.

```
      Cl = Brown  //color for legs
      Circle X,Y,X+LD,Y+LD,Cl,Cl
      Circle X+Sp,Y,X+Sp+LD,Y+LD,Cl,Cl
      Circle X,Y+Sp,X+LD,Y+Sp+LD,Cl,Cl
      Circle X+Sp,Y+Sp,X+Sp+LD,Y+Sp+LD,Cl,Cl
  next
Return
//==================================================================
MoveRobot:
   //left blank for now
Return
//==================================================================
```

FIGURE 11.1 *(Continued)*

the ground with a disinfectant powder that can also be sensed by the robot (perhaps with ultraviolet light).

We will simulate the ability of the robot to sense an already cleaned area by sensing the line drawn by the robot. The method used in real life is immaterial so long as the robot is given the ability to decide if it has already vacuumed the area it is currently over.

You have seen in Chap. 9 how to build a specialized ground sensor system to sense a drop off. In this chapter we will do exactly the same arrangement but instead of looking for a drop off the robot will be sensing for the color drawn by the pen. The value returned by the subroutine will be used to decide whether to do a left or right turn while moving forward, in order to move away from an already vacuumed area. This algorithm modifies

```
MoveRobot:
  m =1
  while true
     F = rFeel()
     if F
       if F&3 and not(F&24)
            m = -1
       elseif F&24 and not(F&3)
             m = 1
       elseif F=4
           m = -1
           if random(10000) < 5000 then m = 1
       endif
       rTurn m*(random(4000)/1000+1)
     endif
     while rBumper()&4
        rturn m
     wend
     gosub ForwardRobot
  wend
return
//==============================================================
ForwardRobot:
   rForward 1
Return
//==============================================================
```

FIGURE 11.2 First-attempt algorithm.

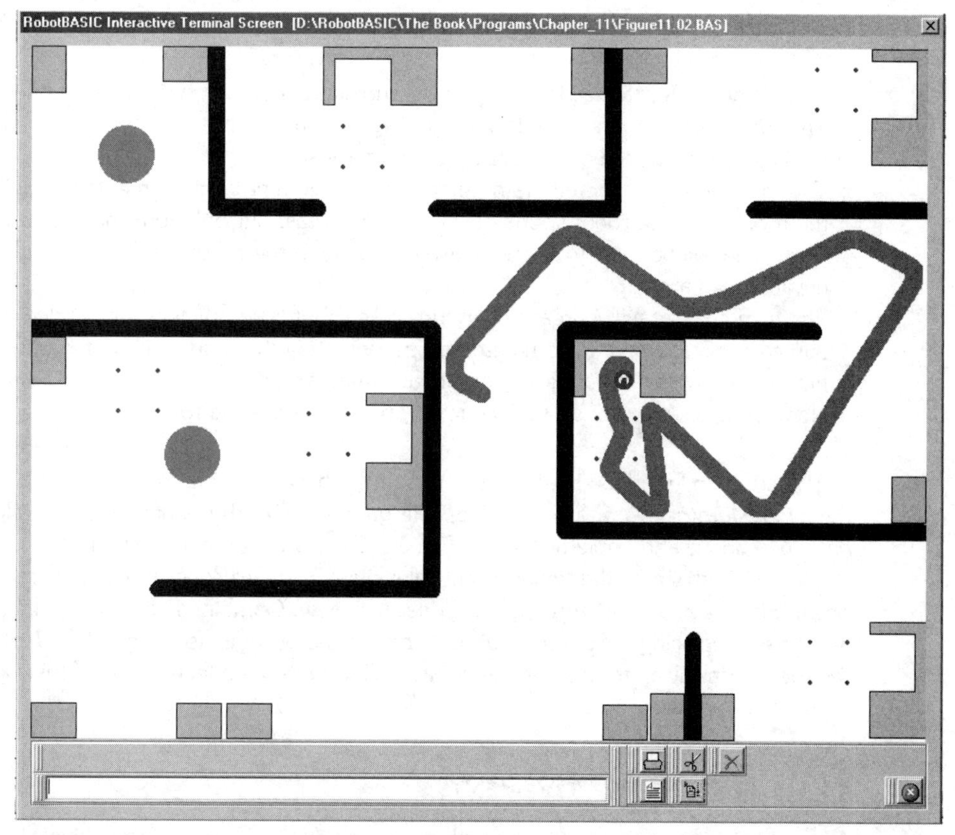

FIGURE 11.3 Result of running the program in Figs. 11.1 and 11.2 combined. Notice the Robot under the desk. It has managed to enter between the chair legs to clean under the desk.

the *ForwardRobot* subroutine to use the *TestSensors* subroutine. The idea is to test to see if the robot is over an area painted by the pen (using color *LnClr*). When the sensors give an indication of where there is color, we make the robot turn a percentage of the time away from a painted area and toward a nonpainted area. This percentage of time will be increased the more time the robot spends over a partially or fully painted area but only up to some maximum amount.

The turn quantity will be increased every time the robot is totally surrounded by a painted area up to some maximum value. If the robot is totally surrounded by a nonpainted area the turn direction will be set to 0. Also, the turn amount will be set back to 1. Additionally the percent of time to do a turn will be reset back to 0.

If the robot is partially surrounded by a painted area the turn direction is set to turn away from the painted area. If the painted area is straight ahead of the robot, or the robot is totally surrounded by a painted area, the turn direction is randomized.

For additional randomness and effectiveness, the percent of time to turn is reset to 0 at a certain percentage of time. This avoids turning in circles forever if the area becomes almost fully covered.

To summarize:

- The robot will test to see if and how it is surrounded by a previously vacuumed area.
- The robot will turn in a direction away from the painted area or in a random direction if it is fully surrounded or the area is straight ahead.
- Turning will only occur a certain percentage of the time. This percentage will increase the more time the robot spends over painted areas, but only to a maximum amount. This value will be reset to 0 on occasion and when the robot is fully surrounded by an unvacuumed area.
- The turn amount will increase every time the robot is completely surrounded by a vacuumed area, but only to a maximum amount. Also this value will be reset to 1 every time the robot is surrounded by an unvacuumed area.
- No turning will occur if an unvacuumed area surrounds the robot.

Place the code in Fig. 11.4a at the top of the base program in Fig. 11.1, just before the label *MainProgram*. These values will greatly affect the behavior of the algorithm described above and implemented by Fig. 11.4b. Try experimenting with them.

Figure 11.4b shows the replacement subroutine *ForwardRobot* that should replace the one in Fig. 11.2. The subroutine *TestSensors* is new. Combine Figs. 11.1, 11.2, 11.4a, and 11.4b (replacing the old subroutines) and run the program (see Fig. 11.5). Notice how the robot turns away from a vacuumed area. Compare the effectiveness of this algorithm to the previous one.

11.1.4 FURTHER IMPROVEMENTS

A possible improvement to this algorithm is to follow the contour of the objects it encounters instead of just avoiding them. We will explore this possibility in the next section on mowing with the robot.

Currently in the *MoveRobot* subroutine (Fig. 11.2) when the robot encounters an obstacle head on it turns in a random direction. If we change this behavior in such a way as to turn away from a covered area instead of a random direction, the robot would be turning to cover more area in less time. Unfortunately, this may cause the robot to get stuck endlessly repeating the same behavior. Some means of detecting the situation will be needed to trigger a different behavior to free the robot (you will see an example of this in Chap. 12).

```
//---Variables
   TurnDirection = 0
   TurnAmount    = 1
   MaxTurnAmount = 3

   ResetPercentTime    = 1
   MaxTurnPercentTime = 30
   TurnPercentTime     = 0
```

FIGURE 11.4a Place this code at the top of the base program of Fig. 11.1.

```
ForwardRobot:
  rForward 1
  //now see if we need to turn too
  gosub TestSensors
  if Sensors = 0  //fully new area
    TurnPercentTime = 0 //recent %time
    TurnDirection = 0  //no turn
    TurnAmount = 1     //reset amount
  else
    //increment %time but to a maximum
    TurnPercentTime = TurnPercentTime+1
    if TurnPercentTime > MaxTurnPercentTime
       TurnPercentTime = MaxTurnPercentTime
    endif
    //turn right unless we need to turn left
    TurnDirection = 1
    if Sensors = 31   //fully old area
      //increment turn amount but to a maximum
      TurnAmount = TurnAmount+1
      if TurnAmount > MaxTurnAmount
         TurnAmount = MaxTurnAmount
      endif
      //random turn direction
      if random(10000) < 5000 then TurnDirection = -1
    elseif Sensors&3 and not(Sensors&24)
      //if painted area on right turn left
      //no need to check for left since turndir is set to
      //right by default
      TurnDirection = -1
    elseif Sensors = 4
      //if painted area is only straight ahead
      //randomize the direction
      if random(10000) < 5000 then TurnDirection = -1
    endif
  endif
  //if not correct percent of time
  //turn off turning
  if random(100000) > 100000%TurnPercentTime
     TurnDirection = 0
  endif
   //reset the %time for turning on occasion
   if random(100000) < 100000%ResetPercentTime
      TurnPercentTime = 0
   endif
   //make the turn if required
   if TurnDirection <> 0
      rTurn sign(TurnDirection)*TurnAmount
   endif
 Return
 //============================================================
 //-- Creates 5 ground sensors
 //-- at +/-90, +/-45 and 0 degrees
 TestSensors:
   Sensors = 0
   for TS_i = 0 to 4
      if rGroundA(90-TS_i*45) = LnClr
         Sensors = Sensors | (2^TS_i)
      endif
   next
 Return
 //============================================================
```

FIGURE 11.4b A better-coverage algorithm.

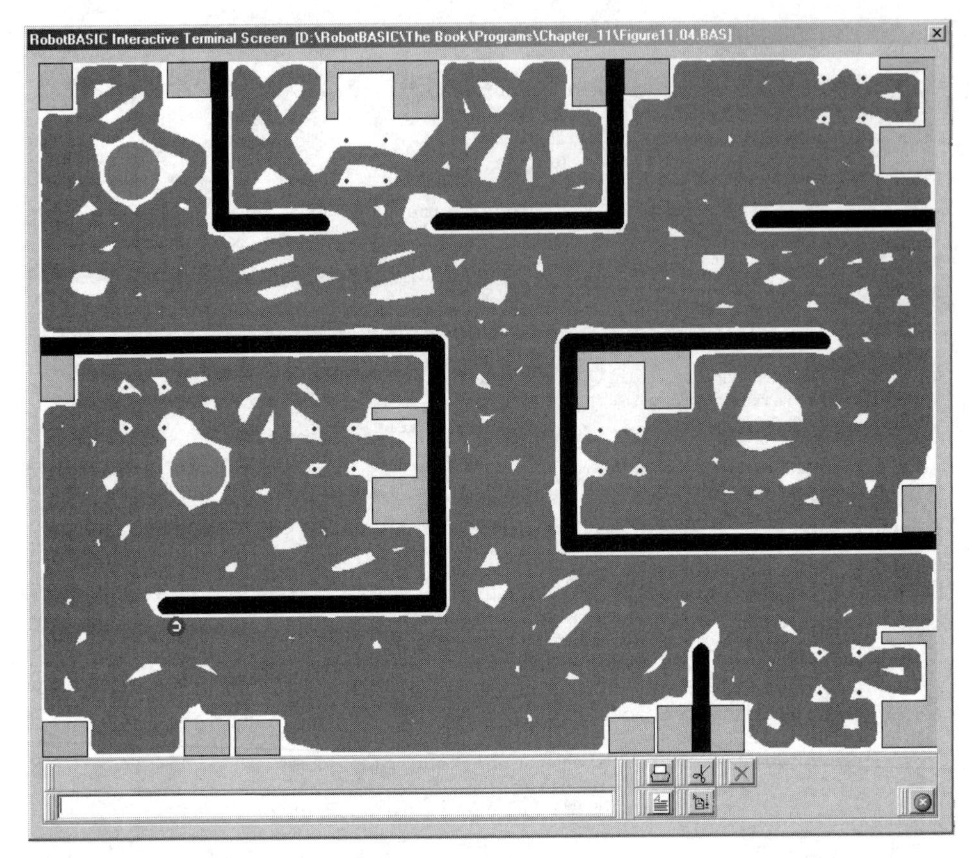

FIGURE 11.5 Nearly full coverage in a much shorter time.

11.2 Mowing Robot

In this section we want to use the robot to mow a lawn. As you have seen in Chap. 9, there are a variety of techniques to confine the robot to an area. We will use the confining method proposed in Sec. 9.1 to limit the robot to the area of the lawn.

We will also use the method in the previous section of this chapter to reduce the likelihood of the robot covering over an already mowed area. Additionally, the methods in Chap. 8 will be used to make the robot cover the perimeter around trees and flowerbeds more efficiently.

With a real lawn-mowing robot there are a variety of methods to test if the robot is over an area that has already been mowed. Sensing the grass height is one way. Another method is to check the load on the mowing motor. If the current being drawn is higher than a threshold value then the grass is still being mowed, but if it is lower, then the blades are not finding resistance while rotating indicating the grass is already cut. A third method might be to measure the amount of clippings going through a discharge system. Whatever method is used in real life, the outcome is that the robot can gauge if it is over an already cut area or not. This will be simulated by using the pen on the robot to leave a trace behind it (using the color *LnClr*) and then sense for this color as we did in the previous section.

11.2.1 THE SPECIFICATIONS

In order to create an effective lawn-mowing algorithm we need the following:

- A confinement algorithm
- A roaming algorithm
- Obstacle avoidance
- Obstacle contour following
- Minimum visiting of already covered areas

We have already covered all of the above behaviors in this and previous chapters. All that is needed in this algorithm is to combine all the behaviors into one program. However, the combination is not a simple matter of placing the routines in the same program. Some of the routines will have to be modified slightly to accommodate the others. Also, the conditions for changing from one behavior to another must be considered carefully.

11.2.2 THE PROGRAM

Notice that the program in Fig. 11.6 has many subroutines that look familiar. Many of the subroutines are ones you have seen before without any change. Others have been modified. We will discuss each routine in turn.

11.2.2.1 *MainProgram, IntitRobot,* and *DrawLawn* The *MainProgram* calls each routine in turn. Since the lawn is light green in color, the robot needs to know that this color is not an obstacle. This is done with the `rInvisible` command.

The variables assigned above the *MainProgram* label should look familiar. The only new value is the variable *BrdrClr,* which is used to assign the color used by the border avoidance logic discussed later. *InitRobot* is the same as in Sec. 11.2, but some variables have been moved up above the main program. This is done to put all variables that you might want to modify (in order to tweak the algorithm) in one place.

The *DrawLawn* routine is simple enough. We make use of the `Data` and `mPolygon` commands to draw the lawn and some tree beds and a flowerbed. The tree beds and flowerbeds will be considered as objects to be avoided.

11.2.2.2 *ForwardRobot* and *TestSensors* These two subroutines are exactly the same as seen in Sec. 11.2. They perform the same actions and are used in the same way as before, study them to review the details.

11.2.2.3 *CheckBorder* This subroutine uses the *TestSensors* routine to check for the border color. First, we swap *LnClr* with *BrdrClr* so that we can check for the border, then the result of the operation is saved in the *Borders* variable. We then swap the colors again to restore the original *LnClr* to what it should be. The variable *Borders,* indicates how the robot is approaching the border. This is similar to the way the robot avoided a drop off in Chap. 9. This subroutine will be used in the routine *MoveRobot* (discussed below). The reason for swapping colors is so that we can use *TestSensors*, which, as you see from the listing, is using *LnClr* as a color to check for. Thus we need to swap *LnClr* with *BrdrClr* so that we can check for the *BrdrClr*. We need to swap back so that the routine will check for *LnClr* again when used by other routines.

```
//---Variables
   TurnDirection      = 0
   TurnAmount         = 1
   MaxTurnAmount      = 3
   ResetPercentTime   = 1
   MaxTurnPercentTime = 30
   TurnPercentTime    = 0
   RR      = 15    //robot's radius
   LnClr   = Cyan  //pen color
   BrdrClr = white //border color
//============================================================
MainProgram:
  gosub DrawLawn
  gosub InitRobot
  gosub MoveRobot
End
//============================================================
InitRobot:
  rlocate 400,300,30,RR
  rInvisible LnClr,LightGreen
  rpen down
  linewidth (RR-2)*2
Return
//============================================================
DrawLawn:
    ClearScr Red
    LineWidth 4
    SetColor white
    Data Lawn; -44,  58,  173,  28,  534,  45
    Data Lawn; 753, 182,  744, 516,  520, 510
    Data Lawn; 402, 574,  401, 574,   60, 552
    Data Lawn;  35, 406,   43,  59,  589, -255
    MPolygon Lawn,lightgreen
    FloodFill 0,0,white

    LineWidth 2
    SetColor black
    Data FlowerBed; -589, 255,  566, 228,  510, 209
    Data FlowerBed;  442, 218,  424, 245,  426, 284
    Data FlowerBed;  459, 296,  500, 281,  519, 280
    Data FlowerBed;  534, 290,  550, 305,  577, 306
    Data FlowerBed;  602, 283,  589, 256,  521,-271
    MPolygon FlowerBed,brown

    Data Trees; 129, 453,  257, 280,  90, 245
    Data Trees; 182, 134,  600, 400, 316, 450
    Data Trees; 391,  94
    DP_D = 50
    for DP_I = 0 to MaxDim(Trees,1)-1  Step 2
         DP_X = Trees[DP_I]
         DP_Y = Trees[DP_I+1]
         Circle DP_X,DP_Y,DP_X+DP_D,DP_Y+DP_D,black,brown
    next
Return
//============================================================
01 MoveRobot:
02   m =1
03   while true
```

FIGURE 11.6 Lawn-mowing program.

```
04      gosub CheckBorder
05      if Borders
06        if Borders & 24 and not(Borders & 3)
07           rTurn (45+random(45))
08        elseif Borders & 3 and not(Borders & 24)
09           rTurn -(45+random(45))
10        else
11           gosub Reverse
12        endif
13      else
14        F = rFeel()
15        if F
16          if F&3 and not(F&24)
17             m = -1
18          elseif F&24 and not(F&3)
19             m = 1
20          else
21             m = -1
22             if random(10000) < 5000 then m = 1
23          endif
24          rTurn m*(random(4000)/1000+1)
25        endif
26      endif
27      if (rBumper()&4) or (rFeel()&12)
28          TurnDir = -1  //left
29          gosub WallFollow
30      elseif (rBumper()&4) or (rFeel()&6)
31          TurnDir = 1  //right
32          gosub WallFollow
33      endif
34      while rBumper()&4
35        rturn m
36      wend
37      gosub ForwardRobot
38    wend
39  return
//============================================================
ForwardRobot:
  rForward 1
  //now see if we need to turn too
  gosub TestSensors
  if Sensors = 0  //fully new area
    TurnPercentTime = 0 //recent %time
    TurnDirection = 0  //no turn
    TurnAmount = 1     //reset amount
  else
    //increment %time but to a maximum
    TurnPercentTime = TurnPercentTime+1
    if TurnPercentTime > MaxTurnPercentTime
       TurnPercentTime = MaxTurnPercentTime
    endif
    //turn right unless we need to turn left
    TurnDirection = 1
    if Sensors = 31   //fully old area
      //increment turn amount but to a maximum
      TurnAmount = TurnAmount+1
```

FIGURE 11.6 *(Continued)*

```
      if TurnAmount > MaxTurnAmount
         TurnAmount = MaxTurnAmount
      endif
      //random turn direction
      if random(10000) < 5000 then TurnDirection = -1
    elseif Sensors&3 and not(Sensors&24)
      //if painted area on right turn left
      //no need to check for left since turndir is set to
      //right by default
      TurnDirection = -1
    elseif Sensors = 4
      //if painted area is only straight ahead
      //randomize the direction
      if random(10000) < 5000 then TurnDirection = -1
    endif
  endif
  //if not correct percent of time
  //turn off turning
  if random(100000) > 100000%TurnPercentTime
     TurnDirection = 0
  endif
  //reset the %time for turning on occasion
  if random(100000) < 100000%ResetPercentTime
     TurnPercentTime = 0
  endif
  //make the turn if required
  if TurnDirection <> 0
     rTurn sign(TurnDirection)*TurnAmount
  endif
Return
//============================================================
//-- Creates 5 ground sensors
//-- at +/-90, +/-45 and 0 degrees
TestSensors:
  Sensors = 0
  for i = 0 to 4
     if rGroundA(90-i*45) = LnClr
        Sensors = Sensors | (2^i)
     endif
  next
return
//============================================================
CheckBorder:
   Swap BrdrClr,LnClr
   gosub TestSensors
   Swap BrdrClr,LnClr
   Borders = Sensors
return
//============================================================
Reverse:
  for i = 1 to (random(10)+10)
    if rBumper() & 1 then break
    rForward -1
  next
  if random(1000) >= 500
    rTurn 90-random(45)
  else
```

FIGURE 11.6 *(Continued)*

```
      rTurn -90+random(45)
    endif
  return
  //=============================================================
  WallFollow:
    rTurn TurnDir*random(150)
    while not rSense()
      while (rFeel() & 6) or (rBumper() &6)
        rTurn TurnDir
      wend
      if rBumper() &4 then return
      rForward 1  // forward always to prevent stall
      if rFeel()=1 or rFeel()=0  // too far from wall or no wall
        while not rFeel() // turn back quickly to find wall again
          rTurn -TurnDir*5
          rForward 1
        wend
      endif
    wend
  return
  //=============================================================
```

FIGURE 11.6 *(Continued)*

11.2.2.4 *Reverse* This subroutine is exactly the same as the one in Chap. 9. It is used to avoid the border if the robot approaches the border head on. You will see how it is used in the *MoveRobot* subroutine.

11.2.2.5 *WallFollow* This routine is similar to what you have seen in Chap. 8. It will be used to follow around the contour of flowerbeds and tree beds. The contours can be followed to the left or to the right depending on the variable *TurnDir* (negative is to the left and positive is to the right). You will see how this is done in the discussion about the *MoveRobot* routine.

The wall-following behavior can go on forever if we do not have a way of stopping. We do this by using the `rSense()` function to see if the robot is going over a painted (already mowed) area. If so, the wall-following behavior will be abandoned. Also the routine is terminated if the robot encounters an obstacle while it is following the contour (a dead end).

11.2.2.6 *MoveRobot* This is the overall coordinating behavior that controls the robot's movement and triggers what other behaviors ought to take place. The idea is to roam around forever. If a border is encountered then avoid it. If an obstacle is encountered then follow its contour.

As in Sec. 11.2 we try to minimize going over a previously mowed area. This is done in the same manner as before with the aide of the *ForwardRobot* and *TestSensors* subroutines (Line 35). Lines 4 to 13 test to see if there is a border violation. If there is no border violation then we test to see if there is an obstacle (the `else` block). The call to the *CheckBorder* subroutine (Line 4) assigns a value to the variable *Borders* that indicates how the border is being approached. If the border is to the left of the robot then it turns to the right and if it is to the right it turns left. The amount of turn is randomized, but no less than 45° and no more than 90°. In any other border combination the robot reverses away from the border and executes a turn of a random amount and direction. This is achieved by using the *Reverse* subroutine.

If the robot does not see a border it will test for an obstacle (Lines 14–25). If an obstacle is to the right it turns left or if it is to the left it turns right, otherwise it turns in a random direction. The amount of turn is randomized, but no more than 4°.

If the robot ever encounters an obstacle head on with the bumpers or the three front infrared (0° and ±45°) sensors (Lines 27–33), then the wall-following behavior is triggered. We decide on how to follow the contour of the obstacle (left or right) depending on which infrared sensors are being triggered. The variable *TurnDir* is set to +1 if a right turn is to be performed or −1 if left, then the *WallFollow* subroutine is called to do the wall-following action. The subroutine *WallFollow* will end if the robot sees that it is going over a painted area while following around the obstacle. This brings the program flow back to Line 34. The routine then tests to see if there is an obstacle still causing the front bumper to close. If so, the robot turns to avoid the obstacle. The turn direction is the same as the last time it was turning to avoid any obstacle detected by the infrared sensors (Lines 34–36). Figure 11.7 shows the result of the program in Fig. 11.6.

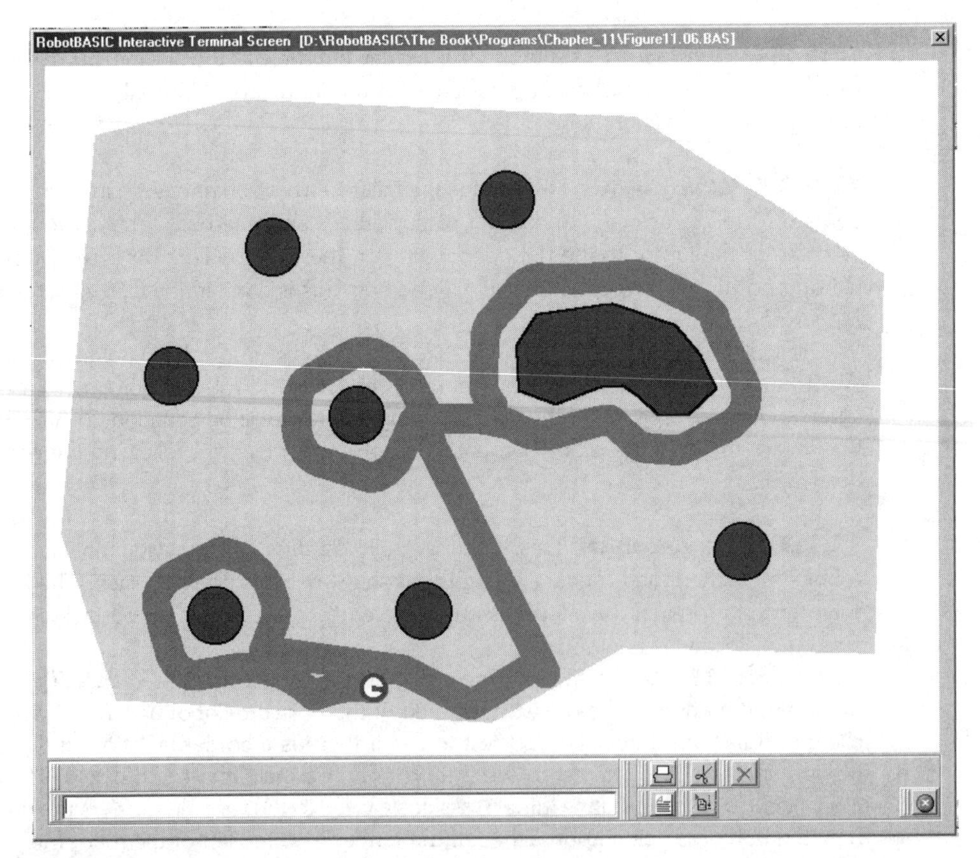

FIGURE 11.7 Lawn-mowing robot.

11.2.3 A SHORTCOMING

In the *DrawLawn* subroutine right after the last `Data` statement add this statement:

```
Data Trees; 600, 450
```

Now run the program and see what happens when the robot encounters a tree. You may have to wait a little while. The robot will try to follow the contour of the new tree bed. However, while it is doing so it will exit the boundary of the lawn. If you notice the tree is right at the border. The robot does not seem to heed the boundary (see Fig. 11.8). Why is that?

The wall-following subroutine does not have any code to prevent the robot from exiting the boundary and if the contour following causes the robot to exit the boundary it will not be stopped from doing so. We need to incorporate boundary checking in the wall-following behavior. Can you do this with all the routines at your disposal? This is an excellent example of how a reasonably well thought out algorithm can fail in an unanticipated situation. Additionally, this situation illustrates the vital role a simulator can play in the research and development stage. The simulator facilitates changing the environment to make it as complex and as varied as possible. Many variations can be tested with ease.

FIGURE 11.8 Notice the robot has exited the boundary on the bottom right hand.

11.3 Further Thoughts

You may have observed certain limitations in the algorithms we have developed, and you may even have pondered certain questions. In this section we will discuss some issues and philosophize about possible solutions.

11.3.1 CONSIDERING THE BATTERIES

In this chapter we have not paid any attention to the battery charge level while the robot was doing its task. This, of course, is not possible in real life. In Chap. 13 we will explore methods for charging the robot's batteries.

Consider the office-sweeping situation. The robot has finally managed to work its way into a section where it has not yet vacuumed effectively. Suddenly, it has to abandon its work and seek a charging station. The station is located in the area that has been effectively vacuumed. The robot goes there and recharges itself. After it gets a full charge, it starts the vacuuming behavior, which now has to start in the already vacuumed area until the algorithm causes it to go to the unvacuumed area. If this takes awhile, the robot may need to recharge again.

You can see that this situation leaves a lot to be desired. One solution is to have multiple recharging stations and allow the robot to recharge itself at the station that is within the area that still has to be vacuumed. This way when it finishes recharging it can start its work in the same area that needs vacuuming.

Another solution is to enable the robot to save the position where it stopped vacuuming. Also it needs to save the path it took to the recharging point. Once it finishes charging it will retrace its path back to the saved position before it starts the sweeping behavior again. This way the time spent outside the uncovered area is minimized.

The algorithms given in this chapter will require some modification to make them work in the situation described above. In later chapters you will see many algorithms that can be used to achieve solutions for the above dilemmas.

11.3.2 LIMITED COVERAGE AROUND OBSTACLES

In both the office and lawn examples you can see that it is hard for the robot to cover areas close to walls and around the contours of obstacles. If you look at Fig. 11.5 you will notice that the robot has vacuumed under the desk on the bottom left, but has left a lot of white space. This is a limitation of the apparatus used to do the vacuuming. We could equip the robot with a specialized nozzle to vacuum in corners and around skirting, but this would be difficult to manage.

Also in Fig. 11.7 you will notice that around flowerbeds and trees the robot was not able to mow the grass effectively. This is also a limitation of real lawn mowers and is why humans use specialized devices for doing the work around such places. The robot could be equipped with such a device, or even a specialized robot could be used to go around behind the first robot to do this job. Notice how using the simulator can help you understand the problems you would face if you actually built a real robot mower.

11.3.3 USING GPS GRIDS

In the algorithms above we tried to minimize time spent over previously covered areas and we employed randomness and some programming techniques to do so. Nevertheless, the robot did spend a lot of its time over previously visited areas. The possibility of this happening becomes progressively higher as more of the area is covered. This means that an increasing percentage of the robot's time is wasted and battery utilization becomes less efficient.

One way to alleviate this problem is to employ multiple robots and assign each a smaller area. However, this would be expensive in hardware. Another way is to use the same robot but divide the entire area into grids. The robot then sweeps each grid in turn, moving from one grid to another after it has finished the work for the one it is currently in.

The grid system does not need to be delineated by any kind of physical devices or barriers. Rather, it would be a set of coordinates saved in the robot's memory. The robot would use its GPS (or LPS) to decide how to navigate from one grid to another and which grids still require visiting. The problem of finding the battery charger when needed would become trivial, since the robot can be given the coordinates of the charging station.

Another approach is to divide the area into small square grids delimited by RFIDs (radio frequency identification devices). The robot can then note in its memory that it is within grid *N* and know that this grid has or has not been vacuumed. Also the robot can have a preplanned procedure for how to move among the grids. RobotBASIC's `rSense()` function (line sensors) can be used to simulate RFID detectors.

11.3.4 A REALITY CHECK

The algorithms in this chapter are experiments and not real solutions. Robots that mow and vacuum are on the frontiers of technology. Certainly, they are a very good idea, but in a real home or office environment there are numerous obstacles that can hinder any robot from effectively vacuuming the floor.

In lawn mowing we have not considered safety issues such as children or animals running in front of the robot. We did not consider the issues of a steep sloped garden, nor gardens with pathways.

There are a multitude of issues to consider in a real world robot that has to tackle such tasks. Some of these problems may be very hard to resolve, but the ideas in this chapter and the simulator can be used to experiment with possibilities.

11.4 Summary

In this chapter you have learned:

- How to combine routines and methodologies from previous chapters to allow the robot to perform useful work.
- How the `rPen` feature can be used to provide visual feedback on the effectiveness of an algorithm. You have also seen how the pen can be used to simulate further functionalities.

- ❑ How the utilization of randomization can improve the effectiveness of algorithms.
- ❑ How the `DrawShape` command can be used to easily draw complex objects on the screen.
- ❑ Further uses of arrays and the `Data` command.

Now, try to do the exercises in the next section. If you have difficulty read the hints.

11.5 Exercises

1. In the algorithm of Sec. 11.1 we did not implement wall following. Add wall-following to the program.

HINT: See Sec. 11.2.

2. The wall-following subroutine in Fig. 11.6 uses the `rFeel()` function to sense the walls. This causes the robot to stay further away from the walls than might be desirable in this application. In Chap. 8 (Sec. 8.4) the `rRange`(*ExprN*) function was used to control the distance from the wall. Change the *WallFollow* subroutine in this chapter to use the one in Fig. 8.8 of Chap. 8.

HINT: Remember you will need to use a method to abort wall-following once the robot has gone around the object.

3. Try to modify *WallFollow* as discussed in Sec. 11.2.3.

HINT: You will need to use the *CheckBorder* subroutine to abandon the routine if *Borders* is not zero.

4. After studying the problems and solutions in this chapter, try to design your own algorithm for handling a mowing or sweeping problem. Perhaps your algorithm could try to mow each new path while slightly overlapping a previous path. Maybe your robot could work in spirals to cover a selected area efficiently. Or perhaps you have a unique idea of your own.

CHAPTER 12

LOCATING A GOAL

In the preceding chapters we had no fixed destination for the robot to go to. The robot just moved around whether randomly, or following a line drawn on the floor, or around an object, but with no final destination in mind. This kind of behavior has been useful in applications such as mowing or sweeping areas that the robot can visit.

There are many applications where the robot will need to go from one point to another. It would be simple enough to make the robot go to a point, as you have seen in Chap. 4, if there are no obstacles in the way. However, if there are obstructions between the robot and its target destination it will become necessary for the robot to circumnavigate the obstructions while making headway toward the target.

In this chapter, we will address two general methods for indicating to the robot where to go:

- Using a marker beacon that hangs over the target position. The robot can see and home in on this beacon.
- Giving the robot a GPS (global positioning system) unit and a compass along with destination coordinates so it can calculate a path to the target and follow it.

Once the robot knows its path it will proceed toward the goal. When it encounters obstacles it will have to momentarily abandon progress toward its destination and deal with the obstruction. We will assume that there is at least one path that can lead from where the robot is to the goal position. We will deal with situations where there might be no path in

Chap. 14. Chapter 15 will deal with the more complex situation of moving from room to room in a typical home or office.

12.1 Using a Beacon

In this section, we are going to mark the desired destination by hanging a beacon above it. Since the beacon is high in the air, the robot is able to see it even if there are objects on the floor between the robot and the goal point. If the beacon is a flashing light, either visible or infrared, a real robot could use circuitry capable of recognizing a particular frequency to detect it.

A robot with a camera aimed slightly upward could detect a beacon of a specified color and even use triangulation to estimate how far it is from the robot. In our simulation, we will assume the robot has a means of detecting a beacon of a specific color using a directional sensor aimed along the robot's heading.

12.1.1 THE ALGORITHM

In order to develop the algorithm, imagine you are the robot. Assume you are in a cluttered room and have limited senses. To make you feel more like the robot, imagine that the beacon is a bright-flashing light. Your eyes are closed so you can't really see, but you can detect the bright-flashing light when it is in front of you.

Your first action would be to turn around slowly until you face the flashing light. You would then move forward toward the light, feeling ahead of you with your hands to make sure you don't bump into something (remember your eyes are closed).

If you bump into an object you try to go around it. Since you can't actually see, this is not a simple task. You could follow around the edge of the object until you think you are around it (you don't have any idea of how big the object is) and then try to face the beacon again. If you repeat these steps over and over, you should eventually arrive at the goal.

The subroutine in Fig. 12.1 shows the implementation of the algorithm discussed above. The routine assumes there are subroutines that can accomplish the required tasks. Each subroutine executes until its task is complete (or the robot has reached the beacon) and then terminates. The loop ensures that the tasks are executed in turn, one after the other repeatedly, until the beacon is found. We will discuss each routine in the following sections.

```
FindBeacon:
  repeat
    gosub FaceBeacon
    gosub ForwardTillBlocked
    TurnDir = 1
    gosub GoAround
  until BeaconFound
Return
```

FIGURE 12.1 This subroutine locates and finds the beacon.

```
MainProgram:
  while true
    gosub SetEnvironment
    gosub FindBeacon
  wend
End
```

FIGURE 12.2 A `while`-loop causes the program to test the algorithm repeatedly.

12.1.2 THE MAIN PROGRAM

The main program sets up an environment with obstacles and then starts the goal-seeking behavior (Fig. 12.2). The subroutine *SetEnvironment* (see below) sets the environment and places the robot and beacon at random positions.

In order to test the algorithm we need to run the program several times to see if any obstacle arrangement can baffle the code and cause the robot to fail to reach the goal. We could do this by manually running the program many times. A better way, though, is to have the main program repeat the sequence of creating a random environment and locating the goal in an endless loop.

12.1.3 CREATING A CLUTTERED ROOM

The subroutine in Fig. 12.3 clears the screen then draws three circles and three squares (experiment with more or less). The size and location of each object is chosen randomly. This makes the environment full of obstacles at random positions that can be hard to circumnavigate.

The robot is located at a random position on the left side of the screen and the beacon (a red circle) at a random position on the right side.

```
SetEnvironment:
  ClearScr
  // Draw three circles and three squares
  for i=1 to 3
    SetColor Black
    LineWidth 4
    x = random(450) + 100
    y = random(300)+100
    size = random(50)+50
    Circle x,y,x+size,y+size
    x = Random(450)+100
    y = Random(300)+100
    size = random(100)+50
    Rectangle x,y,x+size,y+size
  next
  // place robot
  rLocate 25,Random(350)+100
  rInvisible Red
  // place beacon
  bx =750
  by = Random(350)+100
  Circle bx-10,by-10,bx+10,by+10,red,red
Return
```

FIGURE 12.3 Creates a cluttered room and places the robot and beacon at random positions.

12.1.4 FACING THE BEACON

The rBeacon(*color*) function in RobotBASIC is used to locate a beacon of a specified color. It returns zero (*false*) if the beacon is not directly in front of the robot. If the beacon is directly ahead of the robot, the function returns the distance to the beacon. You can consider the number returned as a nonzero number and therefore is equivalent to being *true*. This means that you can use the function to test if the beacon is directly ahead of the robot or not. However the function can also be used to return the distance to the beacon. This can be useful in many situations, especially to determine when the robot has reached the point under the beacon (see later).

The function is usable to look for any color you specify as a parameter. Normally, the robot will see colors on the screen as objects to be avoided. If you want the robot to assume that objects of the beacon color are in the air and thus cannot cause collisions, you need to issue the rInvisible *color* statement listing the appropriate color. This statement tells the robot that the color being used as a beacon is not an obstacle. Figure 12.4 shows how to create a function that turns the robot until the beacon is directly in front of it. The expression not rBeacon(Red) is the same as saying:

```
rBeacon(Red) = false    or    rBeacon(Red) = 0
```

12.1.5 MOVING TOWARD THE BEACON

The subroutine *ForwardTillBlocked*, shown in Fig. 12.5, moves the robot forward until it encounters an object or it reaches the beacon. The expression in the while-loop checks for an obstacle with the three front infrared sensors and the front and side bumpers. The subroutine *CheckFound* determines if the robot has reached the beacon and sets the variable *BeaconFound* to true or false to indicate the current status. If the beacon has been found the while-loop is exited with a Break statement.

```
FaceBeacon:
  while not rBeacon(Red)
    rTurn 1
  wend
Return
```

FIGURE 12.4 This subroutine turns the robot toward the beacon.

```
ForwardTillBlocked:
  while not (rFeel() & 14) AND not (rBumper() & 14)
    rForward 1
    gosub CheckFound
    if BeaconFound Then break
  wend
Return
```

FIGURE 12.5 This code moves the robot forward until it reaches an object or the beacon.

12.1.6 GOING AROUND AN OBSTACLE

If the robot encounters an object, it needs to go around it. It is certainly possible to develop many different ways to go around an object, but we already have one from Chap. 8. All our robot needs to do is follow the edge of the object as if it were a wall. However, if the code from Chap. 8 is used as it was written the robot would just continue to follow around the object forever. We need a way to tell it to stop when it has reached the other side.

The robot has no easy way to determine when it has reached the other side of the object. In fact, if there are other objects close by, the robot might not even be able to get to the other side without causing a collision. An easy solution is to simply let the robot follow the wall for a little while and stop. If you study Fig. 12.1 you will see that if the robot stops too early it will just try to face the beacon again and start over. Obviously, the robot does not have to follow the wall until it gets to the other side; it only has to follow it for a reasonable length of time. The question is "How long is reasonable?"

If the object is small, then a short time is best because we don't want to go all the way around the object. If we always use a short time though it is conceivable that some combination of objects could occur that would *trap* the robot. This might happen if the robot does not go far enough around the object to get a clear (or at least clearer) path to the beacon. In such situations, the robot might simply continue to retrace its steps repeatedly moving toward the goal until blocked, following the wall but not far enough, moving toward the goal again, but essentially in the same situation as before.

One way to solve such a problem is to introduce some randomness into the robot's behavior. If you examine the code in Fig. 12.6 you will notice that it is the same as the code in Chap. 8, but in place of a `while`-loop we are now using a `for`-loop. The `while`-loop in Chap. 8 caused the robot to follow the wall forever. The `for`-loop causes this code to be executed between 20 and 270 times. These numbers were

```
GoAround:
  If BeaconFound Then return
  rTurn -random(150)
  if TurnDir > 0
     FN  = 6
  else
     FN  = 12
  endif
  for i=1 to 20 + random(250)
    while (rFeel() & FN) or (rBumper() &4)
      rTurn -TurnDir
    wend
    rForward 1
    while not rFeel()
      rTurn 5*TurnDir
      rForward 1
    wend
  next
Return
```

FIGURE 12.6 This code follows the contour of an obstacle for a random amount.

chosen experimentally based on the general size and number of objects expected to be in the room.

An `if`-statement at the beginning of the subroutine causes the routine to exit and not attempt to follow a wall if the beacon has already been found. The line after the `if`-statement is very important. It causes the robot to turn away from the wall a random amount (0°–150°). This single statement prevents the robot from being stuck in many situations because the random turning of the robot eventually puts it into an orientation where the sensors allow it to move. Remove the line when you test this algorithm and you will see the robot eventually encounter a situation where it cannot free itself.

12.1.7 DETERMINING IF THE BEACON IS FOUND

The routine in Fig. 12.7 determines if the robot has reached the beacon. Remember the `rBeacon()` function returns the distance to the beacon if it is directly ahead of the robot. If the robot has just faced the beacon and the beacon is less than 20 pixels away then the robot has reached the goal. Notice the use of the function `Within()`. We need to check if the returned value from `rBeacon()` is not zero and also less than or equal to 20, so the parameters for `Within()` are 1 and 20. The variable *BeaconFound* is then set to *true* or *false*, depending on whether the beacon value is within 1 to 20 pixels from the robot. Remember a value of 0 means the robot is not facing the beacon.

12.1.8 A POTENTIAL PROBLEM

Combine all the code from Figs. 12.1 to 12.7 into one file and run the program. The program is almost perfect and executes properly nearly all the time. However, if you let it run for an extended period, eventually the obstacles may be placed in such a combination where it is possible for the robot to find its way into a cavity where it cannot escape. One possible solution to this is shown in Fig. 12.8.

Replace the old *FindBeacon* routine with the new one in Fig. 12.8 and also add the new routine *UnStick* as shown in Fig. 12.8 then run the new program.

The basic premise of these additions to the algorithm is randomness. The new routine *FindBeacon* counts the number of attempts to locate the beacon and after 20 attempts it assumes it must be stuck and executes the subroutine *UnStick,* which executes a series of random turns and moves. Notice that when this happens, the counter is reset to zero so that after another 20 failed attempts the robot will again call *UnStick*. When you run the new program long enough, the robot will appear to get stuck, but if you wait long enough, it eventually frees itself by using this routine.

If a robot must deal with a totally unknown environment, especially if that environment itself is randomly changing, the robot needs to have some randomness built into its behavior. Without randomness there is no way to absolutely ensure that your algorithm will be able to handle the infinite number of possible situations that can occur.

```
CheckFound:
  BeaconFound = Within(rBeacon(Red),1,20)
Return
```

FIGURE 12.7 This code determines if the robot has reached the beacon.

```
FindBeacon:
  cnt=0
  repeat
    cnt=cnt+1
    gosub FaceBeacon
    if cnt <20
      gosub ForwardTillBlocked
      if (cnt=1) and (rFeel()&8) Then cnt=10
      TurnDir = 1
      gosub GoAround
    else
      gosub UnStick
      cnt=0
    endif
  until BeaconFound
Return
//============================================================
UnStick:
  if Random(100)<50 Then rTurn 180
  for i=0 to 100+random(200)
    while not(rbumper()&14)
      rForward 1
    wend
    rTurn Random(8)-3
  next
Return
```

FIGURE 12.8 This *FindBeacon* subroutine replaces the original and adds a new subroutine called UnStick.

12.2 Using a Beacon and Camera

Adding randomness to the code is an effective way to handle unknown situations (i.e., try something new). We can improve the algorithm further if we use the `rLook()` function. If you recall, in Fig. 12.6 we used a `for`-loop to make the robot follow the wall a random distance before trying to face the beacon again. If you run the program you will see that sometimes, when the robot has chosen to follow the wall for a long distance, it will actually pass by the beacon without noticing it. This happens because the robot does not check to see if it has a free path to the goal point while it is following the contour of the obstacle.

If after finishing the wall-following, the robot happens to end up on the same side as the goal (cleared the object), the robot is able to look around, see the beacon, and go to it. Unfortunately, sometimes the robot goes too far around the object and ends up with the object still between it and the goal. If the robot had the means to check for a clear path to the goal point while it was going around the object, it could proceed to the goal at the first opportunity.

The original algorithm worked due to the randomness built into the code and the fact that the code repeats until it succeeds, but the robot does not look very intelligent when the robot goes past the goal without noticing it.

Adding a camera to the robot's sensory input can solve this problem. (The camera sensor differs from the beacon sensor in that it cannot see the goal if there are objects blocking its view). The idea is that while the robot is executing the *GoAround* subroutine, we use

```
GoAround:
  If BeaconFound Then return
  rTurn -random(150)
  if TurnDir > 0
     FN  = 6
  else
     FN  = 12
  endif
  for i=1 to 20 + random(250)
    while (rFeel() & FN) or (rBumper() &4)
      if rLook()=Blue then Return
      rTurn -TurnDir
    wend
    rForward 1
    while not rFeel()
      rTurn 5*TurnDir
      rForward 1
      if rLook()=Blue then Return
    wend
  next
Return
```

FIGURE 12.9 This new routine watches for an object under the beacon as it follows the wall.

the camera to constantly look for the goal. If the goal is seen, the *GoAround* subroutine is terminated and the program flow returns to the *FindBeacon* routine where it will immediately face the beacon and *try* to move to it. The robot will try but it may not succeed because there could be an object partially blocking the path to the goal, even if there is a clear line of sight to it. When this happens the robot simply resumes the wall-following behavior.

Figure 12.9 shows how the `rLook()` function can be incorporated to achieve the above solution. Notice that the camera is looking for a blue object. It cannot look for the beacon directly because the beacon is supposed to be up in the air. The `rLook()` function also cannot see the beacon color (red) since it has been designated as invisible. Also it is useless to look for the beacon anyway since it can always be seen even if there are obstructions between the robot and the position directly below the beacon.

What has to be done is to place a blue object below the beacon (marking the goal). The camera (which faces straight ahead instead of being angled upward like the beacon detector) can then be used to look for this object and if there is a clear line of sight (no objects in the way) the `rLook()` function will see it. Adding the following line to the very end of the *SetEnvironment* subroutine (Fig. 12.3 just before the `Return` statement) achieves this:

```
circle bx-7,by-7,bx+7,by+7,blue,blue
```

There is another improvement that can be made that may give the robot a little more effectiveness. In the code so far we programmed the robot to always follow an object clockwise (*TurnDir* = 1). We can allow the robot to follow the object counter-clockwise by changing *TurnDir* to −1. But how should we decide to do this? We could do it on a random basis, but it is important that the robot continues to follow the wall using the same direction for a reasonable amount of time. See Exercise 2 below for more discussion on how to achieve this.

12.3 Using a GPS and Compass

Using a beacon in the previous section was one way to indicate the desired destination to the robot. Adding a GPS and a compass to the robot's sensory capabilities opens up many additional possibilities for effective navigation of the robot's environment. We have seen this approach in Chaps. 4 and 9 and we will use the GPS in coming chapters.

In this section we will investigate how to use the GPS to seek the goal rather than a beacon. The two routines that will be changed are *FaceBeacon* and *CheckFound*. Previously these routines used the beacon to do their work. They will be changed to use the GPS to accomplish the same action. We will not rename the first routine so as not to have to rewrite all the other code in the other routines. This means that to use the GPS we only have to replace these two routines with the ones listed in Fig. 12.10.

You have seen the code in Fig. 12.10 before in Chap. 4 and it was explained there so no further explanation will be given here. The improvement given in Sec. 12.2 is still possible without change. Also there is no need to change the subroutine *SetEnvironment* because even though the beacon is no longer needed, there will be no harm in drawing it.

It might not be obvious why the number 50 was used in the subroutine *CheckFound*. Remember that the GPS returns the position of the robot's center. The robot has a radius of 20 (by default), and if you look at the code in Fig. 12.3 you will see that we have given the beacon a radius of 10. Additionally, remember that the `rFeel()` "function checks for objects in a robot's radius, away from the perimeter of the robot". In order to prevent the robot from colliding with the blue object at the goal point we assume that the robot has reached the goal if it is 20 + 20 + 10 pixels away.

```
FaceBeacon:
    dx = bx-rGpsX()
    dy = by-rGpsY()
    If dx=0 AND dy = 0 Then Return
    Theta = PolarA(dx,dy)*180/pi()+90-rCompass()
    if Theta > 180 then Theta = Theta-360
    if Theta < -180 then Theta = Theta+360
    rTurn Theta
Return
//============================================================
CheckFound:
  rGPS x,y
  // remember beacon is at bx,by
  if PolarR(x-bx,y-by)<50
    BeaconFound = true
  else
    BeaconFound = false
  endif
Return
```

FIGURE 12.10 Using the GPS to face the goal and check if the robot is close.

12.4 Summary

In this chapter you have:

- Learned how to make the robot reach a goal despite obstacles in the way.
- Learned that navigating from one place to another can be achieved by the use of simple instruments and methods [`rBeacon()`] or by the use of more sophisticated devices (`rGps`)
- Learned how simple behaviors can be linked together to form more complex behaviors.
- Seen how previously developed routines can be slightly modified and incorporated with other routines to achieve new behaviors and solve new challenges.
- Seen how randomness can be used to create unusual environments.
- Learned why randomness should play a role in many robot behaviors.

Now, try to do the exercises in the next section.

12.5 Exercises

1. The routine in Fig. 12.4 makes the robot always rotate clockwise when trying to face the beacon. As we have set up the simulation, we know that the beacon should generally be east of the robot. The robot would look much more intelligent if it could decide which way to turn based on its current compass heading. For example, if the robot's current heading is between 0° and 180° it should turn left when looking for the beacon. Otherwise it should turn right. Even if you implement this idea successfully, the robot may still turn the wrong way sometimes, but it will work better most of the time. Add this behavior to the routine in Fig. 12.4.
2. Change the subroutine *FindBeacon* in Figs. 12.1 or 12.8 so that it reflects the algorithm suggested at the beginning of Sec. 12.2. Decide through experimentation if the robot should perform a fixed number of attempts in succession using the same wall-following direction and then switch direction for the next set of attempts, or if it should decide randomly which way to follow every time its movement is blocked. Another way to decide on the direction is to consider how the robot is approaching the object. If it is to the right then follow to the right and vice versa. Another alternative would be to use the relative position of the beacon as a deciding factor.
3. In the code of this chapter there was a lot of use of the color of the beacon (red) in many lines of the code. If we desire to change the color of the beacon to, say yellow, how many lines would have to be changed? Using the editor you can find and replace all occurrences of red with yellow to accomplish the task. However, it is much better programming practice to set a variable at the top of the program to say *BeaconColor = Red*, and then instead of using the word red in the program code use this variable. This way when you want to change the color you just change one line of code in one place to assure that all lines are changed as desired. This concept applies to any other

usage of constants in your programs. Change the code in this chapter to do this. Parameters that affect the algorithm's responses should be placed in variables at the top of the program. Use these variables in the body of the program in place of the literal numbers. This way you can experiment with different parameters without having to search for them in the body of the program. Do this with the various parameters used in this chapter (an example is the parameters in *CheckFound*).

CHAPTER 13

CHARGING THE BATTERY

No mechanical or electrical device can function without some form of energy to power it. The sensors, motors, and computers in robots are mechanical, electrical, and electronic devices that need power to function. There are many ways an autonomous mobile robot can be powered to be able to do its work:

- An engine that runs on some form of fuel (gasoline, hydrogen, propane, etc.) can generate the mechanical motive force to propel the robot, and also generate electricity for powering the numerous electronics. This kind of robot would only be suitable outdoors where the noise and exhaust fumes would do no harm. However, the fuel supply would have to be replenished sooner or later.
- An umbilical cord that connects the robot to a power source and even a central computer is a possible solution, however, this limits the robot's ability to be autonomous and the cord can create problems.
- Solar energy, in combination with a battery to provide the power for electrical motors and all the electronics, is definitely an ideal solution. This solution is used on robots like the Mars Rover. With this powering method the robot is able to move around forever and would never need to seek a recharging station. However, this solution can become expensive and we may need a hefty solar panel and access to effective light energy (the sun or bright lights).
- The standard method for most cases is a rechargeable battery that powers all the motors and electronics. This battery may be lead-acid, lithium-ion, or any other type that provides a high energy to weight ratio. No matter how efficient these battery types are, they still need to be recharged sooner or later from a source of electrical energy.

In order for the robot in RobotBASIC to simulate a real robot we gave it a method for simulating a battery that discharges and can be recharged. In this chapter we will examine ways of making our robot seek a recharging station whenever it senses that its battery is in need of replenishing.

13.1 The Robot's Battery

By default, our simulated robot does not care about the battery level. It will function regardless of the battery's charge condition. This enables us to not worry about the battery while developing and prototyping solutions for certain problems. However, to make the simulation more realistic, we should consider the battery sooner or later. This is accomplished by ordering the simulated robot to not ignore the battery condition.

Every time you use a sensor [`rBumper()`, `rFeel()`, etc.] or when you issue commands to make the robot move (`rForward`, `rTurn`) the robot's battery discharges a little (motors use twice as much power as sensors in the simulation). To oblige the robot to heed the battery's condition you have to issue the command:

```
rIgnoreCharge false
```

What happens if the robot runs out of battery charge and the command above has been issued? Any commands to make the robot move or turn will cause an error and any functions returning values from sensors will return useless data (see Sec. C.9). If you want the robot to stop heeding the battery level execute the command again with the value *true*.

There are many ways to enable a robot in real life to detect the charge level on its battery. You could, for example, use a digital voltmeter to determine how much the battery voltage drops when the battery is supplying current. Our robot has this ability by using a function that returns the percentage of battery charge remaining. The statement below assigns the remaining percentage charge to a variable *B*.

```
B = rChargeLevel()
```

If the battery is 70 percent depleted, *B* will have the value 30, indicating that there is 30 percent capacity remaining.

Run the program in Fig. 13.1 and leave it running for a while. The program makes the robot move back and forth on the screen and reports its battery charge level. Once

```
MainProgram:
   rLocate 50,200,90
   rIgnoreCharge false
   while true
      XYString 3,3,"Charge=",rChargeLevel(),"%    "
      rForward 700
      rTurn 180
   wend
End
```

FIGURE 13.1 Discharging the battery.

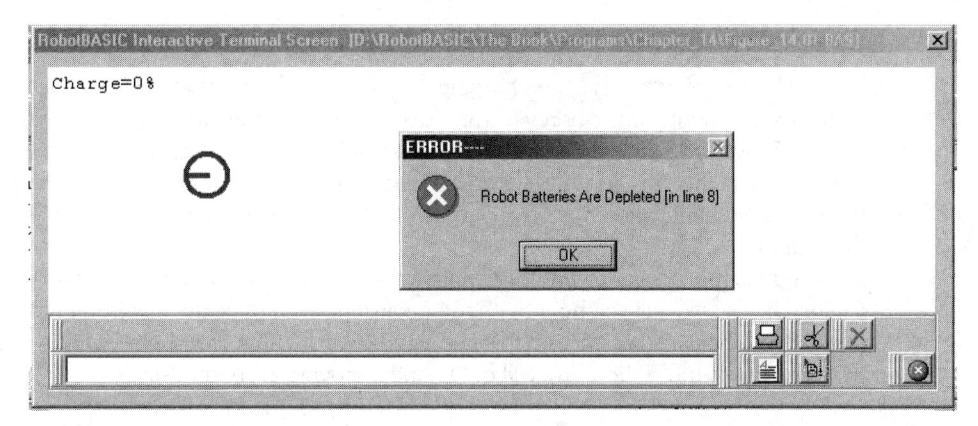

FIGURE 13.2 Battery depletion error.

the battery is depleted the next statement that tries to make the robot move will cause an error as shown in Fig. 13.2.

Real robots will have to connect themselves (or be connected) to a terminal to recharge the battery, and the process may take hours. It serves no purpose for a simulator to simulate hours of recharging. Our robot can be recharged with the command:

```
rCharge ExprN
```

ExprN is an expression that results in a value between 1 and 100. If you pass a value outside these limits RobotBASIC will assume the closer limit.

This command instantaneously recharges the battery to the level you specify. Simulating time delays or having to be at a particular place and orientation can be done programmatically. The following sections will explore methods to accomplish this.

13.2 Real-World Charging

The battery must be removed from many robots in order to charge it. However, it would be more convenient if the robot's battery could be charged by just plugging the robot into some mechanism without having to remove the battery at all.

There are many situations where it would be desirable for the robot to be self-charging. If the robot is to be fully autonomous it certainly ought to be able to:

1. Recognize that its battery needs a charge.
2. Abandon any action it is currently performing to seek a recharging station.
3. Reach the charging station promptly.
4. Orient it self correctly and dock with the charging outlet.
5. Monitor the charge level and wait until the battery is fully recharged.
6. Go back to performing its duties.

13.2.1 FINDING THE STATION

To enable the robot to find the charging station it can be marked with a beacon that is visible to the robot from wherever it is likely to be while performing its duties. Another way is to give the robot the GPS (global positioning system) coordinates of the station. The robot can then use its GPS system to reach the station. In either case, the robot may have to negotiate around obstacles and avoid objects while it is making progress toward the station.

Going from one spot to another while avoiding obstacles and objects was covered in Chap. 12, and we will see how to negotiate around a complicated environment, such as a house or office in Chap. 15. In this chapter we will assume that the robot is in the same general area as the station and will only handle moving around obstacles.

13.2.2 THE CHARGING STATION

There can be many schemes to couple a robot to a charging station; the variety of methods is only limited by your ingenuity. A simple method is to place electrodes on the back of the robot at the same height as corresponding terminals on a battery charging station. In order to make it easy for the robot to make contact with the terminals you could make each terminal a 1-in^2 copper or aluminum plate mounted on a spring. The spring ensures a good connection when the robot backs into the charging station until its rear bumper triggers, letting the spring-tension hold the terminals together.

The 1-in^2 sized terminals would allow the robot a reasonable tolerance when approaching the station. However, it is important that the robot approaches the charger at the correct angle otherwise a proper connection with the charger will not be possible.

13.2.3 ENSURING A PROPER APPROACH ANGLE

There are many ways to make sure the robot approaches the charger at the correct angle. If the robot is equipped with a GPS and a compass it can approach the charging station and orient it self using the GPS and compass. In this chapter however, we are going to assume there is a line on the floor as shown in Fig. 13.3. A beacon is placed over the line so the robot can locate it. Once it reaches the line, the robot follows it until it reaches the charger. Notice how the line is shaped so that the robot will reach the charger regardless of which direction it follows the line.

13.3 The Simulation

In the simulation we are going to let the robot randomly roam around the room (as in Chap. 5). While roaming it will constantly monitor its battery charge level. If the battery level drops below a specified threshold, the robot will abandon roaming to immediately seek a charging station by triggering a goal following behavior (as in Chap. 12). The goal-following behavior locates a beacon serving as a marker for the station. As you have seen in Chap. 12, the robot homes in on the beacon avoiding obstacles by going around them while it is making progress toward the beacon.

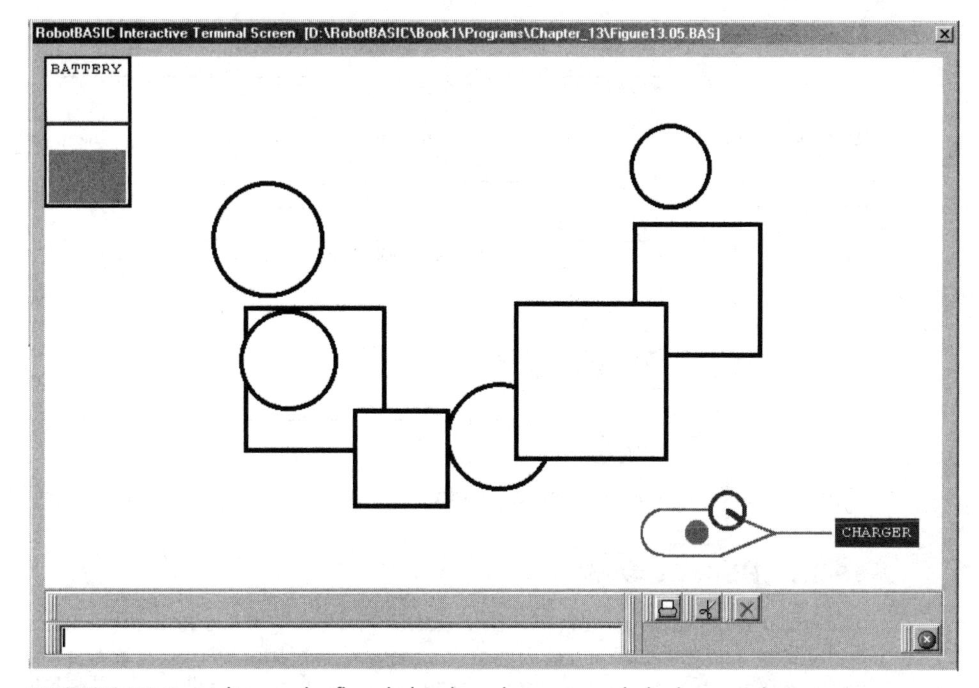

FIGURE 13.3 A line on the floor helps the robot approach the battery-charger at a proper angle.

As the robot approaches the beacon, it needs to check for the line on the floor. When the line is acquired, the robot follows it (Chap. 7) until the charger is reached. Upon reaching the charger, the robot turns around and backs into the charging station for a proper docking procedure and charges the battery. Once the battery is charged, the robot returns to randomly roaming the environment until it needs to find the charger again.

13.3.1 SUBROUTINES HIERARCHY CHART

To achieve the behavior sequence described above we will use many of the subroutines developed in earlier chapters. These routines will be changed slightly to allow for the correct flow of control from one behavior to another. The changes in the routines will be to give them two additional abilities:

1. To recognize when to abort the behavior.
2. To check and/or display the battery charge level.

As your programs become progressively more complicated and subroutines become numerous, the calling sequence becomes intertwined and harder to follow. A diagram of the subroutines hierarchy is an indispensable aide to understanding the overall structure of a program. The chart in Fig. 13.4 is such a diagram.

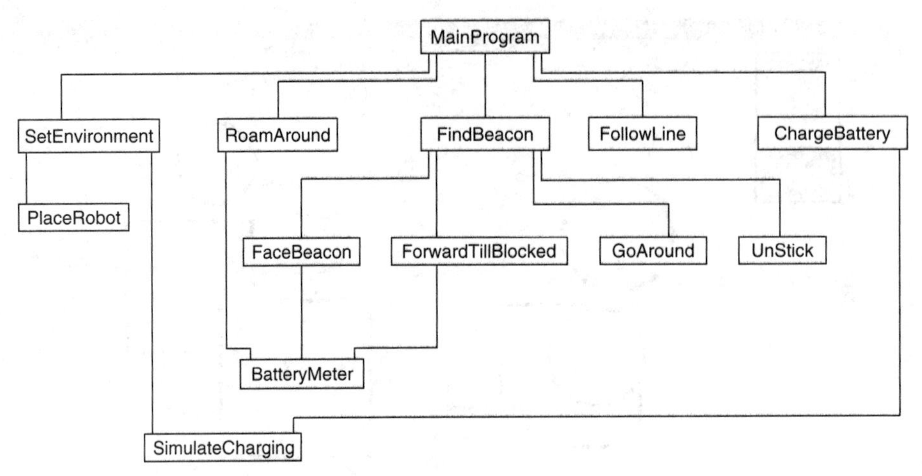

FIGURE 13.4 Subroutine hierarchy chart.

13.3.2 THE PROGRAM

The program is shown in Fig. 13.5. Refer to Fig. 13.4 to see how the subroutines call each other. The chart is not a calling *order;* rather it is hierarchy diagram. The chart shows which routine calls which, but not the order in which they are called. For this information you will either need a flowchart or pseudocode. The language of RobotBASIC can be self-documenting if you use proper names for variables and subroutines so it is almost its own pseudocode. This means that the program is its own documentation.

Refer to the program listing in Fig. 13.5 and review the chart in Fig. 13.4 to understand the program well. You have seen most of the routines in previous chapters. The only new routines are *ChargeBattery*, *SimulateCharging,* and *BatteryMeter*.

The *BatteryMeter* subroutine erases (by drawing a white line) a little of the meter level drawn previously by *SimulateCharging*, which draws a progressively rising colored rectangle to simulate a charge level meter. This is of course to simulate the robot being charged. The actual charging (as explained before) is done by the `rCharge` command and occurs instantaneously. The *ChargeBattery* routine is invoked after the robot has followed the line into the charging station (*FollowLine*). The routine makes the robot turn around so it can back into the charging station and then reverses until its rear bumper closes. The routine then invokes the charging simulation.

At the top of the main program there is a set of constants and variables. The constant *MAX_CHARGE* is set to 100 percent but you can change this number to allow for recharging to a lower level if you want to test the program and do not want to have to wait too long. You can also achieve the same action by raising the *LOW_LEVEL* value. This value is the level at which the robot will start seeking a charging station. The variable *Battery* holds the value of the batteries current charge level. Notice how it is set by the *BatteryMeter* routine. This value will be compared to the *LOW_LEVEL* constant while roaming to determine if it is time to abandon roaming and look for the charger. Most of the subroutines call *BatteryMeter* to update the display and the value *Battery*.

The *FindBeacon* subroutine and its subordinate routines are almost identical to those in Chap. 12. The only difference is that now they call the *BatteryMeter* subroutine to

```
//---Variables & Constants
  LOW_LEVEL    = 65
  MAX_CHARGE   = 100
  BeaconX      = 600
  BeaconY      = 520
  Battery      = 0

//==========================================
MainProgram:
  GoSub SetEnvironment
  While true
    gosub RoamAround
    gosub FindBeacon
    gosub FollowLine
    gosub ChargeBattery
  wend
End
//==========================================
SetEnvironment:
  // Draw four circles and four squares
  for i=1 to 4
     SetColor Black
     LineWidth 4
     x = random(420) + 120
     y = random(250)+60
     size = random(50)+50
     circle x,y,x+size,y+size
     x = random(450)+100
     y = random(250)+100
     size = random(100)+50
     rectangle x,y,x+size,y+size
  next
  //--draw docking line
  SetColor Green
  LineWidth 2
  gotoxy BeaconX+100,BeaconY
  lineto BeaconX+50,BeaconY
  Lineto BeaconX,BeaconY-20
  Lineto BeaconX-50,BeaconY-20
  Line BeaconX-50,BeaconY+20,BeaconX,BeaconY+20
  Lineto BeaconX+50,BeaconY
  Arc BeaconX-70,BeaconY-20,BeaconX-
30,BeaconY+20,DtoR(90),DtoR(180)
  //draw beacon
  circle BeaconX-30,BeaconY-10,BeaconX-10,BeaconY+10,red,red
  //draw charging station
  Rectangle BeaconX+105,BeaconY-
12,BeaconX+180,BeaconY+12,blue,blue
  SetColor White,Blue
  xyString BeaconX+110,BeaconY-10,"CHARGER"
  //--Battery meter
  SetColor Black,White
```

FIGURE 13.5 This program shows how the robot can charge its own battery when it becomes depleted.

```
  LineWidth 3
  Rectangle 0,0,76,130,black
  gotoxy 1,123-LOW_LEVEL
  LineTo 75,123-LOW_LEVEL,3,Brown
  SetColor Black,White
  xyString 5,2, "BATTERY"
  //--charge the robot initially
  GoSub PlaceRobot
  GoSub SimulateCharging
return
//=========================================
PlaceRobot:
  rLocate BeaconX+80,BeaconY,-90
  // designate the beacon and line colors as non-objects
  rInvisible Green,Red
  rIgnoreCharge False
Return
//=========================================
RoamAround:
  while true
    GoSub BatteryMeter
    if Battery < LOW_LEVEL then return
    // forward until an object is found
    while not (rFeel( )&14) AND not (rBumper()&14))
       rForward 1
    wend
    D=random(100)
    for i=0 to random(50)+50
      if D>50
        rTurn 1
      else
        rTurn -1
      endif
    next
  wend
return
//=========================================
FindBeacon:
  cnt=0
  Repeat
    GoSub BatteryMeter
    cnt=cnt+1
    gosub FaceBeacon
    if cnt <20
      gosub ForwardTillBlocked
      // decide to follow left or right
      if (cnt=1) and (rFeel()&8) then cnt=10
      if (cnt<10) or (random(20)<2)
        TurnDir = -1
      else
        TurnDir = 1
      endif
```

FIGURE 13.5 *(Continued)*

```
      GoSub GoAround
    else
      gosub UnStick
      cnt=0
    endif
  Until rSense()&2 // Line Found
Return
//=========================================
FaceBeacon:
  while not rBeacon(Red)
    rTurn 1
  wend
  // Show battery condition
  GoSub BatteryMeter
return
//=========================================
ForwardTillBlocked:
  while not (rFeel() & 14) AND not (rBumper() & 14)
    rForward 1
    if rSense() then break
  wend
  GoSub BatteryMeter
return
//=========================================
GoAround:
  if rSense() then return
  If TurnDir > 0
     FN = 6
  Else
     FN = 12
  Endif
  for i=1 to 20 + random(250)
     While (rFeel()&FN) or (rBumper()&4)
       rTurn -TurnDir
     Wend
     rForward 1
     rTurn TurnDir
  Next
Return
//=========================================
UnStick:
  if random(100)<50 then rTurn 180
  for i=0 to 100+random(200)
     while not(rbumper()&14)
        rForward 1
     wend
     rTurn random(8)-3
  next
Return
//=========================================
```

FIGURE 13.5 *(Continued)*

```
FollowLine:
  while rFeel()=0
    rForward 1
    while rSense() & 1
      rTurn 1
    wend
    while rSense() & 4
      rTurn -1
    wend
  wend
Return
//=========================================
ChargeBattery:
  rTurn 180
  while not rBumper()
    rForward -1
  wend
  GoSub SimulateCharging
Return
//=========================================
SimulateCharging:
  SetColor LightMagenta
  for i= Battery to MAX_CHARGE
   gotoxy 5,125-i
   lineto 70,125-i
   Delay 100
   gotoxy 1,123-LOW_LEVEL
   LineTo 75,123-LOW_LEVEL,3,Brown
  next
  rCharge MAX_CHARGE
  Battery = MAX_CHARGE
Return
//=========================================
BatteryMeter:
   // Show battery condition
   Battery=rChargeLevel()
   SetColor White
   gotoxy 5,120-Battery
   lineto 70,120-Battery
   gotoxy 1,123-LOW_LEVEL
   LineTo 75,123-LOW_LEVEL,3,Brown
Return
//=========================================
```

FIGURE 13.5 *(Continued)*

update the meter display, and also they are made to abandon the behavior once the line on the floor is sensed by the `rSense()` function.

RoamAround is the same as discussed in Chap. 5 but with the added call to *BatteryMeter* to update the display and the value *Battery*. There is also a check to see if the battery charge level is below the *LOW_LEVEL* threshold to be able to abandon the behavior.

13.4 Summary

In this chapter you have:

- Learned about RobotBASIC's simulated battery and how to use it in a program.
- Seen how several of the algorithms from previous chapters can be combined to give the robot the ability to locate a charging station, dock with it, and replenish its battery.
- Seen how to organize a program using a subroutines hierarchy chart.
- Learned how to use programming to simulate real-time actions.
- Seen more examples of how to reuse previously developed behaviors and algorithms with slight changes to achieve new behaviors.

Now, try to do the exercises in the next section.

13.5 Exercises

1. Run the program in Fig. 13.5 to see how it performs. Study the code to see how the algorithms from previous chapters have been modified to work together, passing control from one behavior to another, and how the main program acts as a sequencing manager to invoke the appropriate routines in the correct sequence once a behavior terminates. You may want to change the value in *MAX_CHARGE* to allow for a shorter wait before the battery is depleted.
2. Think of other ways to locate and home in on a charging station. Write a program to demonstrate your ideas.
3. The subroutines *FollowWall* and *FollowLine* in Fig. 13.5 do not call the subroutine *BatteryMeter*. This means that the meter display will not be updated while these routines are running. Change these routines to correct this.
4. What happens if you set *LOW_LEVEL* to a low number (say 10 precent)? If the robot cannot reach the charging station before its battery is fully depleted what happens? If the low-level value is set too low and the robot happens to be in a place where it cannot reach the station without having to go around objects, what situation may occur?

CHAPTER 14

NEGOTIATING A MAZE

Solving mazes is a popular pastime for many people. There are mazes in puzzle books and magazines, and even on placemats in some family restaurants. People love mazes so much that many parks and gardens around the world have hedge mazes. People even build mazes for rats and mice so that they can conduct behavioral experiments. Making our robot solve a maze is an enjoyable challenge. Navigating mazes is also a popular contest in many robotic clubs.

All mazes have a common feature. You start at one point in the maze then try to follow a path that leads to a goal point elsewhere in the maze. The challenge is to reach the goal point without encountering too many dead-ends that oblige us to retrace paths already taken.

There are many types of mazes:

- Line mazes, where you follow a line that has branches that can lead to dead ends.
- Corridor mazes, where you trace a path in the center of a labyrinth of corridors that can lead to dead ends.
- Offices, homes, cities, and highways are also mazes that we negotiate on a regular basis without even realizing it. Every time we travel from one place to another during our daily activities we are actually solving a maze.

In this chapter we will develop algorithms for solving the first two types of mazes. We will only consider mazes with vertical and horizontal lines. In the next chapter we will tackle the more general situations of office and home mazes.

14.1 A Random Solution

If you place a mouse in a maze for the first time, the mouse will have no knowledge of how to exit the maze. It will scurry around taking random turns at junctions. It may encounter many dead ends before it finally happens upon the exit by pure chance. In this section we will develop a simple algorithm that relies on random chance to solve the maze. In the subsequent sections more complicated routines will be developed. This first simple algorithm can be useful in indicating how we can proceed to more intelligent attempts. Additionally, the base program in this section will be used throughout the chapter with changes made only to the routines that need modification to improve the behavior of the robot.

14.1.1 THE PROGRAM

The program of Fig. 14.1 allows the user to place the robot anywhere in a randomly generated line maze, but it has to be over a line. Also, the user is allowed to give the robot an initial heading, but this heading can only be north, south, east, or west. If the robot is placed where its front is not on a line the program will turn the robot to put its front on a line. After placing the robot the user then chooses a goal location. The location must be on a line, but it can be anywhere the user wishes.

After placing the robot and positioning the goal, the program initiates the search. Once the goal is reached the program displays a message and waits for the user to press a key or the right mouse button before repeating the whole action with the same maze as before (see Fig. 14.3). The maze is randomly generated only the first time the program starts. Throughout the program messages are displayed indicating what is happening.

14.1.1.1 *MainProgram* The *MainProgram* is self-documenting and requires no explanation. It also indicates the sequence of actions taken to accomplish the entire process.

The variable *FirstTime* is used by the *PlotMaze* routine to determine whether to generate the maze or use the one previously generated.

14.1.1.2 *DisplayInstructions* This routine displays instructions to the user within a dialog box and then waits for a left mouse click on the OK or Cancel button or pressing the *Enter* or the *Esc* key. The subroutine makes use of the `MsgBox()` function. This function will return the key pressed (OK or Cancel) but no use is made of that information. Read about this function in the IDE help pages.

14.1.1.3 *WaitForMouseOrKey* As the name of the subroutine implies it waits for the user to press any key on the keyboard or the right mouse button.

14.1.1.4 *PlotMaze* This routine generates and plots the line maze. It creates a grid of junctions from which radiate a maximum of four lines. The combination of lines is

```
//-----Variables
    GoalClr = Red
    LnClr = Cyan
//=============================================================
MainProgram:
   FirstTime = true
   gosub DisplayInstructions
   while true
     gosub PlotMaze
     gosub PlaceRobot
     gosub SelectGoal
     gosub SolveMaze
     FirstTime = false
     Message = "Goal Found---Press Any-Key or Right-Mouse"
     Message = Message + "-Button to repeat with same maze"
     gosub DisplayMessage
     Beep
     gosub WaitForKeyOrMouse
   wend
end
//=============================================================
WaitForKeyOrMouse:
  repeat
    readmouse x,y,b
    getkey k
  until k <>0 Or b = 2
return
//=============================================================
DisplayInstructions:
  data IM;"Figure14.01.Bas"
  data IM;"This program creates a random Line-Maze then allows"
  data IM;"you to place the robot anywhere on the maze by"
  data IM;"clicking the Left-Mouse-Button on that position.",""
  data IM;"Keep the mouse button down to make the robot rotate"
  data IM;"to the desired direction. Always make sure the robot"
  data IM;"is facing a line not empty space.",""
  data IM;"Then select any position on the maze to place the"
  data IM;"goal to be found.",""
  data IM;"If there is no connection between the place where"
  data IM;"the robot is and where the goal is then it will not"
  data IM;"be possible for the robot to reach the goal."
  n = MsgBox(IM)
return
//=============================================================
PlotMaze:
  SetColor Black,White
  ClearScr
  if FirstTime then Dim Maze[5,7]
  For i = 0 to 4
    for j = 0 to 6
      S = 0
      If not FirstTime then S=Maze[i,j]
      X =(j+1)*100
      Y =(i+1)*100
      for k = 0 to 3
        if FirstTime and (random(10000)<8000) then S=S|(2^k)
```

FIGURE 14.1 Randomly negotiating a line maze.

```
        if S & (2^k)
          GotoXY X,Y
          dX = 55*round(cos(Pi(k)/2))
          dY = -55*round(sin(Pi(k)/2))
          LineTo X+dX,Y+dY,4,LnClr
        endif
      next
      if FirstTime then Maze[i,j] = S
    next
  next
return
//==============================================================
PlaceRobot:
   Message ="Place Robot"
   gosub DisplayMessage
   while true
     readmouse x,y,b
     if b = 1
       ReadPixel x,y,pc
       if pc = LnClr then break
     endif
   wend
01 //make sure is on center of line
02 xx = x#100
03 if within(xx,96,104)
04     x = (x/100+1)*100
05 endif
06 //make sure is on center of line
07 yy = y#100
08 if within(yy,96,104)
09     y = (y/100+1)*100
10 endif
11 Rx = x
12 Ry = y
13 rLocate Rx,Ry,90
14 rInvisible LnClr
15 repeat
16   delay 400 //allow for too long press
17   readmouse x,y,b
18   if b = 1 then rTurn 90
19 until b <> 1
20 while not rSense()
21    rTurn 90
22 wend
Return
//==============================================================
SelectGoal:
   Message = "Select Goal"
   gosub DisplayMessage
   while true
     readmouse x,y,b
     if b = 1
       ReadPixel x,y,pc
       if pc = LnClr then break
     endif
   wend
```

FIGURE 14.1 *(Continued)*

```
    Gx = x
    Gy = y
    Circle Gx-6,Gy-6,Gx+6,Gy+6,GoalClr,GoalClr
return
//===============================================================
DisplayMessage:
  Rectangle 0,0,800,20,white,white
  Rectangle 0,0,Length(Message)*10,20,Blue,Blue
  SetColor Yellow,blue
  xystring 2,2,Message
  SetColor black,White
Return
//===============================================================
SolveMaze:
  Message ="Searching"
  gosub DisplayMessage
  while true
    S= rSense()
    if S = 0
       rTurn 180
    elseif S &5
       gosub MakeATurn
       rTurn m
    endif
    if rBumper() then break
    rForward 1
  wend
Return
//===============================================================
21 MakeATurn:
22   for MT_i=1 to 20
23     if not (rBumper()&4)
24       rForward 1
25     else
26       m=0
27       return
28     endif
29   next
30   m = 90
31   if rSense() <> 0
32     if random(10000) < 5000
33       m = 0
34       return
35     endif
36   endif
37   if S = 7
38     if random(10000) < 5000 then m = -m
39   elseif S = 6
40     m = -m
41   endif
42 Return
//===============================================================
```

FIGURE 14.1 *(Continued)*

determined randomly. If two adjacent junctions have lines toward each other then the junctions would be connected. We save the maze characteristics in an array *Maze*[ ] to be able to use the maze again. The array holds a binary number (4 bits) to indicate which lines are to be plotted.

The code also takes consideration of whether it is the first time the program is being run which means that the maze should be generated. If it is not the first time the program is run, the maze is plotted from the array *Maze*[ ].

Think of this maze as a city with north-south and east-west streets. Some streets intersect, some are dead ends, and parts of the city may not be reachable from some starting points.

14.1.1.5 *PlaceRobot* This subroutine allows the user to indicate a location (using the mouse) where the robot will be placed. Once the left mouse button is clicked over a position on the screen that also has the color of the line (i.e., it is part of the maze) the robot is placed at that location. If the user keeps the mouse button clicked the robot is rotated (90° at a time) until the user releases the mouse button.

The routine also uses `rInvisible` to set the line color so the robot can use the `rSense()` function to move over the line. The initial robot position is saved in the variables *Rx* and *Ry*. We will not make any use of these in this program, but they will be useful later.

If you review the routine you will notice some lines are numbered. These numbers are used only for the purposes of this discussion. The code in Lines 1 to 22 ensures that the robot center is at the center of the line. Since the line is four pixels wide it is possible for the mouse to have been placed slightly off the center of the line when the user selected the position to place the robot. We need to ensure that the robot is at the center of the line because of the line following routine, which is discussed later.

You should be familiar with the use of the `Delay` command; we use it to make it possible to control the rotation of the robot. Without the delay the mouse input will be too fast for the user to control.

Lines 20 to 22 ensure that the robot's front is over a line. If it is not, the robot is turned until its front is over a line. This ensures that when it starts following the line it will not fail.

14.1.1.6 *SelectGoal* The mouse is used to allow the user to indicate a location on the maze where the goal is to be placed. Once the user clicks the left mouse button a small circle is drawn in red to create an obstacle at the indicated position. This obstacle is used to indicate the final goal of the maze. Once the robot senses this object (think of it as a red cone on the ground) it will have solved the maze.

The position of the goal is saved in the variables *Gx* and *Gy*. These variables are not used in this program, but will be used in later improvements of the algorithm.

14.1.1.7 *DisplayMessage* Most of the subroutines in this program will display a message to the user at the top left corner of the screen. These messages help the user understand what is going on and what is required.

The calling routine sets a variable *Message* with the required text and then calls *DisplayMessage,* which clears the top of the screen and displays the message inside a blue rectangle.

14.1.1.8 *SolveMaze* This routine executes the logic for solving the maze. The robot moves along the lines or turns depending on the result of `rSense()`. The value of `rSense()` is saved in the variable S for multiple usage. If the line sensors do not see the line ($S = 0$) the robot has reached a dead-end and has to execute an about face (180° turn).

If only the front sensor senses the line, the robot goes forward unless it has bumped into the goal. If the goal is found the routine returns back to the main program. If any of the outer sensors are triggered, a junction has been encountered and a turn may have to be made based on the logic in the *MakeATurn* subroutine. No active line following is performed as we did in Chap. 7 because the robot is placed squarely over the center of the line. Since the lines are straight and the robot moves only horizontally or vertically there is no need to check if the robot is coming off the line. The only consideration is to see if the robot is sensing a junction or a dead end. This simplifies the program so we can concentrate on how to solve the maze. Once we understand how to solve mazes the robot can be programmed for a more complicated line-following behavior if you wish.

The algorithm in this initial attempt is simple because the robot simply makes random turns when it encounters a fork in the road.

14.1.1.9 *MakeATurn* At the beginning of this routine the robot is moved forward so that its center is over the junction (Lines 22–29). If, while moving forward, the goal is encountered then the routine terminates and returns to *SolveMaze,* which then returns to the main program.

Once over the junction we have to decide what kind of junction it is. The possible types are shown in Fig. 14.2.

If it is type (2), (4), or (6) then the center line sensor will still detect the line after moving over the junction (Lines 31–36) and `rSense()` will return a nonzero value. In this case we decide whether to continue going forward or turn by random chance (50 percent). If no turn is to be made then the variable m is set to 0 and the routine is terminated.

If a turn is to be made instead of going forward, or if it is a junction of type (1), (3), or (5) [i.e., `rSense()` returns a zero] then we proceed to Lines 37 onward.

The `if`-statements use the S variable to decide what kind of turn it was. S is used because the robot has already moved and the old value of sensors is needed not the current value.

If S equals seven then the junction was either type (1) or (2) and we need to decide whether to turn left or right. A random number is used to make each turn 50 percent of the time (Line 38). If S equals six then the junction is type (3) or (4) and the robot needs

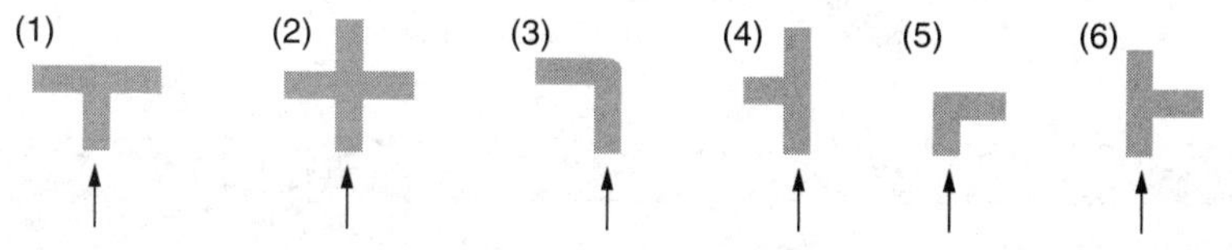

FIGURE 14.2 Types of junctions.

to turn left (Line 40). If *S* is none of these values then the junction is type (5) or (6) and nothing needs to be done since *m* is set to a right turn by default (Line 30).

14.1.2 OBSERVATIONS

This algorithm works, but is unsatisfactory because the robot is no smarter than a mouse wondering aimlessly through a maze. Given the same initial position and goal, the robot may find the goal quickly sometimes and very slowly at others. This inconsistency is due to the robot making totally arbitrary decisions on how to turn.

We will explore various means of improving this algorithm. However, for now, run the program and try various goal and robot positions with different mazes. Observe the behavior of the robot and see if you can think of ways to make the robot act more intelligently.

NOTE: It is possible to place the goal in a section of the maze that the robot cannot reach. Obviously, if you do this, the robot will never be able to solve the maze.

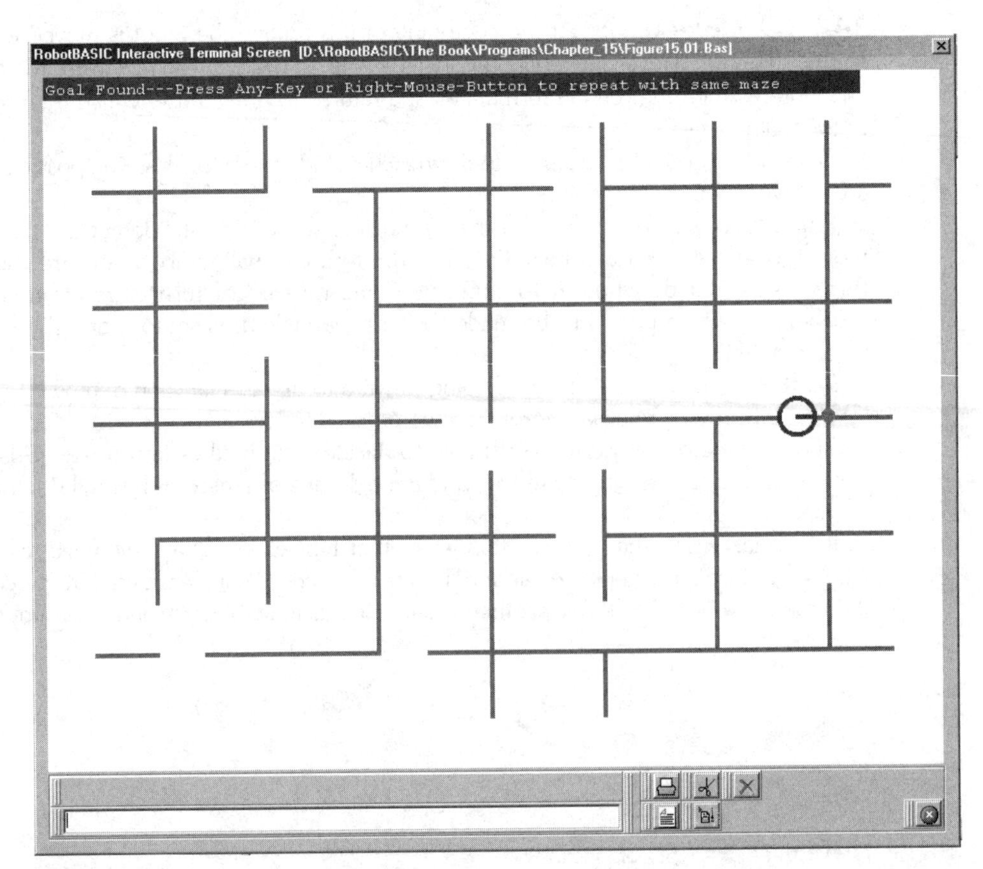

FIGURE 14.3 Solving a line maze.

14.2 A Directed Random Solution

If the robot were a mouse and the final goal a block of cheese then we would expect that the mouse would be able to smell the cheese and tend to take turns that are more toward the direction of the cheese. This may enable the mouse to make *luckier* turns that lead to the goal faster or, depending on the maze, it may cause problems.

The above strategy could get the mouse stuck in a dead-ended section of the maze because it is favoring a direction toward the cheese that has no path to the goal. In practice this simply means that the mouse should generally move toward the smell of the cheese, but it must occasionally make some random choices to prevent getting stuck in a dead-end that faces the cheese.

We can give our robot the same strategy. We will give the robot the ability to *look* for the goal when it is trying to decide how to turn at a junction. Its decision of how to turn will be influenced by the relative direction of the goal and the possible turns that can be made.

The subroutine *MakeATurn* in Fig. 14.4 is a replacement for the one in Fig. 14.1. This new routine implements the logic discussed above. Replace the routine and run

```
01 MakeATurn:
02   for MT_i = 1 to 20
03     if not (rBumper()&4)
04       rForward 1
05     else
06       m=0
07       return
08     endif
09   next
10   m = 90
11   if rSense() <> 0
12      Prcnt = 5000
13      for MT_i = -45 to 45
14          if rLook(MT_i) = GoalClr
15             Prcnt = 500
16             break
17          endif
18      next
19      if random(10000) > Prcnt
20         m = 0
21         return
22      endif
23   endif
24   if S = 7
25     Prcnt = 5000
26     for MT_i = -135 to -45
27         if rLook(MT_i) = GoalClr
28            m = -90
29            Prcnt = 500
30            break
31         endif
32     next
33     if random(10000) < Prcnt then m = -m
34   elseif S = 6
35     m = -m
36   endif
37 Return
```

FIGURE 14.4 A directed random solution.

the program and notice how the robot behaves. There is a definite improvement in its ability to seek the goal. The robot *does* appear to be a lot more intelligent most of the time.

The logic is that we look in the direction of the turn before deciding whether to turn or not. If the goal is in that direction we favor the turn 95 percent of the time. If the goal is not in that direction then we make the turn only 50 percent of the time. The 5 percent randomization prevents the robot from getting stuck in a dead-end that faces the goal. We only take the goal into consideration when trying to decide to turn, no other time. Additionally, we only look for the goal ±45° from the direction of the turn. This means that the goal will not influence the robot's decisions to turn when it is behind the robot, which prevents the robot from getting stuck.

Lines 1 to 10 are exactly as before. Lines 13 to 18 look for the goal ahead of the robot (±45°). Notice that this version of the `rLook()` function is given an angle relative to the robot's heading, otherwise it is the same as you have seen in previous chapters. Read Sec. C.9 for details on this function. If the goal is seen then the robot continues ahead (no turn), but Lines 19 to 22 force a turn anyway 5 percent of the time due to Line 14. If the goal is not seen then the decision to continue ahead or turn is made on a 50 percent basis by Line 12.

In Lines 24 to 33 we decide whether to make a left or right turn on a T or + junction. Lines 26 to 32 look for the goal on the left. If it is found then a left turn is made, but Line 33 forces a right turn 5 percent of the time due to Line 29. Notice that if the goal is not there then 50 percent of the time the turn is changed from right to left due to Line 25.

If the junction is of type (3) or (5) (see Fig. 14.2) then we have to make a turn regardless of the goal. If it is a left we make a left turn, otherwise a right turn is made by default and no further action is required. Notice that the randomness given to the robot to stop it from getting stuck can also cause it to make a wrong turn (5 percent of the time). Also since the randomness is 50 percent if the goal is not a consideration, the algorithm will perform no worse than the fully random one in Fig. 14.1.

Run the program and try out different mazes and different combinations of robot and goal positions. Can you think of ways to improve the algorithm further? We will explore one improvement in the next section.

14.3 A Minimized Randomness Solution

In the previous sections the robot was not given any ability to decide if it has already tried a certain path. When a decision to turn was made, no consideration was given to whether that direction had already been tried. Without this kind of decision-making ability the robot will sometimes retrace the same dead-end route repeatedly; only chance would make it take the correct path (which may take a long time).

An array can be used to store the coordinates of the position of the robot every time it makes a turning decision along with a cumulative value to indicate what decisions it has made. This array can be used every time the robot is over the same spot so that the robot can give more preference to the direction not tried previously. This increases the possibility that the robot will take the correct path. This section presents an implementation of this strategy.

14.3.1 A CORRIDOR MAZE

As mentioned at the beginning of the chapter, there are many types of mazes, so in order to experience how to program the robot in a different kind of maze we will implement the behavior discussed above using a randomly generated *corridor* maze. Different sensors will have to be used, and different logic for moving the robot will be applied, but the principles involved in solving the maze and the logic of handling the elimination of the previously tried paths will be the same regardless of the type of maze used.

The algorithm developed in this section will negotiate the maze the first time taking turns and trying paths while keeping track of which paths lead to dead ends. Once the robot finds the exit, it will be placed back at the beginning to try again. However, this time through the maze, the robot will utilize the information obtained on its first run to avoid all the dead-ends.

A program that implements this strategy is shown in Fig. 14.5. A sample screen shot of the programs output is shown in Fig. 14.6.

14.3.2 THE PROGRAM

The main program calls all the subroutines in turn as needed. The subroutine *DisplayInstructions* is the same as seen before, but with different instruction strings. All the other routines are discussed below.

14.3.3 GENERATING THE MAZE

The *PlotMaze* subroutine in Fig. 14.5 generates a maze with only one path to the goal. Unlike many maze algorithms, it creates cells that are potentially open on all four sides, which results in a more challenging maze. This routine makes use of the *DrawPiece* subroutine to draw a four-cell combination with a random wall. These four-cell combinations are then connected by randomly created doors.

14.3.4 SOLVING THE MAZE

Each time the robot moves to a new cell in the maze, it will first count the number of exits it senses from that cell. There are a maximum of three ways: left, forward, and right (we don't count the way the robot entered). The dimensions of the maze were selected so that the infrared sensors can be easily used to determine where the walls are.

In addition to counting the number of exits in the cell (the variable *cnt*), the robot will also fill the variable *dir* with a value that indicates the direction of the exits. Figure 14.7 shows how the exits are indicated.

Once the values for *dir* and *cnt* are formed the robot has some choices to make. If there is only one exit from the cell, then it takes that exit. If the cell has multiple exits then the robot will check to see if it has visited this cell before (how this is done will be discussed below). If this cell has not been visited, then we save all the information needed later if the cell is revisited (how this is done will be discussed shortly). After saving the data, the robot will try one of the exits. It will always try left first, then forward, and finally right (assuming each are available).

```
// Debugon
// uncomment the above line if you wish to see the robot's
//       "thoughts" as it moves through the maze
//==============================================================
MainProgram:
   gosub DisplayInstructions
   gosub PlotMaze
   gosub SolveMaze
   gosub RetraceMaze
End
//==============================================================
DisplayInstructions:
  data IM;"Figure14.05.Bas"
  data IM;"This program creates a random Corridor-Maze then puts"
  data IM;"the robot at one end. The robot will move through the"
  data IM;"maze until it sees the red wall at the other end.",""
  data IM;"The robot saves all its moves and remembers the ones"
  data IM;"that work. ",""
  data IM;"The second time through, the robot will negotiate the"
  data IM;"the maze perfectly leaving a trail behind it showing"
  data IM;"the path it took."
  n = MsgBox(IM)
Return
//==============================================================
PlotMaze:
  ClearScr
  LineWidth 3
  // draw basic blocks
  y=60
  for j=0 to 3
    x=80
    for i=0 to 4
      gosub DrawPiece
      x=x+119
    next
    y=y+119
  next
  gotoxy 678,536
  lineto 740,536
  lineto 740,362
  SetColor Red
  lineto 678,362
  SetColor White
  LineWidth 4
  // open doors between blocks
  y=60
  for j=0 to 3
    x= 200
    for i=0 to 4
      nn=random(10)
      if (nn<5 and i<4) or j=3
        gotoxy x,y+60
        n=random(10)
        if n<5 then lineto x,y+3
        if n>=5 then lineto x,y+120-3
      else
```

FIGURE 14.5 Reduced randomness solution to a corridor maze.

```
          gotoxy x-60,y+120
          n=random(10)
          if n<5 then lineto x-120,y+120
          if n>=5 then lineto x,y+120
      endif
      x=x+119
    next
    y=y+119
  next
  gotoxy 80,60
  lineto 80,120
  rLocate 108-60,90
  rTurn 90
Return
//================================================================
DrawPiece:
  // draw 4 cell piece at x,y
  rectangle x,y,x+120,y+120
  // draw a random line
  gotoxy x+60,y+60
  n=Random (4)
  if n=0 then lineto x+60,y
  if n=1 then lineto x+60,y+120
  if n=2 then lineto x,y+60
  if n=3 then lineto x+120,y+60
Return
//================================================================
SolveMaze:
  Dim Stack[100,4] //100 records of Compass,Turn,gpsX,gpsY
  StackPtr = 0
  rForward 60
  while rLook() <> Red
    // determine if current cell has more than one exit
    numExits = 0
    dir = 0
    cnt = 0
    if not(rFeel()& 16)
       cnt = cnt+1
       dir = 4 // left side
    endif
    if not(rFeel() & 4)
       cnt = cnt+1
       dir = dir+2 // straight ahead
    endif
    if not(rFeel() & 1)
       cnt = cnt+1
       dir = dir+1 // right side
    endif
    // if there is only one exit, take it
    if cnt = 1
       if dir=4 then rTurn -90
       if dir=1 then  rTurn  90
       rForward 60
       continue
    endif
    if cnt = 0 // no exit, turn around and go back
```

FIGURE 14.5 *(Continued)*

```
       rTurn 180
       rForward 60
       continue
    endif
    // there are multiple exits to this cell
    // see if we have been here before
    gosub CheckBefore
    if yep
debug "been here before"
      // been here before so take the other route
      // first face the saved direction
      while rCompass() <> Stack[StackPtr-1,0]
        rTurn 90
      wend
debug "now facing original direction"
      // must get and resave stack data to elminate wrong turns
      // get original dir
      dir = Stack[StackPtr-1,1]
      // check to see if multiple choices still exist
      if dir=3 or dir=5 or dir=6 or dir=7
         if dir & 4 // left was the first choice taken
            dir =dir & 3  // eliminate first choice
                          // could still be straight or right
         else // check to see if straight was a possibility
            dir = 1 // orig. dir had to be 3
         endif
         Stack[StackPtr-1,1]=dir // save it again
debug "data resaved with last choice removed ",dir
         // now make the next move
         if dir&2   // 2nd choice must be straight
            rForward 60
         else   // or a right turn
            rTurn 90
            rForward 60
         endif
      else // this is at least the second time back here
           // but only one choice remains
           // means we have taken all paths and came back
debug "no choices left"
         // get original heading
         while rCompass() <> Stack[StackPtr-1,0]
           rTurn 90
         wend
         rTurn 180 // go back
         rForward 60
         // remove this node
         StackPtr = StackPtr - 1
debug "bad path - choice romoved from stack"
      endif
    else
      // never been here before and multiple exits
      // Save compass, possible turns, gpsX, gpsY
      Stack[StackPtr,0] = rCompass()
      Stack[StackPtr,1] = dir
      Stack[StackPtr,2] = rGPSx()
```

FIGURE 14.5 *(Continued)*

```
      Stack[StackPtr,3] = rGPSy()
      StackPtr = StackPtr+1
      // assume two exits so take first one
      //(start with left, then CW)
      if dir & 4
        rTurn -90
        rForward 60
      else // forward MUST be first option
        rForward 60
      endif
    endif
  wend
Return
//=============================================================
CheckBefore:
  yep = false
  if StackPtr = 0 then return
  dX = rGPSx()-Stack[StackPtr-1,2]
  dY = rGPSy()-Stack[StackPtr-1,3]
  if PolarR(dX,dY) < 10 then yep = true
Return
//=============================================================
RetraceMaze:
  // now do it again - but no mistakes
  x=rGPSx()  // save current position
  y=rGPSy()
  rLocate 108-60,90 // put robot back in the beginning
  rTurn 90
  SetColor White // and erase old one
  circle x-20,y-20,x+20,y+20
  rPen Down,LightGreen
  rForward 60
  StackPtr = 0
  while rLook()<>Red
    // determine if current cell has more than one exit
    numExits = 0
    dir = 0
    cnt = 0
    if not(rFeel()& 16)
      cnt = cnt+1
      dir = 4 // left side
    endif
    if not(rFeel() & 4)
      cnt = cnt+1
      dir = dir+2 // straight ahead
    endif
    if not(rFeel() & 1)
      cnt = cnt+1
      dir = dir+1 // right side
    endif
    // if there is only one exit, take it
    if cnt = 1
       if dir=4 then rTurn -90
       if dir=1 then  rTurn  90
       rForward 60
       continue
    else
```

FIGURE 14.5 *(Continued)*

```
        // take the exit on the stack
        dir = Stack[StackPtr,1]
        if dir & 4
           rTurn -90
        elseif dir&2
           // do nothing
        else
           rTurn 90
        endif
        rForward 60
        StackPtr = StackPtr+1
     endif
   wend
Return
//==============================================================
```

FIGURE 14.5 *(Continued)*

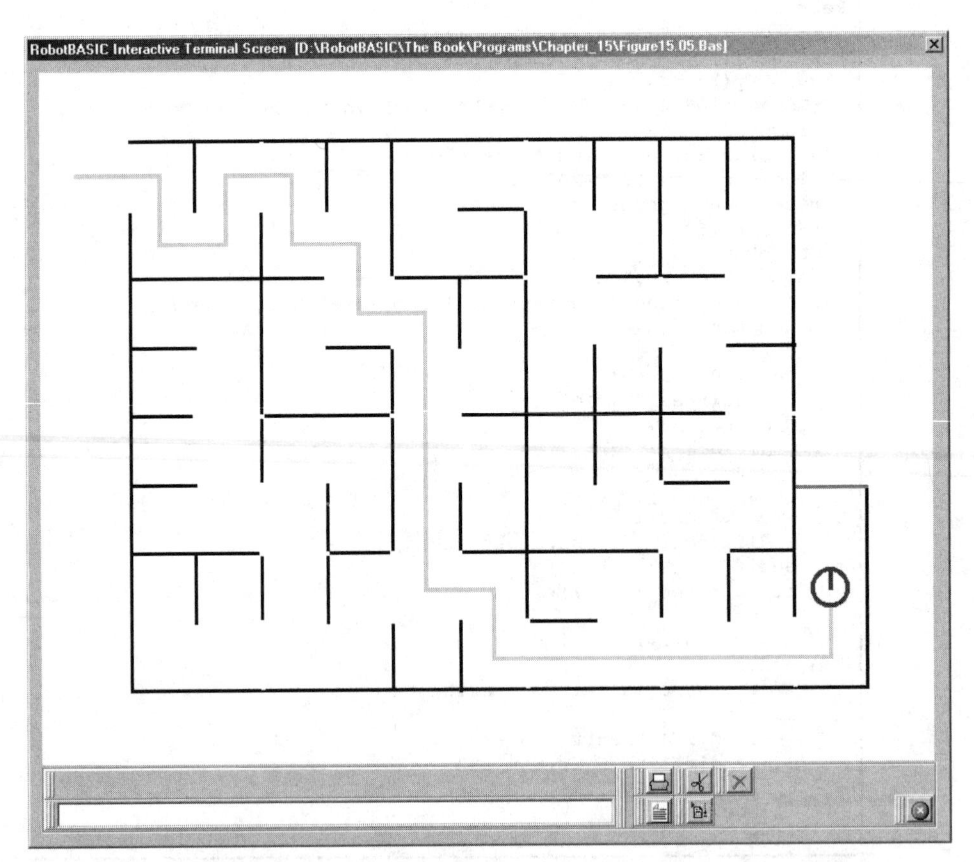

FIGURE 14.6 Corridor maze solving.

dir (decimal)	Binary equivalent	Meaning
0	000	no exits
1	001	exit only to right
2	010	exit only forward
3	011	exit forward or right
4	100	exit only left
5	101	exit left or right
6	110	exit left or forward
7	111	all three exits available

FIGURE 14.7 The variable *dir* holds a code for the possible exits from the current cell.

If the robot determines that it has been to this cell before, it will retrieve the saved information. There are four items that are saved:

1. The GPS (global positioning system) coordinates of the cell, which will be used to determine if the cell has been previously visited (two values the *X* and the *Y* coordinates).
2. The compass heading the robot had when it entered the cell.
3. The value *dir* discussed earlier.

Once it has retrieved the saved data, the robot orients itself so it has the same heading that it did when it originally entered the cell. It then looks at the value of *dir* to see what the possible exits are. The robot knows (since it is revisiting this cell) that the exit it took did not work (or it would not be back to this cell). The robot needs to mark this exit as *not usable*, and it can do so by simply removing the first *one* bit (starting at the left most bit) in the binary number *dir*. Remember, the exits are tried starting from the left, which is why the first *one* in dir represents the choice last made. After the bit is removed, all of the data is resaved to be used again if the cell is revisited.

The robot then looks at the new value of *dir* and uses the first *one* bit found to decide which path to try this time. If the value of *dir* is 0, then the robot knows that all the possible paths have been tried (and all have failed). When this happens, the robot turns around and goes back from where it came. It also needs to forget the information about this cell since it will not be coming back here again. It might not be obvious that the robot will not be back here again so let's examine this idea.

The whole premise of this algorithm is that the robot methodically tries every possible option available to it when it visits a cell that has multiple exits. Each failed try eliminates that option from the future choices available to the robot. When all exits have failed, the robot returns to the last cell with multiple exits and tries the next exit available there. Either one of the choices in that cell will work, or the robot will go backward again until it eventually finds a cell that has an exit that does work.

If the robot repeatedly executes the above logic, it will eventually get out of the maze. In our example a red line is placed at the end so the robot can use the camera to determine when it has completed its task. The data saved by the robot is placed in an array called *Stack*. To better understand the principle of a computer stack, think of a stack of dinner plates. When you want to add a new plate to the stack, you place it on the top of the stack. When you want to get a plate from the stack, you take the one at the top of the stack. This structure can be implemented in our program using an array. A variable is used to point to the array position currently being used. This variable (*StackPtr*) starts with a value of zero. Each time data needs to be put on the stack it should be stored at the position indicated by *StackPtr*. After the data is stored, the pointer is incremented so that it always points to the next available position in the array.

When the last data stored in the array is needed, we retrieve it from position *StackPtr-1*. When all the exits have failed, we discard the data for the cell by decrementing the value of *StackPtr*.

Once the maze has been solved, the array *Stack* will contain only the data (the choices) that worked for each of the multi-exit cells in the correct path through the maze. If the robot tries to run the maze again, it simply has to pull the next piece of data from the *Stack* each time a cell with multiple exits is encountered. The data pulled will indicate which choices are left in the variable *dir*. The first choice indicated is the last choice the robot made for this cell, and consequently must be the correct choice.

The subroutine *SolveMaze* in Fig. 14.5 accomplishes all the above logic. The code is well documented and follows the algorithm described above.

The subroutine *CheckBefore* in Fig. 14.5, determines if the current cell has been previously visited. This is accomplished by saving the robot's GPS coordinates the first time a cell with multiple exits is visited. Later when the robot wishes to see if it has been here previously, it uses this subroutine to compare its current GPS coordinates with the ones saved. Since the robot may not be in exactly the same place as it was previously, the distance between the two points is calculated using the pythagorean theorem [using the `PolarR()` function]. If this distance is less than 10 pixels, we assume that the cell has been visited before.

14.3.5 RENEGOTIATING THE MAZE

As mentioned earlier, once the robot solves the maze, the program uses the subroutine *RetraceMaze* to place the robot back at its original starting position so it can try again. On the second attempt, the robot uses the saved data to make the correct turn each time it encounters a cell with multiple exits. Also, on the second run through the maze, the robot uses its pen to highlight the path it takes.

14.3.6 EMBEDDED DEBUG COMMANDS

If you examine the code in Fig. 14.5 you will see that there are numerous `Debug` commands inside the code. None of these commands will execute unless you remove the comment symbols from the first line in the program. Do this and then execute the program. As the robot progresses through the maze the debug window will pop-up at appropriate situations and tell you what the robot is doing. When this happens analyze the displayed information and click Step to proceed. Hopefully you will find this debugging exercise helpful in understanding the algorithm discussed in this section.

14.4 A Mapped Solution

When we decide to go from one place to another on the highways, we do not just get in the car and drive in the general direction of our destination taking roads that might lead there. Rather we consult a map and plan a route through the maze of roads and then follow that route.

Our robot can be programmed to do the same thing if it is given a map of the maze. The robot can then consult the map to plan a route from the start position to the goal position. The robot would then follow the planned route. This way the robot can determine ahead of time if it is impossible to reach the goal from where it is and not waste time if it cannot.

14.4.1 MAPPING THE MAZE

In order to achieve this algorithm a GPS and a map will be needed. How can we represent this map in the most efficient way for the robot? Since the computer generates the maze randomly, the computer must generate the map as well. Highway and road maps are represented on computers and GPS devices as a list of junctions and a connectivity list showing how the junctions are connected. In computer science this type of data structure is called a **graph.**

There are various types of graphs that differ in how the connection information is represented. Directional graphs represent one-way connections. Weighted graphs represent the distance of connections. Simple graphs are just a connection list where a connection does not have a weight and is bidirectional. This is the type of graph we will use for our maze map.

The array in Fig. 14.8 is a representation of the graph shown below it. A 1 in the array means that there is a connection. A 0 means no connection. Notice that the matrix is symmetrical, which means that if junction 1 is connected to junction 2 then there is also a connection from junction 2 to junction 1. If the graph is not bidirectional then it is not necessarily symmetrical. Also, notice that all the numbers are 1. If the graph is weighted then the numbers represent the length of the connections (like in roads).

We will do the same and represent our maze as an array of connections. This will allow the algorithm to figure out how to go from one place to another using this connectivity array. The graph array will be generated automatically from the array *Maze*[ ].

The elements in each row of the array represent the nodes that have a direct connection to the node with the same number as the row. So, for instance, if you look at row 2 you know that from node 2 you can go directly to nodes 1 and 3, but not node 4. In this example there is no connection from the node to itself (in other scenarios this may not be the case). You can see that the array is a true representation of the diagram below it. In summary, if *Graph[I,J]* = 1 we know there is a direct path from node *I* to node *J*, if it is 0 it means that there is no direct path.

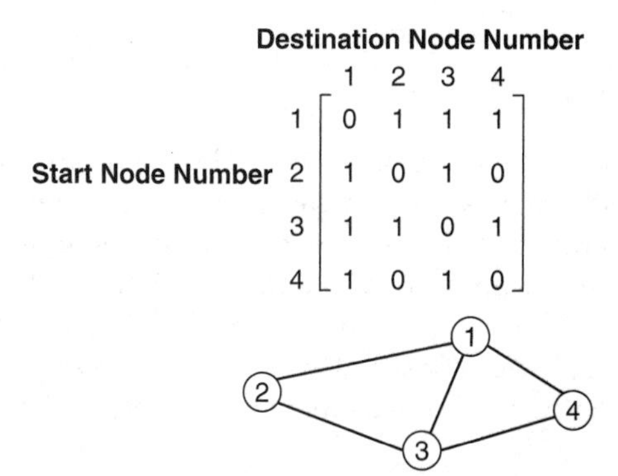

FIGURE 14.8 The graph shown describes the maze below it.

If you look at the diagram (or array), you can see that there is no direct path between 2 and 4. However it is still possible to go from 2 to 4 by following a path 2, 1, then 4 or 2, 3, then 4. In a much more complex graph it might not be so obvious if there is a path, and whether it is the shortest.

Humans are amazing at being able to glance at diagrams like the mazes you saw in the previous sections (see Fig. 14.6 and 14.3) and figuring out in seconds if there is a possible pathway between two points on the maze. We are also very adept at finding a path (not always the shortest) just by looking. A robot is not so fortunate.

Using the above setup, computer routines can be developed to *traverse* the array of connections to determine *a path*, or even the *optimal* path. We will see how to do this in the routines we will develop in the following sections.

14.4.2 THE PROGRAM

The base program in Fig. 14.1 will be used to demonstrate how maps can improve the robot's navigational abilities. We will give only the new and modified subroutines in Fig. 14.9.

The new main program section is not very different from before, except the messages are changed to reflect the fact that the robot now knows if it cannot reach the goal. The variable *Found* is set to true or false by the *SolveMaze* subroutine.

All subroutines not listed in Fig. 14.9 are the same as in Fig. 14.1.

14.4.3 CREATING THE MAP'S GRAPH

The routine *BuildMapGraph* creates the graph of the maze as discussed above. The array *MapGraph*[ ] will hold a 35 × 35 grid of numbers allowing it to represent a maze with up to 35 junctions. Each junction in the maze is given a number (0–34). The array will have the value 0 at the position [*i*, *j*] if there is no connection between junctions *i* and *j*. If there is a connection then the value will be 1. As you can see, since the maze is a two-way maze the position, [*i*, *j*] is the same as [*j*, *i*] (i.e., the matrix is symmetrical).

The way to check if a junction is connected to another is by using the values stored in the *Maze*[ ] array which was created by the *PlotMaze* subroutine.

`If`-statements are used to determine if a line radiates from the junction (north, south, east, and/or west) and if there are lines emanating from any adjacent junction at the same time. If there are, then the junctions are connected, otherwise they are not.

14.4.4 SOLVING THE MAZE

The routine *SolveMaze* calculates the junction nearest to the goal (*GoalNode*) and the junction nearest to the robot (*RobotNode*) using calls to the subroutine *CalcNodeNumber*. The procedure for calculating the nearest node takes into account that there must be a road between the goal or robot and that nearest node. *BuildMapGraph* is called when the program is run for the first time. The routine then calls the *FindPath* subroutine (discussed below) which creates a path (if one is possible) in the array *Stack*[ ] starting at the *RobotNode* and ending at the *GoalNode* showing what intermediate nodes to take to create the path.

```
MainProgram:
   FirstTime = true
   gosub DisplayInstructions
   while true
     gosub PlotMaze
     gosub PlaceRobot
     gosub SelectGoal
     gosub SolveMaze
     FirstTime = false
     if Found
        Message = "Goal Is Found"
     else
        Message = "Goal Is Unreachable"
     endif
     Message = Message+"---Press Any-Key or Right-Mouse"
     Message = Message + "-Button to repeat with same maze"
     gosub DisplayMessage
     Beep
     gosub WaitForKeyOrMouse
   wend
End
//===============================================================
BuildMapGraph:
  Dim MapGraph[35,35]
  MConstant MapGraph,0
  for i = 0 to 4
     for j = 0 to 6
        S = Maze[i,j]
        NN = i*7+j
        if (S&1) and (j<6)
           if Maze[i,j+1]&4
              MapGraph[NN,NN+1] = 1
           endif
        endif
        if (S&2) and (i>0)
           if Maze[i-1,j]&8
              MapGraph[NN,NN-7] = 1
           endif
        endif
        if (S&4) and (j>0)
           if Maze[i,j-1]&1
              MapGraph[NN,NN-1] = 1
           endif
        endif
        if (S&8) and (i<4)
           if Maze[i+1,j]&2
              MapGraph[NN,NN+7] = 1
           endif
        endif
     next
  next
Return
//===============================================================
SolveMaze:
   //get Node Nearest to Goal
   Tx = Gx
   Ty = Gy
   gosub CalcNodeNumber
   GoalNode  = NodeNumber
```

FIGURE 14.9 Negotiating a mapped maze.

```
   //Get Node Nearest to Robot
   rGPS Tx,Ty
   gosub CalcNodeNumber
   RobotNode = NodeNumber
   if FirstTime then gosub BuildMapGraph //Build Graph
   if RobotNode <> GoalNode
      gosub FindPath //Search Graph
   else
      Found = True
      Sp = 0
   endif
   if Found
     if Sp > 0
        For I = 0 to Sp  //for each node on path
           Dnn = Stack[I,0] //destination node number
           x = (Dnn#7 + 1)*100
           y = (Dnn/7 + 1)*100
           gosub GotoPoint
        next
     else
         x = (RobotNode#7+1)*100
         y = (RobotNode/7+1)*100
         gosub GotoPoint
     endif
     // go to the actual goal after reaching its
     // nearest junction
     x = Gx
     y = Gy
     if not rBumper() then gosub GotoPoint
   endif
Return
//==============================================================
CalcNodeNumber:
   //Input----Tx and Ty
   //Output---NodeNumber
   Tj=Tx/100+Round(frac(Tx/100.0))
   if Tj = 0 then Tj = 1
   if Tj > 7 then Tj = 7
   Ti=Ty/100+Round(frac(Ty/100.0))
   if Ti = 0 then Ti = 1
   if Ti > 5 then Ti = 5
   NodeNumber = (Ti-1)*7+Tj-1 //Node Number
   NodeX      = Tj*100
   NodeY      = Ti*100
   dX = NodeX-Tx
   dY = NodeY-Ty
   if abs(dX) > 35
      ReadPixel NodeX-sign(dX)*12,NodeY,Pc
      if Pc <> LnClr then NodeNumber = NodeNumber-sign(dX)
   endif
   if abs(dY) > 35
      ReadPixel NodeX,NodeY-sign(dY)*12,Pc
      if Pc <> LnClr then NodeNumber = NodeNumber-sign(dY)*7
   endif
Return
//==============================================================
FindPath:
   Dim Stack[35,2]
```

FIGURE 14.9 *(Continued)*

```
    Dim Visited[50]
    MConstant Visited,0
    //push on the stack
    Sp = 0
    Stack[Sp,0] = RobotNode
    Stack[Sp,1] = 0
    Found = false
    while true
      i = Stack[Sp,0]
      j = Stack[Sp,1]
      Visited[i] = 1
      if MapGraph[i,j] = 1 and not Visited[j]
         Sp = Sp+1  //push on the stack
         Stack[Sp,0] = j
         Stack[Sp,1] = 0
         If j = GoalNode //Goal found
            Found = true
            Break
         endif
      else
         while true
            Stack[Sp,1] = Stack[Sp,1]+1
            if Stack[Sp,1] > 34
                Sp = Sp-1 //pop the stack
                if Sp >= 0 then continue
            endif
            break
         wend
         If Sp < 0 then break //no more nodes on the stack
      endif
    wend
Return
//================================================================
GotoPoint:
    dx = x-rGpsX()
    dy = y-rGpsY()
    if dx=0 AND dy = 0 then return
    Theta = PolarA(dx,dy)*180/pi()+90-rCompass()
    if Theta > 180 then Theta = Theta-360
    if Theta < -180 Then Theta = Theta+360
    rTurn Theta
    Distance = Round(PolarR(dx,dy))
    For GP_I = 1 to Distance
      if rBumper() & 4 then break
      rForward 1
    next
Return
```

FIGURE 14.9 *(Continued)*

If a path is found the routine uses the information in the created *Stack*[ ] array to follow the path. The goal may not be exactly over a junction and thus the routine causes the robot to make the extra moves necessary to reach the goal from the closest junction.

If the robot's junction and goal's junction are the same then there is no need to call the *FindPath* routine and we bypass following a path, but we do forward the robot toward the nearest junction and then to the goal which may be a little further than the actual junction.

Moving the robot to any point is achieved using the *GotoPoint* subroutine. You have encountered this routine in Chap. 4 and other chapters; it has not been modified.

14.4.5 FINDING A PATH

The routine *FindPath* analyzes the array *MapGraph*[ ] to search for a path from the *RobotNode* to the *GoalNode*. Upon returning from the routine, if a path exists, the array *Stack*[ ] will hold the path. Also the variable *Sp* will hold the length of the path and the variable *Found* will be *true* if a path was found (*false* otherwise).

The algorithm of this routine does not find *the shortest path;* rather it finds *a path.* This path may not be the optimal path or even the most intelligent one. However the algorithm is simple to follow, so we examine it first.

In computer science this principle is called a *depth-first search*. There are other ways to traverse a graph (like a breadth-first search), but depth-first is easier to understand. The idea is to begin at the start node and look in the graph array for the first node that has a connection, then search that node for connecting nodes and so on until the end of nodes or the goal node is reached. If the goal node is not yet reached, we go back one node up to search for the next connected node and so on.

This algorithm will find a path if one exists. However, you can see that the path found may not be optimal due to the method of depth-first searching. If breadth-first searching (with some more logic) is used we would be able to find the optimal path (see next section).

14.4.6 THE OPTIMAL PATH

This algorithm simulates a real-life situation where a GPS system holds a map of the city and knowing where you are and your destination it tells you the shortest path through the city to your destination. Many of these GPS devices can also calculate the fastest path and can reroute you if there is a traffic accident. If you think of our maze as a plan view of a city with north-south and east-west roads then our robot looks like it is traveling through the city.

Figure 14.10 shows a replacement for the *FindPath* subroutine in Fig. 14.9. This new routine implements a breadth-first search through the graph. This method will find the shortest path. The strategy is to look at the start node and then look at *all* the nodes that have a direct connection to it. These nodes are put in a queue to be considered in the same manner as the start node. This process is repeated until we reach the last reachable node of the last reachable node and so on. While doing this we build a list of immediate predecessors for all the nodes.

Think of a queue as a line of people waiting to be served by a teller operator. The first person at the head of the queue will be processed, while a new person joins the queue at its tail. A queue is a first-in-first-out structure, while a stack is a last-in-first-out structure. You can see how the different structures enable the breadth-first and the depth-first searches of the graph.

At the end of the traversal of the graph, we have a list showing what is the immediate predecessor for each node. A predecessor to a node is the node that leads to the node with only one hop while going from the start node to the goal node. You can see how this list can be useful in building a shortest path. All that is needed is to start at the goal node,

```
FindPath:
   Dim Queue[35]
   QHead = 0
   QTail = 0
   Dim Visited[35,3]
   MConstant Visited,999999
   Queue[QTail] = RobotNode //add to queue
   QTail = QTail+1
   Visited[RobotNode,0] = 1  //visited
   Visited[RobotNode,1] = 0  //distance
   Visited[RobotNode,2] = -1 //predecessor
   while Visited[GoalNode,0]<>1 and QHead < QTail
      v = Queue[QHead]  //deque
      QHead = QHead+1
      for w = 0 to 34
        if MapGraph[v,w] = 1 and Visited[w,0] <> 1
           Visited[w,0]=1
           Visited[w,1] = Visited[v,1]+1
           Visited[w,2] = v
           Queue[QTail] = w //add to queue
           QTail = QTail+1
        endif
      next
   wend
   if Visited[GoalNode,0]  = 1
      Found = true
      dim Stack[35]
      Sp = Visited[GoalNode,1]
      Stack[Sp] = GoalNode
      For FP_i = Sp-1 to 0
        Stack[FP_i] = Visited[Stack[FP_i+1],2]
      next
   else
      Found = false
   endif
Return
//==============================================================
```

FIGURE 14.10 Shortest path-finding routine.

go to its predecessor, then the predecessor's predecessor, and so on until we reach the start node. The path from the start node will be the reverse of this list, and it will be the shortest path. The proof of this assertion is a subject in computer science and is too complex to be considered here, but you will be able to see for yourself that it is a true assertion when you run the program. We wrote the new *FindPath* to use the same variables as before. In the routine of Fig. 14.9 a stack was used during the Depth-First search and the array *Stack*[ ] and its pointer *Sp* were set with the required path and its length as a result of the search. In this algorithm we do not use a stack and we end up creating a path list by reversing the travel from the goal node to the start node, visiting the predecessor of each node. This list should be called, for example, *PathList*. However this means we would have to change too many routines. Therefore we will maintain the same naming as before and use the array *Stack*[ ] to hold the path and *Sp* will be the length of that path, and both will be usable in *SolveMaze* as before along with *Found*. The routine also expects *GoalNode* and *RobotNode* to be set as before.

14.5 Final Thoughts

In this chapter the mazes were vertical and horizontal lines or tight corridors; this simplified the techniques for controlling the robot's movement. We concentrated more on the strategies for negotiating the maze. What if the lines were not straight north/south-east/west lines? What if the corridors were wide enough that the robot could not see all the exits at once with its infrared sensors? What if the forks in the lines were not 90° turns, rather more or less, so instead of T(or +) junctions we have Y junctions, or N junctions, and so on? These are problems that must be solved in many real-world situations.

In the next chapter we will devise techniques for negotiating a home or office environment. You can think of these environments as corridor mazes, but the corridors are too wide and the robot has to avoid obstacles and find the doorways.

The techniques of this chapter have applications in many situations other than maze solving. The world is a maze. We humans are so adept at negotiating mazes that we don't even know when we are doing it. It is very satisfying to gain insight into our own thought processes. Trying to program a robot to do what we take for granted gives us an introspective view into our *intelligence*.

The mapping and path-searching techniques we have introduced here are applicable to many situations with real-world significance such as:

Airline scheduling
Delivery truck routing
Process management
Communications networks
GPS systems
Cartography
Artificial intelligence (AI)
Database systems

As you can see, robotics is a very useful as well as entertaining pursuit.

14.6 Summary

In this chapter you have:

- ❑ Learned how to randomly generate line and corridor mazes.
- ❑ Explored various strategies for solving a maze.
- ❑ Seen how directed randomness *can* improve a totally random behavior.
- ❑ Seen how arrays can be used to store a history of actions, enabling the robot to *learn* from its past attempts.
- ❑ Learned to use arrays to generate a map for the robot to consult before trying to solve the maze.

- Learned how to use some advanced computer science structures to give the robot better artificial-intelligence capabilities.
- Seen how debugging statements can be used to trace the "thought-processes" of the robot.

Now, try to do the exercises in the next section. If you have difficulty read the hints.

14.7 Exercises

1. Devise or research different methods to generate random mazes and incorporate them in the programs of this chapter.
2. In this chapter a line maze is generated the first time the program runs. Modify the program to allow the user the option of generating a new maze, or using a maze that has already been saved (giving the file name), or to save the current maze (giving a file name).

HINT: Study the commands `Mwrite`, `Mread`, and `Input` in Sec. C.7. Use these along with a means of presenting the options to the user and accepting a choice from the user.

3. In this chapter we developed four methods for solving the line maze. Write a program that allows the user to select which strategy to use. This way the user can try all strategies on the same maze and can compare and contrast the strategies against a common reference maze. Combine this and the result of Exercise 2 in a single program.

HINT: Present the user with a menu within the main program loop. See the `AddButton`, and `GetButton` commands.

4. In Sec. 14.3 an algorithm was devised to check if a path has been taken so as not to go down that path again. Combine this logic with the logic of the routine given in Fig. 14.4 to make the robot `rLook()` for the goal while avoiding dead-ends.
5. The algorithms in this chapter made the robot move in straight lines and forwarded or turned fixed amounts. With real-life robots it is difficult to have the robot turn or move accurately. There are two ways to solve this problem. One way is to use advanced electronic sensors and motors. The other way is to use software to ensure that the robot did indeed execute the required action.

   To simulate this, RobotBASIC has a command `rSlip` to allow the simulated robot to behave in a random fashion. Study this command in Sec. C.9 and use it in any of the programs of this chapter. Observe how this causes havoc to the robot's behavior. You need to modify all the forwarding and turning commands to allow for this. Implement the required code.

HINT: Think about this. For more ideas and example programs see Chap. 15.

6. When the program in Fig. 14.5 executes, you will notice that the robot makes a lot of seemingly unnecessary turns. It does this because the robot turns to the same heading that it had when it originally visited a cell. After that turn it then makes the correct turn relative to that position. Instead of doing all this turning, it would make more sense to use a little mathematics to determine the proper direction and then turn to that position directly (and turning either left or right based on which would be shorter). Letting the robot "think" about a problem rather than solving it with brute force not only makes the robot look smarter, it uses less battery power and speeds the robot's travel through the maze. This concept is discussed in much more detail in Chap. 16. Modify the program to make it more efficient as described above.

HINT: See the *GotoPoint* routine in Fig. 14.9.

7. The logic of the routine *FindPath* in Figs. 14.9 and 14.10 can be hard to follow. If you place some Debug statements at the right places you will be able to observe how the program traverses the graph to find a path from the start node to the goal node. Do you know where to place these statements and what their contents should be?

HINT: See the program in Fig. 14.5 for an example of how this is achieved.

CHAPTER 15

NEGOTIATING A HOME OR OFFICE

In previous chapters you learned about RobotBASIC's programming environment and its internal commands and functions, which we used to give our robot a repertoire of useful behaviors. Later chapters combined many behaviors to achieve more complex compound behaviors.

We now have enough tools and experience to be able to design a large project. We will show the necessary steps for designing a useful robot. Many of the skills we have developed so far, along with the toolbox of routines and behaviors we have implemented will be used. We will also show how to organize code so that other programmers can understand and modify it if we wish to share it with others.

Here are a few examples of how an autonomous mobile robot can be used to convey a load from point to point:

- Mail, documents, or other articles (e.g., a coffee machine) can be carried from one room to another in an office or home environment.
- Food items and trays can be carted back and forth between the kitchen and customer tables in a restaurant.
- Boxes or crates in a warehouse or depot can be transferred to and from trucks.

- A robotic arm or other form of manipulator can be transported to where it can perform useful work.
- A laboratory of instruments and analyzers can be moved around hostile or inaccessible environments.

All of these examples have one thing in common: the robot must move from point A to point B (perhaps via point C) and wait for further commands. In this chapter we will design a robot to simulate an office messenger that navigates between rooms on the office floor according to user requests.

15.1 The Design Process

Before doing any coding the specifications of the project should be outlined. Once the requirements are known, we analyze what tools and processes are needed to achieve the specified requirements. Often the tools needed to create a project have been designed in previous projects. These tools may have to be adapted to work together and/or made to function in a slightly different manner. The process of determining what tools need to be developed from scratch and which ones can be adapted from previous projects is an essential step in the design process.

A complex task should be divided into a set of less complex subtasks. If these subtasks are complex, then they too should be divided. This process is repeated until you end up with a set of sub and sub-sub tasks that are easy to tackle and design. Some tasks have inputs and no outputs, some have outputs and no inputs, but more often they have both inputs and outputs. The task operates on the inputs in a certain fashion and generates outputs. It is extremely important to be absolutely clear as to what inputs a task requires and what outputs it is going to generate. Knowing the inputs and outputs of a task is part of understanding how it achieves its functionality.

In fact you can consider each subtask as a project in its own right and apply the entire design process to it as you would the main project. Design is an iterative recursive process where you may have to repeat the same process on subtasks and repeat the entire process multiple times until you arrive at a working design. When the design process appears to be complete and you start coding, flaws in the design that cause you to repeat the entire process will become apparent. The final product is influenced as much by the designer's experience and preferences as by the tools available to achieve the requirements.

15.2 An Office Messenger Robot

We will develop a program that allows a user to command the robot to move from one room to another in an office environment. When it arrives at its destination, the robot waits for further commands. If the robot does not receive any further commands within a certain time it will go to a charging station and dock with the outlet. The robot will also return to this station when its battery discharges below a certain threshold.

15.2.1 THE OFFICE SPECIFICATIONS

In an office environment it is reasonable to assume that there is a network of computers that allow each employee to communicate with the robot when required. For this project we will assume that the office is a single floor and robot friendly (i.e., no stairs or steps). We will also assume that the doors are always open or that they open automatically when the robot approaches (see Exercise 1).

We will limit the robot to moving around the office, along a network of paths that guarantees access to all rooms in the office. Also, we assume that any obstacles in the robot's path will be transient objects (such as people) that will eventually move out of the way. Additionally, when the robot is instructed to move to a room it will move to a predetermined spot in that room (see Exercise 2).

The office will have an area designated for the robot's maintenance and recharging. This area will not be available as a destination point in the command program. The robot will go there on its own whenever it requires a charge or has no pending tasks to accomplish.

To maintain simplicity, the command system will not allow a user to issue a request if the robot is still obeying a previous command (see Exercise 3).

Further specifications may arise while designing the individual subroutines and will be discussed as they become needed. Remember, the design of each subroutine should be approached as if it were a project on its own except that consideration must be made for how it will interact with the overall system. We will discuss what tools are to be used and limitations to be tackled in the design of each subroutine. The entire project will be a set of subroutines (as you have seen before).

15.2.2 THE MAIN PROGRAM AND SUBROUTINES HIERARCHY CHART

As an overall procedure for achieving the above requirements we will display instructions to the user, then draw a simulation of the office environment, initialize and place the robot in the office. Finally, we will use an endless loop to wait for a command from a user and execute it. Figure 15.1 shows this process in code. The set of variables above the

```
//---Variables
   TimeOutLimit    = 60
   LowChargeLimit  = 20
   ChargeDelay     = 2000
   RobotSize       = 15
   LnClr           = Cyan
   BcnClr          = Red
   SlipValue       = 0
//=============================================================
MainProgrm:
  gosub DisplayInstructions
  gosub DrawOffice
  gosub MapOffice
  gosub PlaceRobot
  while true
    gosub WaitForCommand
    gosub MoveToRoom
  wend
End
```

FIGURE 15.1 The *MainProgram.* All subroutines will be listed in separate figures.

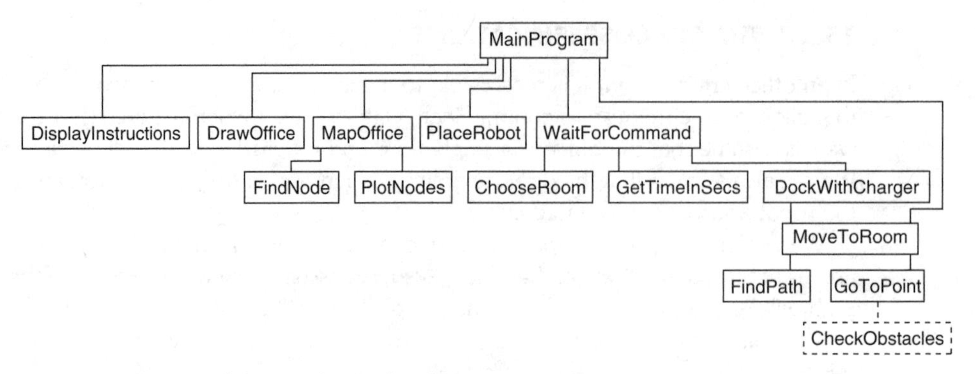

FIGURE 15.2 Subroutines hierarchy chart.

MainProgram label are parameters that affect the operations of the various subroutines and will be discussed as they become needed. They are placed here so they can be easily located for modification.

In order for the robot to move around the office efficiently and effectively a map of the office is needed. This is achieved by the subroutine *MapOffice,* which will be discussed in its own section later.

A subroutines hierarchy chart can be extremely useful for documenting programs that may have to be modified in the future. The chart in Fig. 15.2 resulted when the design process discussed in Sec. 15.1 was applied. Notice that the subroutine *CheckObstacles* is dotted. This routine is not implemented in the first version of the program. However, the routine is required for later improvements of the system. To avoid drawing a new diagram later, it is incorporated here for completeness.

The placement of the subroutines' names is made to minimize crossing lines and to emphasize the hierarchy of calling. When you study Fig. 15.2 carefully, along with the program's listing, you will find that the program's actions are easier to follow. A well documented and annotated listing shows the order of calling the subroutines, and the hierarchy chart keeps you from becoming lost in the depths of nested calls.

Another thing to look for when you study a hierarchy chart is *helper* routines. For example, the subroutine *MoveToRoom* is used by two subroutines. This means that this routine is a utility routine and can be considered as part of a library of utilities. *GotoPoint* is not used by more than one routine but you know from previous chapters that this utility routine has been used many times in various programs.

15.2.3 THE USER INTERFACE

The subroutine *DisplayInstructions* in Fig. 15.3 displays a screen with instructions detailing the actions of the simulation. It makes use of the `MsgBox()` function to display the text in a dialog box and wait for a left mouse click on the OK or Cancel buttons or pressing the *Esc* or *Return* keys. No use is made of the information returned by the function indicating which key or button was pressed.

The *DisplayInstructions* routine could have been a simple set of `Print` statements as you have seen in other chapters. However, it is time, at this advanced stage, to start using

```
//============================================================
//--- Subroutine DisplayInstructions
//--- Inputs  : none
//--- Outputs : none
//--- Calls To: none
//---
//--- displays the instructions in a window and an OK and Cancel
//--- buttons and waits for mouse click on the button or the
//--- Space bar or Enter or Esc
//--------------------------------------------------------------
DisplayInstructions:
   data IM;"Office Messnger"
   data IM;"This program simulates an office messenger robot."
   data IM;"Pressing ""m"" or ""M"" or right-mouse-button brings"
   data IM;"up a command menu.",""
   data IM;"The menu allows you to command the robot to go to"
   data IM;"any room in the office. It also shows stats on the"
   data IM;"robot's current position and the battery level.",""
   data IM;"The robot will go to that room and await further"
   data IM;"instructions. If you do not command it with a new"
   data IM;"room to go to within 60 seconds, or if the battery"
   data IM;"charge level goes too low, the robot will go to"
   data IM;"the charging station and wait for more instructions"
   data IM;"while charging."
   n = MsgBox(IM)
Return
//==============================================================
//==============================================================
//--- Subroutine ChooseRoom
//--- Inputs  : Nodes[], RobotNode
//--- Outputs : GoalNode
//--- Calls To: none
//---
//--- this routine displays a menu of rooms to choose from
//--- once the user selects a room number it sets GoalNode
//--- to the number chosen.
//--- The routine also displays the battery charge level and
//--- the name of the room where the robot is currently
//--- situated when the routine is called. It uses the Nodes[]
//--- array to get the name of the node.
//--------------------------------------------------------------
ChooseRoom:
  if not VType(CR_t)
     CR_t = true
     data CR_btns;"&Cancel","&BreakRoom","Office &1"
     data CR_btns;"Office &2","Office &3","Office &4"
     data CR_btns;"Office &5","&HallWay 1","Hall&Way 2"
  endif
  SaveScr
  Rectangle 55,55,755,555,black,black
  Rectangle 50,50,750,550,cyan,cyan
  ERectangle 52,52,750,550,2,white
  setcolor white,cyan
  xytext 160,80,"Select the Room To Go To:","",15,fs_Bold
  xytext 140,400,"Battery =        Robot is at:","",15,fs_Bold
  setcolor black,white
```

FIGURE 15.3 User interface subroutines.

```
xytext 260,400,Format(rChargeLevel(),"##0%"),"",15,fs_Bold
xytext 493,400," "+Nodes[RobotNode,0]+" ","",15,fs_Bold
for CR_I = 0 to MaxDim(CR_btns,1)-1
   AddButton CR_btns[CR_I],300,110+CR_I*25,150
next
while true
  GetButton CR_btn
  if keydown(kc_Esc) then break
  if CR_btn <> "" then break
wend
RestoreScr
for CR_I = 0 to MaxDim(CR_btns,1)-1
  RemoveButton CR_btns[CR_I]
next
GoalNode = RobotNode
if CR_btn <> "" and CR_btn <> CR_btns[0]
  for GoalNode=1 to MaxDim(CR_btns,1)-1
    if CR_btn = CR_btns[GoalNode] then break
  next
  GoalNode = GoalNode-1
endif
Return
//==============================================================
```

FIGURE 15.3 (*Continued*)

the additional functions and commands that RobotBASIC provides. The subroutine *ChooseRoom* in Fig. 15.3 demonstrates the use of many powerful commands to create an advanced user interface as you can see in Fig. 15.4. You should always refer to the Appendices and IDE help pages for more in-depth information regarding new functions and commands.

In a real-life system we would need a program that runs on a computer network that allows office workers to command the robot. The user interacts with the robot through a PC program and chooses the room to which the robot must go. The PC would have a wireless connection (perhaps through the network) to the robot to specify the desired action. This functionality is simulated by the subroutines *WaitForCommand* (discussed later) and *ChooseRoom*.

The *ChooseRoom* subroutine uses user-interface functions and commands to create a menu for the user to choose which room the robot must go to. The routine also displays the status of the battery and the room where the robot is currently located.

The routine makes use of the command `xyText` to display the text in a bigger bolded font. Also notice the use of the command `ERectangle`. The function `KeyDown()` is another way to check for a keypress. In this case it is used to check for the *Esc* key. The commands `AddButton`, `GetButton`, and `RemoveButton` are used to create a set of buttons that act as a menu for the user to choose the room to go to. The buttons are activated with the `AddButton` command and deactivated (when no longer needed) using the `RemoveButton` command.

Notice the use of the `VType()` function. It is used to create a little trick. The first time through the routine the variable *CR_t* would not have been created yet and the function will return 0. This makes the flow enter within the `if`-block and the first action taken is to assign a *true* to *CR_t*, which effectively creates the variable. Thus the second time

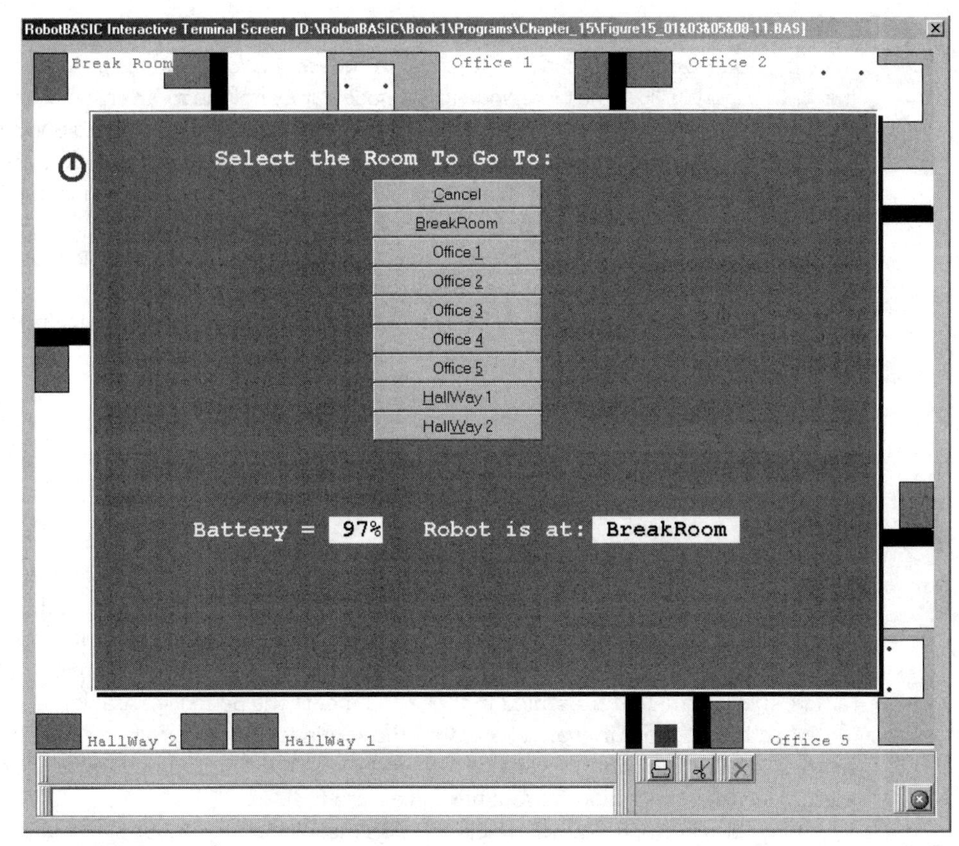

FIGURE 15.4 Choosing a room to go to.

through the subroutine the function will return 105 which is not 0 and thus the `if`-block will not be entered. This prevents the array *CR_btns* from being defined again.

Notice that the routine saves the screen by using the command `SaveScr` and then restores it by using `RestoreScr`. The user interface window is drawn over the office simulation screen. Thus, if we save the screen before we draw the command menu and then restore it after the user chooses a room, the screen will be the same as it was before the command menu window was drawn over it. This makes the command menu appear as if it was a window that opened and closed without affecting the graphics on the screen underneath it.

The routine also displays the robot's current location in the window. The robot's current node number is kept in the variable *RobotNode*. The routine can display this number; however, it is more user-friendly if this number is translated to words displaying the name of the room the robot is in. This is achieved by indexing into an array containing the names of the rooms. This is the array *Nodes*[ ] created in the subroutine *MapOffice*.

The action of the subroutine is to wait for the user to select a button using the mouse or by pressing *Alt+* the underlined character in the button's title. The user is also allowed to press the *Esc key*. The routine will stay in a loop waiting for one of the buttons to be pressed or the *Esc* key. Pressing the *Esc* key or the Cancel button keeps the robot where it is (i.e., the *GoalNode* variable is made to be equal to the *RobotNode* variable). Pressing any of the other buttons will set the *GoalNode* to the node number of the room chosen. Since these rooms have been arranged to be the first eight rooms in the array of nodes and since the array indexing starts at 0 (not 1) then the button's position in the array of button names less one will be the node number of that room. Of course, if the room chosen is the same as the robot's current location, then the robot will remain where it is.

The ultimate outcome of the *ChooseRoom* subroutine is that the value *GoalNode* is set to the desired destination in the office layout.

If you look at the listing in Fig. 15.3 you will see that above the subroutine name there is a set of comment lines that summarize:

1. The inputs the routine requires.
2. The outputs the routine gives.
3. The subroutines the routine calls.
4. The action of the routine.

This method of documenting your work is vital if you are developing nontrivial systems. You will be thankful you spent the time and effort documenting your work in this manner if you ever need to modify the system after the elapse of some time, or if other people will be doing so. Also, if you are going to collect your routines into a library of tools to be reused, the information presented in these comments will be indispensable.

Notice how temporary variables within the routines are named. For example, in the subroutine *ChooseRoom* the variable *CR_I* is used in the `for`-loop instead of *I* or *i* (*CR_* is used due to the fact that the routines initials are CR).

All variables in RobotBASIC are global. This means that all variables are available anywhere in the code. If you have a subroutine that uses a temporary variable (e.g., *i*) and then it calls another subroutine that also uses a variable with the same name (*i* again), then all sorts of problems will arise.

If you are going to set any temporary variables privately within a routine use this naming convention to make it unlikely that the above problem will occur. This is vital when your programs become complicated with subroutines calling other subroutines and so on.

15.2.4 DRAWING THE OFFICE AND PLACING THE ROBOT

The *DrawOffice* subroutine in Fig. 15.5 is similar to the one seen in Chap. 11. The office has been modified a little to have an area dedicated to the robot's charging station (see Fig. 15.6). The various rooms in the office are labeled for the convenience of knowing which room is which, when choosing the room to go to.

The *PlaceRobot* subroutine (Fig 15.5) places the robot so that it is already docked with the charging station. Also, it sets the list of invisible colors and makes the robot heed the battery charge level. The command `rSlip` will be explained later in the chapter.

The last line initializes the variable *RobotNode*. This variable is important for the operations of the system. The robot's initial position is at the charging station. This is designed to be the last node in the array of nodes (see later sections). Thus we set the value to *NodesCount* − 1. *NodesCount* is a variable that holds the number of nodes in the office map (see later sections). The reason we subtract 1 is due to array indexing, as you already know.

```
//============================================================
//--- Subroutine PlaceRobot
//--- Inputs  : BcnClr,LnClr,RobotSize,SlipValue,NodesCount
//--- Outputs : RobotNode
//--- Calls To: none
//------------------------------------------------------------
PlaceRobot:
  rLocate 558,565,0,RobotSize
  rInvisible LnClr
  rSlip SlipValue
  rIgnoreCharge false
  RobotNode = NodesCount-1
Return
//============================================================
//============================================================
//--- Subroutine DrawOffice
//--- Inputs  : none
//--- Outputs : none
//--- Calls To: none
//------------------------------------------------------------
DrawOffice:
  ClearScr
  LineWidth 15
  Data Walls;-165,140,165,0,-357,245,0,245,-590,513,590,600
  Data Walls;-165,140,255,140,-360,140,517,140,-530,513,530,600
  Data Walls;-644,140,797,140,-517,140,517,0,-474,245,699,245
  Data Walls;-474,246,474,419,797,419,-357,247,357,470,113,470
  MPolygon Walls

  Cabinet_H = "rrrddddllluuuu"
  Cabinet_V = "dddllllluuurrrr"
  Desk_H    = "rrrrrrrrrrddddlllluuuulllllddddluuuuu"
  Desk_V    = "dddddddddlllllluuuurrrruuuuullllurrrrr"
  linewidth 1
  //Desks & Cabinets Locations
  Data Furniture; "CH",478,0,"CH",597,559,"CH",769,370
  Data Furniture; "CH",0,252,"CH",0,0
  Data Furniture; "CV",564,0,"CV",40,569,"CV",214,569
  Data Furniture; "CV",169,569,"CV",156,0
  Data Furniture; "DV",348,300,"DV",800,496,"DV",800,0
  Data Furniture; "DH",481,252,"DH",259,0
  //Draw them
  for I = 0 to MaxDim(Furniture,1)-1 step 3
     if Furniture[I] = "CH" then ss = Cabinet_H
     if Furniture[I] = "DH" then ss = Desk_H
     if Furniture[I] = "CV" then ss = Cabinet_V
     if Furniture[I] = "DV" then ss = Desk_V
     DrawShape ss,Furniture[I+1],Furniture[I+2],10
  next
  //Shade them
  Data FF_Cabinets; 10,-17, 10,-271, 488,-21
  Data FF_Cabinets;782,-395, 612,-585, 140,-19, 544,-24
  Data FF_Cabinets; 22,-584, 151,-586, 198,-580
  MPolygon FF_Cabinets,darkgray
  Data FF_Desks;337,-34, 776,-80, 323,-388, 565,-274
  Data FF_Desks;772,-580
  MPolygon FF_Desks,gray
  //Tables
  Circle 59,69,109,119,darkgray,darkgray
  Circle 118,329,168,379,darkgray,darkgray
```

FIGURE 15.5 DrawOffice and PlaceRobot.

```
  //Charger Station
  Rectangle 558-10,554+35-10,558+10,554+35+10,blue,blue
  //Chairs
  Data Chairs;275,27,699,16,500,319,245,316,723,512,75,279
  for I = 0 to MaxDim(Chairs,1)-1 step 2
     X = Chairs[I]
     Y = Chairs[I+1]
     Sp = 35  //leg spacing
     LD = 4   //leg diameter
     Cl = Brown  //color for legs
     Circle X,Y,X+LD,Y+LD,Cl,Cl
     Circle X+Sp,Y,X+Sp+LD,Y+LD,Cl,Cl
     Circle X,Y+Sp,X+LD,Y+Sp+LD,Cl,Cl
     Circle X+Sp,Y+Sp,X+Sp+LD,Y+Sp+LD,Cl,Cl
  next
  data labels; 34,  0,"1=Break Room", 370,  0,"2=Office 1"
  data labels;579,  0,"3=Office 2"  , 590,253,"4=Office 3"
  data labels;241,253,"5=Office 4"  , 652,584,"6=Office 5"
  data labels;224,584,"7=HallWay 1" ,  50,584,"8=HallWay 2"
  setcolor LnClr
  for I=0 to MaxDim(labels,1)-1 step 3
     xystring labels[I],labels[I+1],labels[I+2]
  next
Return
//============================================================
```

FIGURE 15.5 *(Continued)*

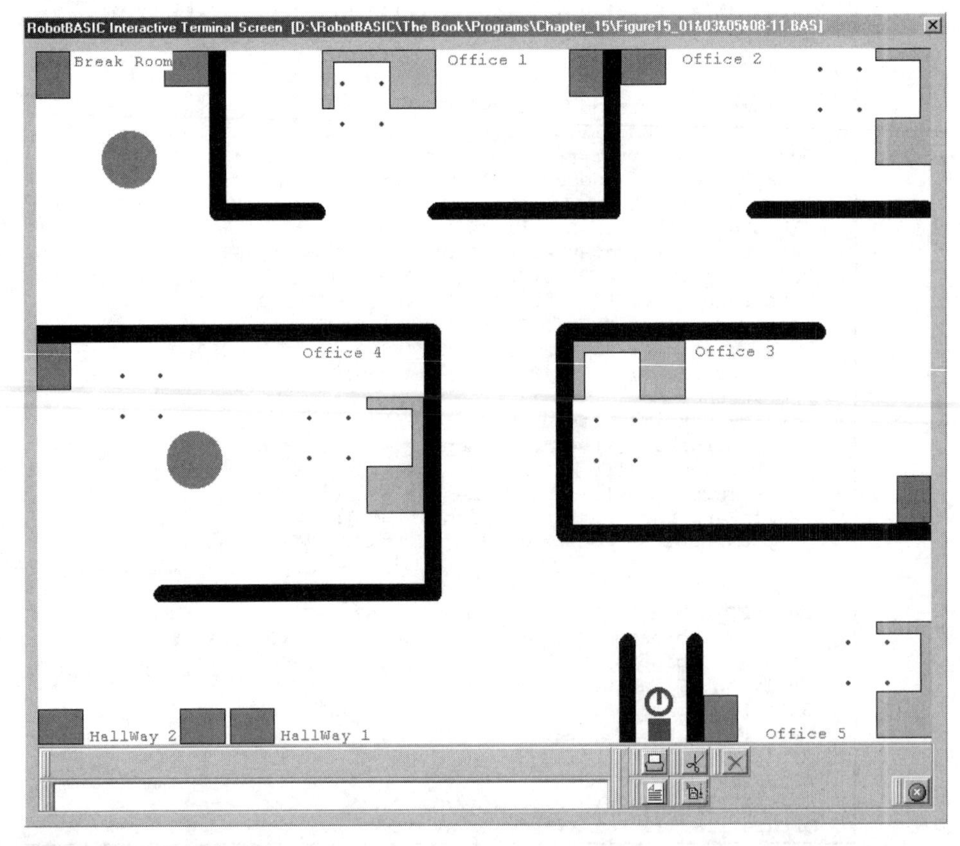

FIGURE 15.6 The office plan.

The parameters *LnClr, BcnClr, SlipValue,* and *RobotSize* are set at the top of the program. They are placed there for easy changing if it becomes necessary, as you will see in the case of *SlipValue* in later sections. *NodesCount* is set by the subroutine *MapOffice* (see later sections).

15.2.5 MAPPING THE OFFICE

If you look at Fig. 15.6 you will see that the office plan is complex enough to baffle the robot if it did not have a planned route from one room to another. It would be impossible to make the robot go to a desired room if it did not have a means of knowing which room is which and how to get there. One solution is to give our robot a map of the office.

You have seen how to do this in Chap. 14, where we used a *graph* to represent the interconnectivity of the various junctions in the maze. The office environment is really a type of maze. The only difference is that there are no lines to follow and the robot can move in any direction, not just horizontally and vertically as before.

Once the map of the office is created (in the form of a graph—see Chap. 14) the process of making the robot move around the map should be similar to what we have seen in Chap. 14. The trick is in how we create the graph.

If you examine Fig. 15.7 you will see that we have drawn a network of *virtual* paths that guarantees the robot access to all the areas of the office. Notice the nodes are not

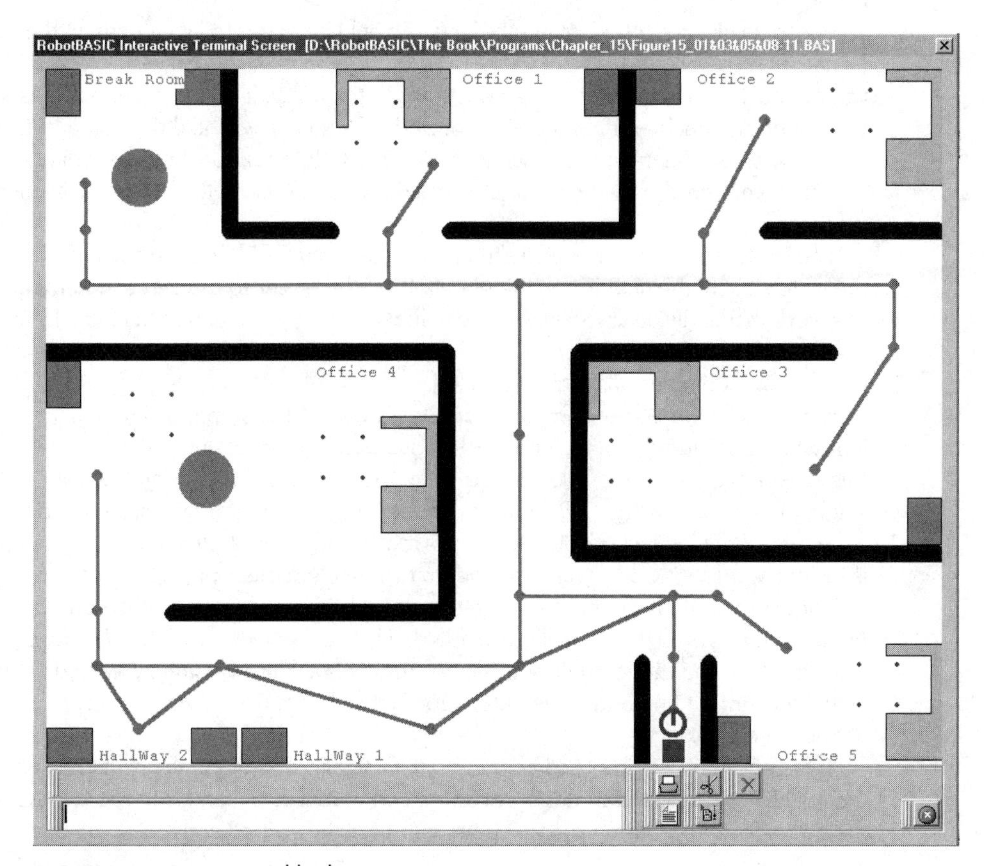

FIGURE 15.7 A virtual highway.

just at intersection points. Some nodes are required as intermediary positions that guarantee the robot the ability to reach the next node without having to negotiate around walls. Also, some nodes are added to minimize the distance between nodes for reasons that will become clear in later sections.

In each room there is a designated spot that the robot will reach and wait for the office employee to interact with it. The robot will not move around the office to any other point. However, if you desire the robot to be able to reach more spots in a room you can design more nodes for it to be able to navigate to (see Exercises).

You can now see how we can map the office. We give each of the nodes in Fig. 15.7 a number and then create an array showing which node is connected to which, just as we have done in Chap. 14. The robot will always move from one node to another. In the command program, simulated by the *ChooseRoom* subroutine, we only allow the user to specify certain nodes that happen to be within each of the rooms. There is no way in our program for the user to command the robot to go to any of the intermediate nodes but we could have done so if we wished. In this example though, there is no logical reason for the robot to stop at any of the intermediate nodes.

Notice that the charging room is not on the list of destinations that the user can command. However, the charging room is a node that the robot will want to go to by itself whenever it is idle for an extended time or when the battery charge drops below a certain level.

We could give each node a number, but it would be better to label them with meaningful names such as "Office 1" or "Office1_D" (for the door) so we can modify and add nodes without having to worry about their positions in the array. This effectively creates a *database* of nodes. Each entry in the array of nodes (*Nodes*[ ]) will hold the name of the node and its *x, y* coordinate in the office (see Fig. 15.8). In a real office there would be a reference point and the *x, y* coordinate would be in reference to that point in inches or centimeters.

The array *Nodes*[ ] is created indirectly via the array *OF*[ ]. We use a set of `Data` statements to specify the data for all the nodes. It is a lot easier to use `Data` statements to do this work than individually assigning the value of each array element in *Nodes*[ ]. We then use the command `MCopy` to copy the contents of *OF[ ]* into the previously dimensioned (to the correct dimensions) array *Nodes*[ ].

Remember, the `Data` command creates a one-dimensional array. This is convenient when entering the data but is not convenient for accessing the data later on. For this reason we create the two-dimensional array *Nodes*[ ] and copy the data from the single-dimensional array *OF[ ]* into it. The `MCopy` command ensures that the correct elements from *OF*[ ] are copied to the correct position of *Nodes*[ ]. So long as the data is in sets of three, `MCopy` will copy the first three elements from *OF*[ ] into the first row in *Nodes*[ ] and then the next three elements into the next row, and so on until all the *NodesCount* rows of *Nodes*[ ] are filled. Notice how the variable *NodesCount* is calculated. As mentioned before in Chap. 9, this is a more versatile way to do this than actually counting by hand. If we later modify the list of nodes we won't have to worry about the count.

Two constraints are imposed on the array. First, all the nodes at the beginning are nodes that will be on the list of nodes that can be commanded. Second, the last two nodes are the nodes for the charging room. The reasons for the second constraint will be discussed

```
//================================================================
//--- Subroutine MapOffice
//--- Inputs   : none
//--- Outputs : MapGraph[], Nodes[], NodesCount
//--- Calls To: FindNode
//---
//--- The Nodes[]array is a data base with name of the
//--- node and the x,y coordinates of it.
//----------------------------------------------------------------
MapOffice:
  //Rooms
  Data OF;"BreakRoom" , 35, 99, "Office1" ,345, 83
  Data OF;"Office2"   ,641, 44, "Office3" ,685,346
  Data OF;"Office4"   , 45,351, "Office5" ,660,500
  Data OF;"HallWay1"  ,343,570, "HallWay2", 82,570
  //Doors
  Data OF;"BreakRoom_D", 35, 139, "Office1_D",305, 141
  Data OF;"Office2_D"  ,585, 142, "Office3_D",757, 240
  Data OF;"Office4_D"  , 45, 468, "Office5_D",597, 455
  //Corridors
  Data OF;"Corridor1" , 35, 186, "Corridor2" ,305, 186
  Data OF;"Corridor3" ,420, 186, "Corridor4" ,585, 186
  Data OF;"Corridor5" ,757, 186, "Corridor6" ,420, 515
  Data OF;"Corridor7" , 45, 515, "Corridor8" ,420, 455
  Data OF;"Corridor9" ,558, 455, "Corridor10",420, 316
  Data OF;"Corridor11",155, 515
  //Charging Station
  Data OF;"ChargeRoom_D",558, 508, "ChargeRoom",558,554
  NodesCount = MaxDim(OF,1)/3
  Dim Nodes[NodesCount,3]
  MCopy OF,Nodes
  Data Edges;"BreakRoom","BreakRoom_D","Corridor1","Corridor2"
  Data Edges;"Office1_D","Office1",-1
  Data Edges;"Office2","Office2_D","Corridor4","Corridor5"
  Data Edges;"Office3_D","Office3",-1
  Data Edges;"Corridor3","Corridor10","Corridor8","Corridor6"
  Data Edges;"HallWay1","Corridor11","Corridor7",-1
  Data Edges;"Office4","Office4_D","Corridor7","HallWay2",-1
  Data Edges;"Office5","Office5_D","Corridor9","ChargeRoom_D"
  Data Edges;"ChargeRoom",-1
  Data Edges;"Corridor2","Corridor3","Corridor4",-1
  Data Edges;"HallWay2","Corridor11","Corridor6","Corridor9"
  Data Edges;"Corridor8",-1
  Dim MapGraph[NodesCount,NodesCount]
  MConstant MapGraph,0
  NName = Edges[0]
  gosub FindNode
  FNode = NodeNumber
  for MO_I = 1 to MaxDim(Edges,1)-2
    if IsNumber(Edges[MO_I])
       NName = Edges[MO_I+1]
       gosub FindNode
       FNode = NodeNumber
       MO_I = MO_I+2
    endif
    NName = Edges[MO_I]
    gosub FindNode
    TNode = NodeNumber
    if FNode >= 0 and TNode >= 0
```

FIGURE 15.8 Mapping the office.

```
        MapGraph[FNode,TNode] = 1
        MapGraph[TNode,FNode] = 1
      endif
      FNode = TNode
    next
    //gosub PlotNodes
Return
//=============================================================
//=============================================================
//--- Subroutine FindNode
//--- Inputs  : NName, NodesCount, Nodes[]
//--- Outputs : NodeNumber
//--- Calls To: none
//---
//--- If Node is not found the NodeNumber will be -1
//--- otherwise it will be the number of the node in the array.
//-------------------------------------------------------------
FindNode:
   NodeNumber = -1
   for FN_I = 0 to NodesCount-1
      if NName = Nodes[FN_I,0]
        NodeNumber = FN_I
        Return
      endif
   next
Return
//=============================================================
//=============================================================
//--- Subroutine PlotNodes
//--- Inputs  : MapGraph[], Nodes[]
//--- Outputs : none
//--- Calls To: none
//---
//--- Uses the arrays Nodes[] and MapGraph[]
//--- to plot a network of the nodes and their connections
//--- using the line color.
//-------------------------------------------------------------
PlotNodes:
  for PN_I =0 to NodesCount-1
    SN_X = Nodes[PN_I,1]
    SN_Y = Nodes[PN_I,2]
    Circle SN_X-5,SN_Y-5,SN_X+5,SN_Y+5,LnClr,LnClr
    for PN_J = 0 to NodesCount-1
      if MapGraph[PN_I,PN_J]=1
        Gotoxy SN_X,SN_Y
        LineTo Nodes[PN_J,1],Nodes[PN_J,2],3,LnClr
      endif
    next
  next
Return
//=============================================================
```

FIGURE 15.8 *(Continued)*

later in the chapter. The reason for the first constraint is to make it easy to find the nodes that will be used on a regular basis. Also, the way we designed the user interface (*ChooseRoom*), assumed that the button number −1 is the node number. This could not have been possible without the first constraint.

The first constraint can be removed if we allow the user to choose the room by name. We then search the database for that name to determine what node number it is. However, the user interface as designed is easy and functional. You will see that we do search the database for a different reason. The subroutine *FindNode* facilitates this. This subroutine goes through the array *Nodes*[ ] looking for the name set in the variable *NName*. If it is found the variable *NodeNumber* is set to the position in the array where that node name is, otherwise it is set to −1. The routine also needs to know the number of nodes in the array; this is the value *NodesCount* set by the *MapOffice* routine.

The reason we search the database is due to the way we specify how the nodes are connected. Rather than saying that node *n* is connected to node *m* and repeating this for all nodes. We create an array (*Edges*[ ] using `Data` statements) that specifies paths by listing the nodes on the paths by name. This makes it easy to modify this connectivity list and we won't have to remember which node is which number since the names are used. Notice how the list of paths is terminated by the number −1 indicating the end of the list. Each list is a path from the first node to the last node in the list. This means that the first node is connected to the next and the next to the one after that and so forth. Notice how some nodes are members of multiple lists.

Edges[ ] along with *Nodes*[ ] are used to create *MapGraph*[ ]. This array is, as described in Chap. 14, a two-dimensional square array of 1s and 0s. A 1 at position [*I*, *J*] implies that node *I* is connected to node *J* and since we have a bidirectional network we will also expect to have a 1 at position [*J*, *I*]. A 0 at the position implies that there is no connection between these nodes.

Notice that the last line before the `Return` statement is commented out. This line is a call to the subroutine *PlotNodes*. This subroutine uses *Nodes*[ ] and *MapGraph*[ ] to plot the nodes (as little circles) and the paths between them using the color defined in *LnClr*. In normal usage of the simulation you won't need to see this plot, but it may become necessary in debugging or visualizing the system. Figure 15.7 was generated by uncommenting the line.

The number of nodes was chosen to make the robot's job of moving between them easy, without the possibility of getting close to walls or being obstructed by corners and doorframes. Experiment with less or more intermediate nodes and see how the system fairs. Also you may want to see what happens if you remove some of the multiple paths between the "Hallways" and "Offices 4 and 5." These paths ensure that the robot will take the shortest route depending on which direction it is coming from. If you remove these redundant paths the robot will look less intelligent in the way it goes from one point to another in that area. We have chosen to make the robot move in the center of the corridors. This can be easily modified by changing the placement of the nodes. This kind of interaction demonstrates the value of a good design. Had we not designed the mapping process to be so generic we would not have been able to change such things without a great deal of work.

Notice how the *FindNode* routine is designed to return a meaningful value if the node name is not found in the database. This value is utilized in the *MapOffice* routine while it is creating *MapGraph*[ ] from the list of edges (*Edges*[ ]). If you mistype the name of a node while specifying the edges the routine will not fail, it will just ignore that edge. Of course this means that the graph will not be correct but you will soon find this out when you run the system or if you examine the graph's data.

15.2.6 WAITING FOR A COMMAND

As we have specified, the robot will wait until a user commands it to a particular room. The subroutine *WaitForCommand* (Fig. 15.9) will wait until the user presses the right mouse button anywhere on the screen or the "m" or "M" button on the keyboard. This indicates to the routine that the user desires to command the robot. The routine will then

```
//==============================================================
//--- Subroutine WaitForCommand
//--- Inputs   : LowChargeLimit
//--- Outputs  : GoalNode
//--- Calls To: GetTimeInSecs,DockWithCharger,ChooseRoom
//---
//--- this routine uses ChooseRoom to set the variable
//--- GoalNode which indicates what room the robot is to go.
//--------------------------------------------------------------
WaitForCommand:
   if rChargeLevel()<LowChargeLimit
     StartTime=-TimeOutLimit
   else
     gosub GetTimeInSecs
     StartTime = Tm
   endif
   repeat
     gosub GetTimeInSecs
     if Tm-StartTime > TimeOutLimit
        gosub DockWithCharger
        StartTime = 99999
     endif
     readmouse xx,yy,b
     getkey k
     k = char(k)
     if b = 2 or k="m" or k="M"
       gosub ChooseRoom
       break
     endif
   until false
Return
//==============================================================
//==============================================================
//--- Subroutine GetTimeInSecs
//--- Inputs   : none
//--- Outputs  : Tm
//--- Calls To: none
//---
//--- this routine sets Tm to the current minutes and seconds
//--- in seconds. It is used like a stop watch.
//--------------------------------------------------------------
GetTimeInSecs:
  Tm  = Time(1)
  TmH = ToNumber(Substring(Tm,1,2))
  TmM = ToNumber(Substring(Tm,4,2))
  TmS = ToNumber(SubString(Tm,7,2))
  Tm  = TmH*3600+TmM*60+TmS
Return
//==============================================================
```

FIGURE 15.9 Waiting for commands.

invoke the *ChooseRoom* routine we have discussed before and then return to the main program. The outcome of all this is that the variable *GoalNode* is set according to the user's command. The main program will then cause the robot to move to the indicated node, as we will see later.

The routine will also execute two other important tasks. The first task makes the robot move to the docking station if there is no user command within a certain period of time. This is achieved by calling the subroutine *GetTimeInSecs* that starts a timer and initiates a stopwatch action. When the time period elapses without any user input the routine calls *DockWithCharger* (see later section).

The second task causes the robot to move to the charging station if the battery charge level is below a certain threshold. This is done by checking the charge level using the `rChargeLevel()` function. When the value drops below a predefined threshold the *StartTime* is set to a value that causes an immediate time out. This causes a call to the subroutine *DockWithCharger* as above.

The routine *GetTimeInSecs* sets the variable *Tm* to the value of the current time converted to seconds. So if the time is 11:23:30 the variable *Tm* will be set to the value $11 \times 3600 + 23 \times 60 + 30$ (41010). This is achieved by dissecting the string value returned by the function `Time(1)` and converting it to numbers and applying the formula. The outcome is the same as noting the time but the value is in seconds. If you save the value *Tm* in a variable *StartTime* and then call the subroutine again at a later time and subtract *StartTime* from the new *Tm* value you will get the elapsed time in between calls. This is the same action as using a stopwatch.

WaitForCommand does the above and compares the time to a specified value *TimeOutLimit*. If the elapsed time exceeds the time out limit the routine invokes *DockWithCharger* and sets the start time to a value that will not cause a time out again, until the subroutine is called again.

15.2.7 EXECUTING THE COMMAND

The command process culminates in specifying a *GoalNode* to go to. The node where the robot is situated is specified as the charge room node upon starting the program, and is always maintained in the variable *RobotNode* as the robot moves. These two variables are necessary for the actions of the subroutine *FindPath*. This subroutine is called by the routine *MoveToRoom* (Fig. 15.10) to find a path of nodes to move from the *RobotNode* to the *GoalNode*.

The subroutine *MoveToRoom* calls *FindPath* (Fig. 15.10) to create *PathList*[ ] and its length *PathLength*. This action is only performed if the destination node and the robot's node are different (i.e., the robot is not at the destination). Once *PathList*[ ] is created and is a valid list (*PathLength* > 0) the routine iterates through the list to make the robot *GotoPoint* to the nodes one at a time. You have seen the *GotoPoint* subroutine many times already in various chapters. It assumes that the robot has a compass and GPS system.

In this section it is assumed that the robot has an accurate compass and GPS (global positioning system) or even better an LPS (localized positioning system). Later sections will discuss how to move the robot without the use of these two specialized and sometimes limiting instruments.

The *FindPath* subroutine is the same as the one discussed in Chap. 14 for finding the shortest path between two nodes. The routine is modified slightly from the listing in

```
//==============================================================
//--- Subroutine MoveToRoom
//--- Inputs   : RobotNode,GoalNode
//--- Outputs  : none
//--- Calls To: FindPath,GotoPoint
//---
//--- causes the robot to move to the goal node from where
//--- it is. It calls FindPath to get the shortest path.
//--------------------------------------------------------------
MoveToRoom:
   if RobotNode <> GoalNode
      gosub FindPath
      if Found and PathLength > 0
         for MTR_I = 1 to PathLength  //for each node on path
            Dnn = PathList[MTR_I,0] //destination node number
            x = Nodes[Dnn,1]
            y = Nodes[Dnn,2]
            gosub GotoPoint
         next
         RobotNode = GoalNode
      endif
   endif
Return
//==============================================================
//==============================================================
//--- Subroutine FindPath
//--- Inputs   : RobotNode,GoalNode,MapGraph[],NodesCount
//--- Outputs  : PathList[],PathLength
//--- Calls To: none
//---
//--- this routine searches the MapGraph[] for a shortest path
//--- between the RobotNode and GoalNode.
//--------------------------------------------------------------
FindPath:
   Dim Queue[NodesCount]
   QHead = 0
   QTail = 0
   Dim Visited[NodesCount,3]
   MConstant Visited,999999
   Queue[QTail] = RobotNode //add to queue
   QTail = QTail+1
   Visited[RobotNode,0] = 1  //visited
   Visited[RobotNode,1] = 0  //distance
   Visited[RobotNode,2] = -1 //predecessor
   while Visited[GoalNode,0]<>1 and QHead < QTail
      v = Queue[QHead]  //deque
      QHead = QHead+1
      for w = 0 to NodesCount-1
        if MapGraph[v,w] = 1 and Visited[w,0] <> 1
           Visited[w,0]=1
           Visited[w,1] = Visited[v,1]+1
           Visited[w,2] = v
           Queue[QTail] = w //add to queue
           QTail = QTail+1
        endif
      next
   wend
```

FIGURE 15.10 Moving the robot.

```
   if Visited[GoalNode,0]  = 1
      Found = true
      Dim PathList[NodesCount]
      PathLength = Visited[GoalNode,1]
      PathList[PathLength] = GoalNode
      For FP_I = PathLength-1 to 0
        PathList[FP_I] = Visited[PathList[FP_I+1],2]
      next
   else
      Found = false
   endif
Return
//==============================================================
//==============================================================
//--- Subroutine GotoPoint
//--- Inputs  : x,y
//--- Outputs : none
//--- Calls To: none
//---
//--- this routine turns the robot towards the location x,y
//--- and then moves the robot there. If there is an object
//--- in the way it terminates before reaching the point.
//--------------------------------------------------------------
GotoPoint:
   dx = x-rGpsX()
   dy = y-rGpsY()
   if dx=0 AND dy = 0 then return
   Theta = PolarA(dx,dy)*180/pi()+90-rCompass()
   if Theta > 180 then Theta = Theta-360
   if Theta < -180 Then Theta = Theta+360
   rTurn Theta
   Distance = Round(PolarR(dx,dy))
   For GP_I = 1 to Distance
     if rBumper() & 4 then break
     rForward 1
   next
Return
//==============================================================
```

FIGURE 15.10 *(Continued)*

Chap. 14. The change is only in naming. In Chap. 14 we returned the path list in the array *Stack*[ ] and its length in *SP*. This was to maintain compatibility with the naming in the previously developed any-path search. Here we do not have to do this and the name of *Stack*[ ] has been changed to *PathList*[ ] and *SP* to *PathLength*, for more appropriate names. Otherwise the subroutine executes exactly the same logic as in Sec. 14.4. As a summary, the routine carries out a breadth-first search through *MapGraph*[ ] to find the shortest path between the *RobotNode* and *GoalNode*.

15.2.8 RECHARGING THE BATTERY

The *DockWithCharger* subroutine (Fig. 15.11) causes the robot to move to the charging room by calling *MoveToRoom*, with *GoalNode* set to the node number of the charging room, which is the value *NodesCount* − 1. Remember we made the charge room node and its door node the last two nodes in the *Nodes*[ ] array for precisely this purpose. Of

```
//============================================================
//--- Subroutine DockWithCharger
//--- Inputs   : NodesCount,ChargeDelay
//--- Outputs  : none
//--- Calls To: MoveToRoom
//---
//--- makes the robot move to the charge room. Once there it
//--- turns around and docks with the charging station with
//--- its back.
//------------------------------------------------------------
DockWithCharger:
    GoalNode = NodesCount-1
    gosub MoveToRoom
    rTurn 180
    while not(rBumper()&1)
      rForward -1
    wend
    delay ChargeDelay
    rCharge 100
Return
//============================================================
```

FIGURE 15.11 Charging the robot.

course, we could have used the subroutine *FindNode* to locate the node if it was not known to be the last node.

Once at the charge room node, the robot turns around 180° and reverses until the back bumper is closed. This is the same docking procedure we used in Chap. 13 but without the use of a line to follow. This is because we use the GPS and compass to situate the robot at the proper position and orientation to achieve a docking. This is accomplished by locating the charge room node and the node before it (the door) at the correct positions to lead the robot straight into the charging station where all that is needed for docking is to turn 180° and reverse a short distance.

Finally the robot simulates charging by delaying *ChargeDelay* milliseconds and using the `rCharge` command.

15.3 A Reality Check

When you combine Figs. 15.1, 15.3, and 15.5 with 15.8 to 15.11 and run the program you will see the system working perfectly well. The robot goes from room to room as commanded and the recharge docking procedure works as expected. However, this is not a real situation. The simulation in Sec. 15.2 is unrealistic in two ways:

1. Rarely do robots go forward the amount commanded without slipping or turning. Most robots use gears that have play due to manufacturing tolerances and wear and tear. Also wheels tend to slip and motors tend to behave differently, all resulting in inaccurate movement of the robot. So the movement you expect is not what you get.
2. Compasses and GPSs are relatively expensive and generally not as accurate as desired. Affordable electronic compasses are often only accurate to ±5° not 1° as we would

like. Most GPS systems are accurate to a few yards at best, not inches as we would like, and most GPS receivers will not work indoors.

You can use encoders and feedback control to help ensure that the robot moves as expected, and there are ways to make an LPS that can be accurate to very small resolutions and work indoors. However, these solutions can be expensive and difficult to implement.

Can we devise solutions in software rather than hardware (or just simple hardware) that would make our algorithm resilient to these deficiencies? We will explore three potential solutions in the next section and discuss further possibilities in the section after that.

15.3.1 COUNTERACTING MOTOR SLIP WITH A GPS AND COMPASS

A real robot has a certain amount of randomness inherent in its movement. Wheels slip slightly on the floor. The friction associated with wheel-bearings and gears varies with temperature, wear, grease quantity, and so on. No two motors have exactly the same specifications. This means that no matter how precise you try to make a real robot, it will never be able to repeat its actions exactly unless you use sophisticated electronics and control methods to counteract the problem.

RobotBASIC has the ability to simulate the above limitations. If you issue the command `rSlip` the robot will add a 2 percent error when turning and forwarding. If you include an argument for this statement, you can specify the percent of error you want. You can specify a 10 percent error, for example, if you use `rSlip 10`. See Sec. C.9 for more information.

To examine how this problem affects the algorithm in Sec. 15.2, a variable at the top of Fig. 15.1 called *SlipValue* has been set. We also have the statement `rSlip` *SlipValue* in the *PlaceRobot* subroutine in Fig. 15.5. Notice that *SlipValue* is set to 0. Try changing this value to 5, 10, 20, 50, and 70 to see how it affects the robot.

The algorithm in Sec. 15.2 assumed an ideal robot that moves as commanded. When we simulate a real robot though, the algorithm fails. Can the algorithm be modified to resolve this issue?

In order to resolve the problem we need to understand *why* the algorithm fails. Of course it failed because the robot did not move as expected, but this is not the *direct* reason for failure. If you do a search on the code of the entire program you will find that the commands `rTurn` and `rForward` are used in only two subroutines: *GotoPoint* and *DockWithCharger*.

The direct reason the algorithm fails is due to the fact that the robot does not arrive at the expected intermediate node. The *GotoPoint* subroutine calculates the heading to turn then turns the robot to that heading. It does not check that the robot actually turned to exactly that heading. Also the algorithm calculates the distance to the point and forwards the robot that amount. It again does not check that the robot actually moved that distance without turning or slipping.

The indirect reason for the failure is motor/wheel slip, but the direct reason is that the algorithm does not verify that the robot actually arrived at the desired coordinates. So all we need to do is modify the algorithm of *GotoPoint* to *ensure* that the robot actually arrives at the desired point.

If you study the code in Fig. 15.12 you will see the required modification. The idea is to stay in a loop that turns the robot toward the destination and then forwarding one pixel, repeating the action until it actually gets there. The robot may turn and move more or less than the required amount, but the loop will ensure that *eventually* it gets there by constantly turning and forwarding toward the goal point. The above actions counteract any slip in the robot's movements, but they still require an accurate GPS and compass. We will address this in the next section.

Another shortcoming in the previous algorithm is that if an obstacle is encountered the robot will never get to the point because it abandons the routine. We shall simulate transient obstacles that move into the robots path by using the mouse. If you press the left mouse button with the cursor in front of the robot, the code in the subroutine *CheckObstacles* will make the robot think that there is an obstacle. This routine also simulates the robot flashing a beacon (you can also make it beep see Exercise 8). The robot will not move until the obstacle moves out of the way and then it will continue on its way. Try this by clicking the mouse and keeping it clicked in front of the robot.

The other routine that must be changed is *DockWithCharger*. The old routine assumed that by turning 180° the robot would have its back facing the charging station. However, the robot may not turn 180°. If there is an error of a few degrees there would be no problem, but this cannot be guaranteed unless we modify the code to ensure that the robot actually turns to face north. This is achieved by making the robot continue to turn in a loop until it actually faces north. The final reversing may still cause problems but if we design the charging room node to be close to the station then the likely error will be within allowable tolerances. Remember that reversing is done until the rear bumper is closed so the robot will always get there.

The routines in Fig. 15.12 are replacements for the ones in Figs. 15.10 and 15.11. *CheckObstacles* is a new routine.

15.3.2 NO GPS OR COMPASS (SLIP IS CORRECTED BY HARDWARE)

Can we still know the robot's position and heading if we had no compass or GPS? If the robot has no slip or any slip is removed by hardware using feedback control and good wheel encoders ensuring that the robot moves as expected, we can effectively create a *software* GPS and compass system.

This is actually the principle behind a navigation system used by aircrafts, ships, and even spaceships. This system is called inertial navigation system (INS). INSs rely on gyroscopes oriented in the three axes of movement. The gyroscopes give a method for the system to measure forces on all three axes. Using calculus and mathematics the translation of the craft can be calculated quite accurately. If you keep track of these translations you are able to calculate the craft's position in three dimensions at all times. The system generates small errors that can accumulate over time but if the reported position is checked against a known reference occasionally, a correction can be applied to maintain the system within acceptable tolerances.

We can simulate an INS system using software. If instead of calling `rForward` or `rTurn` we call a subroutine that simulates keeping track of all translations, we can simulate an INS which can return the robot's heading and x, y coordinates.

```
//==============================================================
//--- Subroutine GotoPoint
//--- Inputs  : x,y
//--- Outputs : none
//--- Calls To: CheckObstacles
//---
//--- this routine turns the robot towards the location x,y
//--- and then moves the robot there. If there is an object
//--- it waits in a loop until the object goes away. Also the
//--- routine keeps trying to turn and move the robot until it
//--- definitely reaches the goal point.
//--------------------------------------------------------------
GotoPoint:
  while true
    dx = x-rGpsX()
    dy = y-rGpsY()
    if dx=0 AND dy = 0 then return
    Theta = PolarA(dx,dy)*180/pi()+90-rCompass()
    if Theta > 180 then Theta = Theta-360
    if Theta < -180 Then Theta = Theta+360
    rTurn Theta
    gosub CheckObstacles
    if not Obstructed then rForward 1
  wend
Return
//==============================================================
//==============================================================
//--- Subroutine CheckObstacles
//--- Inputs  : none
//--- Outputs : Obstructed
//--- Calls To: none
//---
//--- checks for obstacles, also allows for simulated transient
//--- obstacles indicated by the left mouse button kept pushed
//--- down in front of the robot.
//--- sets Obstructed to true or false.
//--------------------------------------------------------------
CheckObstacles:
    Obstructed = true
    CO_B = rBumper()&4
    readmouse xx,yy,bb
    if bb=1 or CO_B
       rGPS Rx,Ry
       if PolarR(xx-Rx,yy-Ry) <= RobotSize+2  or CO_B
          //flash a beacon on the robot
          circle Rx-3,Ry-3,Rx+3,Ry+3,BcnClr,BcnClr  //on
          delay 60
          circle Rx-3,Ry-3,Rx+3,Ry+3,white,white    //off
          delay 40
          return
       endif
    endif
    Obstructed = false
Return
//==============================================================
//==============================================================
//--- Subroutine DockWithCharger
//--- Inputs  : NodesCount,ChargeDelay
```

FIGURE 15.12 Slip resilient moving.

```
//--- Outputs : none
//--- Calls To: MoveToRoom
//---
//--- makes the robot move to the charge room. Once there it
//--- turns around and docks with the charging station with
//--- its back. It keeps turning until the robotís heading is
//--- definitely north.
//-----------------------------------------------------------------
DockWithCharger:
     GoalNode = NodesCount-1
     gosub MoveToRoom
     while true
       dA = -rCompass()
       if dA = 0 then break

      if dA > 180 then dA = 360-dA
      if dA < -180 then dA = 360+dA
      rTurn dA
    wend
    while not (rBumper()&1)
      rForward -1
    wend
    delay ChargeDelay
    rCharge 100
Return
//=================================================================
```

FIGURE 15.12 *(Continued)*

You do not need gyroscopes or much mathematics if you have a very good encoder system on the robot's wheels that accurately records the amount the robot has moved and turned. This would result in a simpler INS that relies only on encoding the amount of turns on each of the wheels of the robot.

We will do this in this section, however, we will not be able to do this simulation if there is any slip. Similarly, in the real world we would need accurate wheel encoders to minimize slip to near zero if we wished to use this method for keeping track of our robot's location. Also there are electronic 3-axis accelerometers that can be used in a similar fashion to the 3-axis gyroscopes mentioned above. See Sec. 15.4 for other ideas on how to counteract slip using this method.

Figure 15.13 has two new subroutines *ForwardRobot* and *TurnRobot* that do the work of moving the robot and maintaining the simulated INS parameters. The *PlaceRobot* routine is a replacement for the one in Fig. 15.5. It has a few extra lines to initialize the simulated INS parameters. *DockWithCharger* and *GotoPoint* are replacements for the ones in Fig. 15.12. These new routines now use the INS rather than `rCompass()` and `rGPS`. Do not forget to set the variable *SlipValue* to zero.

Note that the subroutine *CheckObstacles* (Fig. 15.12) uses the GPS to find the distance from the mouse pointer to the robot. This is only to *simulate* obstacles in the robots way. Using the GPS in this routine can still be done because it does not impact real-life movement or position determination and therefore does not need to be changed.

```
//==============================================================
//--- Subroutine PlaceRobot
//--- Inputs   : BcnClr,LnClr,RobotSize,SlipValue,NodesCount
//--- Outputs  : RobotNode,RobotX,RobotY,RobotHeading
//--- Calls To: none
//--------------------------------------------------------------
PlaceRobot:
  rLocate 558,565,0,RobotSize
  rInvisible LnClr
  rSlip SlipValue
  rIgnoreCharge false
  RobotNode    = NodesCount-1
  RobotX       = 558
  RobotY       = 565
  RobotHeading = 0
Return
//==============================================================
//==============================================================
//--- Subroutine GotoPoint
//--- Inputs   : x,y
//--- Outputs  : none
//--- Calls To: CheckObstacles, ForwardRobot, TurnRobot
//---
//--- this routine turns the robot towards the location x,y
//--- and then moves the robot there. If there is an object
//--- in the way it continues to wait until the obstacle moves
//--- away then resumes moving towards the point. It also uses
//--- the new subroutines ForwardRobot and TurnRobot inplace
//--- of rForward and rTurn to maintain an INS position.
//--------------------------------------------------------------
GotoPoint:
    dx = x-RobotX
    dy = y-RobotY
    if dx=0 AND dy = 0 then return
    Theta = PolarA(dx,dy)*180/pi()+90-RobotHeading
    if Theta > 180 then Theta = Theta-360
    if Theta < -180 Then Theta = Theta+360
    TurnAmount =  Theta
    gosub TurnRobot
    for i=1 to round(PolarR(dx,dy))
       gosub CheckObstacles
       if not Obstructed
          ForwardAmount = 1
          gosub ForwardRobot
       else
          i = i-1
       endif
    next
Return
//==============================================================
//==============================================================
//--- Subroutine DockWithCharger
//--- Inputs   : NodesCount,ChargeDelay
//--- Outputs  : none
//--- Calls To: MoveToRoom, ForwardRobot, TurnRobot
//---
//--- makes the robot move to the charge room. Once there it
//--- turns around and docks with the charging station with
//--- its back. It keeps looping to ensure a North heading
//--- Also it uses ForwardRobot and TurnRobot inplace of
//--- rForward and rTurn to maintain an INS position
//--------------------------------------------------------------
```

FIGURE 15.13 Simulating an INS but no slip is allowed.

```
DockWithCharger:
    GoalNode = NodesCount-1
    gosub MoveToRoom
    TurnAmount = 180
    gosub TurnRobot
    while not (rBumper()&1)
      ForwardAmount = -1
      gosub ForwardRobot
    wend
    delay ChargeDelay
    rCharge 100
Return
//==============================================================
//==============================================================
//--- Subroutine ForwardRobot
//--- Inputs  : ForwardAmount,RobotHeading,RobotX,RobotY
//--- Outputs : RobotX,RobotY
//--- Calls To: none
//---
//--- Moves the robot the ForwardAmount (+ or -) and
//--- maintains track of the robot's INS position
//--------------------------------------------------------------
ForwardRobot:
   rForward ForwardAmount
   Angle = 90-RobotHeading
   if ForwardAmount < 0
     Angle = Angle-180
     ForwardAmount = abs(ForwardAmount)
   endif
   if Angle <= -180 then Angle = Angle+360
   if Angle > 180 then Angle = Angle -360
   Angle = Angle*pi()/180
   dX = CartX(ForwardAmount,Angle)
   dY = CartY(ForwardAmount,Angle)
   RobotX = RobotX+dX
   RobotY = RobotY-dY
Return
//==============================================================
//==============================================================
//--- Subroutine TurnRobot
//--- Inputs  : TurnAmount,RobotHeading
//--- Outputs : RobotHeading
//--- Calls To: none
//---
//--- Turns the robot the TurnAmount (+ or -) and
//--- maintains track of the robot's INS heading
//--------------------------------------------------------------
TurnRobot:
   rTurn TurnAmount
   RobotHeading = RobotHeading+rounddn(TurnAmount)
   if RobotHeading < 0 then RobotHeading = RobotHeading+360
   if RobotHeading > 359 then RobotHeading = RobotHeading-360
Return
//==============================================================
```

FIGURE 15.13 *(Continued)*

15.3.3 RESILIENCE AGAINST SLIP USING BEACONS

The two previous algorithms showed how to make the robot more practical, but as seen, even these techniques have their limitations. Let's look at an entirely different way for the robot to deal with its environment.

We need a means of knowing when the robot arrives at the desired commanded point. If we install a collection of beacons in the ceiling of the office above each node (see Fig. 15.7) and we give the robot a remote control that can switch individual beacons on and off (using a coded signal), we can have a means for the robot to be able to home in on a desired location as in Chap. 12. The robot moves along a designated path by switching the beacons on one at a time along the path. After locating and moving to a beacon, it switches the beacon off and turns on the next one and continues doing this in sequence until the final goal is reached. The robot is always guaranteed to reach the beacon by this homing algorithm. The overall behavior of the robot (moving to each node) is the same as before but the robot responds correctly without a compass or GPS and with no expensive slip counteracting hardware. This idea is implemented with the code in Fig. 15.14.

In the *DockWithCharger* routine we use the beacon at the door of the charging room to orient the robot to north. Set *SlipValue* to some number other than zero (20 is a good test) and replace the subroutines in Fig. 15.12 with the ones in Fig. 15.14. Note that the subroutine *CheckObstacles* (Fig. 15.12) uses the GPS to find the distance from the mouse pointer to the robot. This is only to *simulate* obstacles in the robots way. We continue to use the GPS in this routine because it does not impact our goal of creating a practical robot.

15.4 Further Thoughts

The office or home robot discussed in this chapter is a very practical and useful application and is definitely achievable. All the hardware required by the simulated system can be implemented with current technology. Some of this technology can be expensive, but the beacon algorithm proposed in Fig. 15.14 is easily attainable with a modest budget. Those with a larger budget and access to more sophisticated electronics can implement the more complex but more robust alternatives using a GPS or INS. Perhaps some company will realize the value of the beacon system described in this chapter. We suspect many hobbyists would purchase a set of remote controlled beacons (or beacons that emit a coded identifier) if they were available at a reasonable price. Such a product would make a useful home/office robot very feasible.

The INS alternative required no slip as implemented in Fig. 15.13. However, we can achieve some tolerance for slip if we could ensure a method for updating the robot with a corrected position and heading at, say, each node in the path network (see Fig. 15.7). We also may have to redesign the path network to include more nodes and thus more frequent updates and less distance during which the robot can incur errors. We can place some form of identifying device at the nodes [perhaps magnetic encoded pulses or radio frequency identification tags (RFID)], which the robot can use to identify the node it is currently at. Combined with the *Nodes*[ ] array the robot can update its *x, y* position. To update the heading the node can be given a way of telling the robot where north is and the robot can correct its assumed heading value. If you look at Fig. 15.7 you may get an idea of

```
//================================================================
//--- Subroutine GotoPoint
//--- Inputs   : x,y
//--- Outputs  : none
//--- Calls To: CheckObstacles
//---
//--- this routine turns the robot towards a beacon
//--- and then moves the robot there. If there is an object
//--- in the way it waits until the obstacles moves away and
//--- then resumes moving. The routine will continue to try to
//--- move and turn towards the beacon until it gets there
//--- regardless of errors that may occur due to slip..
//----------------------------------------------------------------
GotoPoint:
  circle x-5,y-5,x+5,y+5,BcnClr,BcnClr //turn beacon on
  while true
    dA = 0
    if not rBeacon(BcnClr)
      for dA = -90 to 90
        if rLook(dA) = BcnClr then break
      next
      dA = dA+sign(dA)*2
    Endif
    rTurn dA
    if within(rBeacon(BcnClr),1,10) then break
    gosub CheckObstacles
    If not Obstructed then rForward 1
  wend
  circle x-5,y-5,x+5,y+5,white,white //turn beacon off
  rForward RobotSize+15
Return
//================================================================
//================================================================
//--- Subroutine DockWithCharger
//--- Inputs   : NodesCount,ChargeDelay
//--- Outputs  : none
//--- Calls To: MoveToRoom
//---
//--- makes the robot move to the charge room. Once there it
//--- turns around and docks with the charging station with
//--- its back. The routine guarantees a North heading by
//--- continuing to orient itself toward the beacon at the
//--- door of the charging station room.
//----------------------------------------------------------------
DockWithCharger:
    GoalNode = NodesCount-1
    gosub MoveToRoom
    x = Nodes[GoalNode-1,1]
    y = Nodes[GoalNode-1,2]
    circle x-5,y-5,x+5,y+5,BcnClr,BcnClr //turn beacon on
    while not rBeacon(BcnClr)
      rturn 1
    wend
    circle x-5,y-5,x+5,y+5,white,white //turn beacon off
    rturn 4     //turn to the center of the beacon
    while not (rBumper()&1)
      rForward -1
    wend
    delay ChargeDelay
    rCharge 100
Return
//================================================================
```

FIGURE 15.14 Counteracting slip using beacons.

making the robot follow lines to counteract any slip and the need for using a GPS. This is definitely achievable and with some encoding at each node, the robot can always know at which node it is, and thus be able to navigate around the office. However, a line on the floor of the office may not be desirable. One way to resolve this is to have an electric wire in the floor that replaces the visible line, or a line painted with invisible paint that can only be seen by specialized sensors.

The ideas presented in this chapter are aimed at achieving a realistic project but also at invoking a thought process for tackling limitations in hardware and how to counteract these limitations with software. The design process is a very important step in achieving a versatile robust system that can be easily modified, and adapted to various environments. You saw how the overall project needed only minor changes and/or additions to be able to handle quite disparate hardware limitations. With a little effort up front you can save major headaches later in the lifespan of the project.

15.5 Summary

In this chapter you have:

- Learned how to design an ambitious project using all the tools and skills developed in this book.
- Explored various strategies for achieving the same objective with varied hardware limitations.
- Seen how software can be used to circumvent the physical limitations of the robot.
- Learned how the `rSlip` command can add realism to a simulation.
- Seen how specialized commands can be used to achieve a nice GUI.
- Seen the value of commenting and annotating your code to make it reusable and supportable by others.
- Learned how using a database of information can be a more versatile means of defining a system than hard coding cryptic information.
- Seen more examples using graphs, queues, and arrays.

Now, try to do the exercises in the next section. If you have difficulty read the hints.

15.6 Exercises

1. You saw how in Fig. 15.14 the robot could command beacons to turn on and off. In the project of this chapter we assumed the doors were either open or would open automatically when the robot approaches. Modify the code to make simulated doors open and close based on commands from the robot.
2. When we designed the network of nodes we made each room have a node inside the room and a node at the door. If you give each room multiple nodes the robot would be able to reach more spots in the room. Change the subroutine *MapOffice* to have more nodes in all or some rooms and then change *ChooseRoom* to allow the user to choose these new locations. This could be useful if, for example, there are several people sharing an office and you want the robot to deliver a package to one of them.

3. We designed the project so that the robot could not accept new commands while executing a current one. In real life this can be achieved by giving the user a message when trying to command the robot. However, it would be a lot more versatile if the robot could accept multiple commands. These commands can be placed in a queue and the robot can execute them one at a time. Modify the system to allow for this.
4. In Fig. 15.12 we developed a simulated transient obstacle and made the robot flash a warning beacon while waiting forever for the obstacle to go away. This is actually quite a satisfactory behavior in this type of application, but it would not be acceptable in other scenarios (e.g., mobile laboratory). Develop other ways the robot can be made to behave when obstacles are encountered and implement this new behavior.

HINT: Refer to wall-following in Chap. 12.

5. In the subroutines *DockWithCharger* in Fig. 15.14 we ensured that the robot faced north, then we reversed the robot until it docked with the station. However, we did not allow for slip while doing so. We assumed that the distance was short by ensuring the charge room node is close enough to the station. Develop other ways to make the docking procedure more slip resilient.

HINT: See Chap. 13.

6. In the subroutine *GetTimeInSecs* (Fig. 15.9) we start a stopwatch by saving the current time value in seconds and then subtracting it from a later time value in seconds. What may occur if the time happens to be 23:59:59 when we start the timer and we check the elapsed time say 30 seconds later? Can you allow for this with a more robust subroutine?

HINT: Thirty seconds later the time would be 00:00:29. How does this affect the new time value? Check out the functions `Date()` and `Timer()`. Why is it that the use of the command `Delay` would not work?

7. Study the commands `Mwrite, MRead, WriteScr,` and `ReadScr` in Sec. C.7. Can you share your office plan and graph with other users without having to give them all the code that creates the office and graph? Write modifications to *DrawOffice* and *MapOffice* to allow for using an already created file. Also change them to allow for saving all the necessary arrays and screen.

HINT: `Mwrite/WriteScr` save your arrays and screen files to give them to a user who then uses `MRead/ReadScr` to recreate the arrays and screen without having to know the details of how they were created. The arrays of importance are *Nodes*[ ] and *MapGraph*[ ], also the counter *NodesCount* has to be recreated after reading the arrays from the files.

8. In the subroutine *CheckObstacles* in Fig. 15.12 we made the robot flash a beacon to warn that it is being obstructed. The command `Sound` in RobotBASIC allows you to make a sound of a certain frequency for a certain duration using the speaker of the PC. Implement a siren sound along with the flashing beacon in the subroutine *CheckObstacles*. See App. C.

HINT: Replace the two `Delay` commands with `Sound` commands, each with a different frequency.

PART 4

GOING FURTHER

Part 4 explores the exciting fields of artificial intelligence and adaptive control. We introduce some of the concepts and offer a thought-provoking program that creates a robot mimicking a living creature with biological needs. The program uses an interesting variation on adaptive behavioral control where there is no direct instruction to the robot as to *where* to seek satisfaction for its biological needs. Nevertheless, due to the use of the concept of learning and adapting through association, the robot soon learns how to seek a quick path to locations where it can satisfy its needs.

Part 4 also shows how to translate the algorithms developed through simulation so they can execute on a real-world robot built with affordable parts from Parallax, Inc. and other vendors. We also show how to utilize Bluetooth transceivers to allow RobotBASIC programs running on a PC to communicate directly with and control real-world robots and other computers.

One chapter discusses the issues involved in creating and participating in contests and proposes that RobotBASIC can provide an exciting new concept for contest organizers and contestants.

Another chapter explores why using RobotBASIC and this book in the classroom can be of value for both the teacher and student during the teaching and learning processes.

CHAPTER 16

TRUE INTELLIGENCE: ADAPTIVE BEHAVIOR

In all the chapters so far, we have developed algorithms to make our robot react to its environment while achieving specific tasks. The approach used was to give the robot specified parameters for how to react to specific sensory inputs. On many occasions the robot was given randomness so it could vary its behavior sufficiently to avoid getting stuck in unanticipated dead-end situations.

When the robot became stuck in a situation where it could not continue doing its work, randomness eventually created the right combination of parameters to enable the robot to escape. However, the robot had no way of learning from its experience. If the robot encountered a similar situation at a later time it had no means to recall and reapply the same parameters.

If we give the robot a memory and the ability to record the parameters that made it succeed as well as the ones that led to it becoming stuck, the robot should be able to adapt its future behavior so that it can avoid the bad situations and favor the good ones. An example of this kind of behavior was seen in Chap. 14, where the robot learned from the first pass through the corridor maze, how to negotiate the maze perfectly the second time. To achieve this we used memory (array) to save past behavior in order to influence future behavior.

16.1 Adaptive Behavior

In a normal control algorithm the robot observes the environmental conditions using sensors. These values are compared to a set of desired values that have been fixed by the programmer. Depending on the deviations of the sensory values from the desired ones, the algorithm will determine a course of action that is translated into a set of commands to actuators (e.g., motors) that manipulate the robot and/or environment (see Fig. 16.1).

The above would result in an environmental change, which then affects the sensory inputs. The algorithm continues in a loop, responding to changes due to the actuators of the robot and/or external environmental factors. The outcome is that the robot will get progressively closer to the desired state. How quickly and how efficiently the robot reaches the desired state, and how quickly it responds to a disturbance in the conditions depends on the control algorithms used.

The field of control is a specialized and exciting field of study and can be very mathematical. The methods for determining the parameters of the algorithm are the subject of a well-established discipline in engineering that uses complex mathematics and calculus to optimize these parameters. The analysis to determine the parameters usually takes into account a certain range of environmental criteria but this range is often fixed and limited.

If, instead of fixing the parameters forever, the robot is given another feedback loop mechanism that serves to automatically modify the parameters in memory (and keep the parameters that work best), the robot would be able to adapt to a wider range of changing environmental situations.

The diagram in Fig. 16.2 shows the modified adaptive feedback loop. Inputs from the decoded sensory data are stored in memory along with the actions that the robot took in response to those inputs. The memory is then consulted whenever the robot is to take

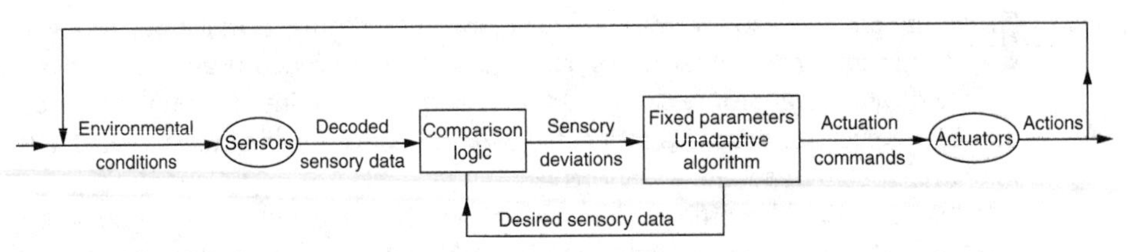

FIGURE 16.1 Unadaptive feedback loop.

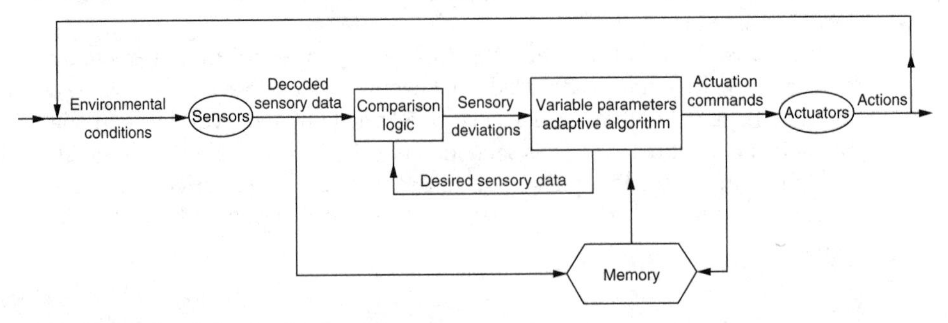

FIGURE 16.2 Adaptive feedback loop.

action to determine if a favorable or detrimental actuation was experienced in the past, and thus modify the parameters and logic of the algorithm if needed.

A programmer can also affect the robot's behavior by filling the memory with appropriate data, effectively *teaching* the robot certain responses and behaviors. An example of this was given in Chap. 15 where a map was provided to enable the robot to plan an optimal path through the office.

Let us consider how adaptive behavior algorithms can be applied to two algorithms we have encountered in previous chapters.

16.1.1 ADAPTIVE WALL-FOLLOWING

In Chap. 8 a contour-following algorithm used a ranging sensor to determine how far the robot was from the wall. The robot turned toward the wall when it got too far from it and away from the wall when it became too close. Various parameters affected the performance of the algorithm.

One of the important parameters was the distance (*RangeLimit*) we wanted the robot to stay from the wall. In another algorithm the parameter *TurnAmount* defined how many degrees to turn when the robot became too close or too far.

If we want the robot to be able to adjust the values of these variables by itself, it would need a means for evaluating how the changes affect its performance. Generally, we want the robot to try to keep as close to the wall as possible *without* hitting it. An adaptive program might try smaller and smaller values for *RangeLimit* until the bumper sensors indicate that it is getting too close.

When a working value for *RangeLimit* is found, the value is only valid for the current wall. If the contour of the wall changes, the value for *RangeLimit* will need to change to maintain an optimum performance. The robot would periodically try to lower the value of *RangeLimit* to see if the current wall can be followed more closely. Anytime collisions occur the robot would increase the value of *RangeLimit*.

If the wall being followed has sudden turns and protrusions, the distance to the wall would change quickly. The robot would detect such situations and adjust the value of *TurnAmount* thus turning toward (or away from) the wall more to keep up with the sharper turns. If the value becomes too high (or low) causing collisions, the robot would adjust the value (and perhaps *RangeLimit* too) accordingly.

The algorithm may have to keep changing the values mentioned above at a constant rate, or may consider a varying rate depending on parameters such as the amount and rate of change of the distance to the wall, and/or the accumulated error amount in the distance. This is called an adaptive-proportional-integral-differential control (APID).

16.1.2 ADAPTIVE LINE-FOLLOWING

In Chap. 7 we developed several algorithms for following a line. In general, all of the algorithms moved the robot along a line at a steady pace. This is not necessarily the most efficient approach.

A car driving along a winding road, for example, may speed up in the straight sections and slowdown when the road curves. Slowing down allows for more time to read the sensory data and analyzing it in more detail before responding becomes required. Regardless of what we want the robot to do when the road curves, it has to be able to determine *when* the road curves.

One way to detect if a line is curving is to equip the robot with more than three line sensors [see Sec. C.9 for details on `rGroundA()`]. When following a relatively straight line, only sensors near the robot's current heading would detect the line. As the line curves, the outer sensors start to trigger. We need to specify how the robot should react when the outer sensors are triggered. There are many options for how we can make the robot react. One possibility is to make the robot turn more sharply. Another is to make the robot slowdown.

Instead of *specifying* what the robot should do when the line curves, imagine a robot that can *decide on its own* how much to turn when it is in such situations. Previously, we simply guessed the required turn amount when sensor data showed that the robot was veering from the line. If the guess was too high the robot turned too much and lost the line. If the guess was too low the robot was not able to stay on the line when the line turned sharply. Our solution has been to test the program on a typical line from the expected environment and manually adjust the amount of turn until the robot performs satisfactorily.

Instead of programming the robot to turn 2°, for example, we could tell it to turn *TurnAmount* degrees where *TurnAmount* is a variable. The robot could be programmed to automatically *try* different values for *TurnAmount* and see what happens. The robot would have to be able to detect when it loses the line and then adjust *TurnAmount* and try again. Once the robot finds an acceptable value for *TurnAmount* it would use this value from then on, but this concept can be improved further.

We can program the robot to *continually* alter the value of *TurnAmount* based on its situation. Of course, a robot that can, in essence, program itself, needs a way to evaluate its own performance. This means that the robot must be given the means to determine when it has lost the line, and a way to find the line again so that it can try again with the adjusted parameters.

16.2 How to Define Intelligence?

Robots that adapt their behavioral rules and parameters are definitely more intelligent than those that behave in a predetermined manner. The question is, are they truly intelligent? The answer depends on how you define intelligence. Most people would argue that the robot is *not* truly intelligent because it is not making decisions the way human beings make them.

16.2.1 HUMAN INTELLIGENCE

Many factors affect how humans make decisions. Certainly memories affect the decision process. If an action causes pain it is less likely to be repeated in the future. On the other hand, actions that create pleasurable outcomes are more likely to be repeated.

The environment also affects the decision-making process. At the very least, the environment limits the range of choices. Even the food you eat affects your actions. It is obvious that normal body chemistry would make you less likely to eat something sweet if you have just eaten a large bowl of ice-cream, but other effects might be less obvious.

Sugar in your blood stream, for example, might make you choose to nap instead of exercising.

16.2.2 INTELLIGENCE THROUGH ASSOCIATION

Actions that create pleasurable outcomes are more likely to be repeated. One question that should come to mind is how do humans determine what is pleasurable. Certainly, human biology imposes many factors. All newborn babies find cold and hunger an unpleasant experience. Conversely, food and warmth are deemed pleasurable. The brain commits to memory many associations with these biological factors as a baby grows to adulthood.

There are also indirect associations. If a mother provides warmth and food for her baby, she will be associated with pleasure and thus is placed on the baby's *good* list. It is not hard to imagine that things that are associated with the mother would also be considered pleasurable. The behavior described above is deceptively simple, yet amazingly effective. In general, it means that a baby learns to achieve pleasure not only from things that directly give pleasure, but also from things that are *associated with* things that give pleasure. These associations, along with the current environmental conditions, control our behaviors. Early associations are straightforward but as the baby matures, creating the lists becomes more complicated. New situations and actions are often associated with things on both the good and bad lists. This means that many situations are not interpreted as strictly pleasurable or painful. Consequently, future choices are not just black or white, right or wrong. When our brain tries to analyze the choices of this nature we refer to the process as making a value-judgment.

The diagram in Fig. 16.3 shows a feedback loop that depicts how humans react and behave. Notice the loop is not really that different from the one in Fig. 16.2. The real difference is in the way that memory affects the sensory comparison process. Human perception is affected by memory as well as by the actual state of the environment and sensory organs. This is why we often *perceive* erroneously even when we sense correctly. This concept also explains why some people make some poor decisions in life.

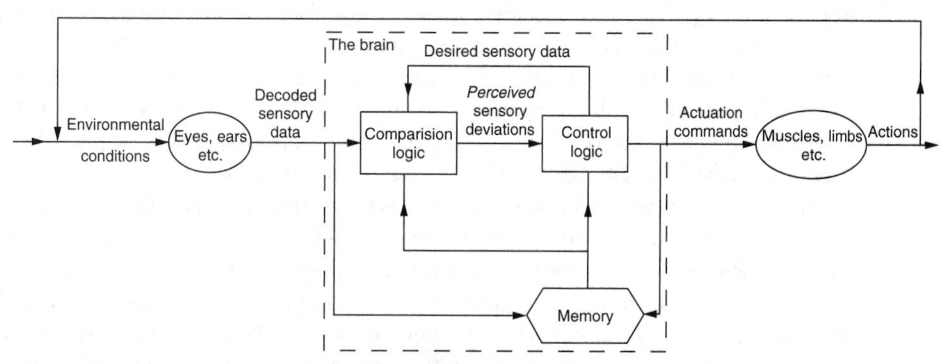

FIGURE 16.3 Human adaptive behavioral algorithm.

16.3 Adaptation through Association

A robot can be programmed to make choices using associations as described in the previous section. Rigorously addressing this problem would certainly result in a reasonably intelligent machine, but a rigorous approach could easily be the subject of an entire book. For now, we will be satisfied to create a relatively simple program to demonstrate that the principles in fact work.

As with biological life-forms, our mechanical creature must have some innate tendencies and abilities. These *instincts* will be used to evaluate situations in order to build the association memories that directly affect the choices made by the robot. The following discussion describes the general algorithm.

The robot must have some built-in desire to roam its environment. It should have some natural curiosity to make it tend to try to go to places it has not visited recently. The robot's initial movements have to be random, but there should be some tendency to keep moving in the same direction once a direction has been chosen.

16.3.1 I FEEL PLEASURE I FEEL PAIN

The robot must have some built-in way to differentiate between pain and pleasure. Our robot will encounter pain in several ways. Areas in its environment will contain briars where entering them will be interpreted as pain by the robot. The environment will also have a fireplace. Getting too close to the fire will cause the robot pain. Finally, if the robot's movement causes it to collide with objects it will feel pain. One of our goals is to make the robot learn *on its own* to avoid the fire, the briars, and collisions.

Defining pleasure is a little more complicated. Our robot will need to eat because it gets hungry, need to sleep because it gets sleepy, and need to play because it gets bored. Satisfying these needs will create pleasure. Roaming through the environment takes energy so the robot will become hungrier. Roaming without purpose will increasingly bore the robot making it want to play, which increases the robot's need for food and sleep. Eating also makes the robot sleepy. In the program, counters will keep track of each of these needs and will be incremented based on the robot's actions. These counters simulate a living creature's biological needs.

When a need exceeds a threshold value, the need becomes a motivator that affects the robot's choices. The need with the largest value will be the only one that currently affects the robot's behavior. This will cause the robot to make choices to attempt to satisfy the current motivator. Once a need is satisfied the next need will become the motivator. If two or more needs exceed the threshold simultaneously, the one with the largest value will be applied. If two or more needs reach the maximum, a priority will be applied. Hunger will have the highest priority followed by sleepiness and then boredom.

Humans don't *exactly* behave this way. We do prioritize objectives, but sometimes we may make use of an opportunity that presents itself even though we were not seeking it. For instance if you are sleepy and are on the way home to go to sleep when a friend calls and invites you to an interesting party, with some nice people, you may postpone the sleep drive to go and play. Of course, if the need to sleep is large enough, it will overshadow other needs and desires. See Exercise 5 for a discussion on how this concept may be implemented in the simulation of this chapter.

16.3.2 ENVIRONMENTAL FACTORS

The robot will not be able to get satisfying sleep unless it is near the fire to stay warm but not so near that it feels pain. There will be a garden area in the environment that provides food if the robot finds it. There will also be an activity area that provides some form of stimulation to relieve boredom. When our robot is "born" it will know nothing about these areas.

In the beginning of the robot's life it will roam aimlessly around the environment. Whenever an action causes it pain, the robot will save that action and the related situation in its memory so that it can avoid the pain in the future. Likewise, whenever the robot feels pleasure (needs are satisfied) it will save the action and the situation that led to the pleasure.

Furthermore, once an action/situation has been associated with pleasure, then actions that lead to that situation will also be saved as being associated with pleasure. This is very similar to the scenario described earlier where objects associated with a baby's mother are considered to be pleasurable because she was associated with food and warmth.

16.4 Implementing the Algorithm

Figure 16.4 shows a program that implements the algorithm described in the previous section. If you are running the program for the first time read the instructions and press *Enter* or click the OK button to continue. You will see the environment shown in Fig. 16.5. Do not press the Cancel button (or *Esc* key) if you have never run the program before, since there are no memories for the robot to reload. However, once there are memories you can always stop the program and run it at a later date and have the robot recall its previous experiences (by pressing the Cancel button or *Esc* key). Notice how the *RestoreMemory* subroutine uses the `FilExists()` function to avoid loading the memory files if they do not exist.

The robot starts out with all of its needs at maximum value as indicated by the bar graphs on the right side of the screen in Fig. 16.5. The robot begins exploring its environment and if it bumps into objects, gets too close to the fire, or wanders into a briar patch, the appropriate information will be saved to the *bad* memory list. Eventually, the robot will find the garden and eat. This will satisfy the hunger need, which will be indicated by the meter on the right of the screen gradually reducing to zero as the robot eats. The action and the environmental conditions that led to the food will be saved in the *good* memory. Notice that when the program is running, two counters at the top of the screen show the number of items stored in the *good* and *bad* memory lists.

In the beginning, the bad memory will fill quickly. The good memory, on the other hand, will expand very slowly because the robot will have a hard time finding situations that satisfy its needs. Remember, the robot has no knowledge of how to go to places that cause it to be happy. The situations it randomly encounters in the environment, along with the built-in biological factors and previous memories, determine what is stored in the robot's memory. Ultimately, this memory will control the robot's behavior and its personality.

```
//---Constants
  //--Action
  NOACTION = 0
  EAT      = 1
  SLEEP    = 2
  PLAY     = 3
  EXPLORE  = 4
  SAVE     = 5
  RETREAT  = 6
  RESPOND  = 7

  //--Status
  HUNGRY   = 0
  SLEEPY   = 1
  BORED    = 2

  //--Feeling
  PAIN     = 1
  PLEASURE = 2

  //--Headings
  NORTH    = 0
  EAST     = 1
  SOUTH    = 2
  WEST     = 3

  NEEDS_THRESHOLD = 85
  EAT_AREA        = 8
  SLEEP_AREA_1    = 20
  SLEEP_AREA_2    = 26
  PLAY_AREA       = 0
  PAIN_AREA_1     = 11
  PAIN_AREA_2     = 29
  PAIN_AREA_3     = 25
//=================================================================
//===============================================================
MainProgram:
  GoSub DisplayInstructions
  GoSub InitializeSimulation
  if not Key then  GoSub RestoreMemory
  GoSub ComeToLife
End
//===============================================================
//===============================================================
//--- Subroutine RestoreMemory
//--- Inputs  : none
//--- Outputs : Memory[],BadList[],MemPtr,BadPtr
//--- Calls To: none
//---
//--- Reads the arrays Memory[] and BadList[] from files
//--- created by previous runs.
//---------------------------------------------------------------
RestoreMemory:
   if FilExists("MemoryGood") and FilExists("MemoryBad")
      MRead Memory,"MemoryGood"
```

FIGURE 16.4 This program creates a robot with biological needs and the ability to learn how to satisfy them.

```
      MRead BadList,"MemoryBad"
      MemPtr = Memory[99,0]
      BadPtr = BadList[99,0]
   endif
Return
//=============================================================
//=============================================================
//--- Subroutine ComeToLife
//--- Inputs  : none
//--- Outputs : none
//--- Calls To: CheckMemory,DisplayAction,DoMovement
//---           CheckBadList
//--- Makes the robot do SOMETHING either based on memory of
//--- past experiences or some new random choice. The robot's
//--- memory is altered as the robot encounters pain and
//--- pleasure.
//-------------------------------------------------------------
ComeToLife:
  while true
   gosub CheckMemory
   if HaveResponse
     Action = RESPOND
     gosub DisplayAction
     Movement = Memory[HaveResponse,2]
     gosub DoMovement
   else
     if random(50)>=20 then Movement = Random(4)
     gosub CheckBadList
     if not Bad
       Action=EXPLORE
       gosub DisplayAction
       gosub DoMovement
       if Status = PAIN
         // Save to bad list
         BadList[BadPtr,0]=LastCell
         BadList[BadPtr,1]=Movement
         BadPtr = BadPtr+1
       elseif Status = PLEASURE
         //Save to Memory if NOT there already
         AddIt = True
         if MemPtr>0
           for i=0 to MemPtr-1
            if Memory[i,0]=CurNeed and Memory[i,1]=LastCell
              AddIt = False
              break
            endif
           next
         endif
         if AddIt
           Memory[MemPtr, 0] = CurNeed
           Memory[MemPtr, 1] = LastCell
           Memory[MemPtr, 2] = Movement
           if (MemPtr=25) or (MemPtr=50) or (MemPtr=75)
             Memory[99,0]=MemPtr+1
             BadList[99,0]=BadPtr
             MWrite Memory,"MemoryGood"
             MWrite BadList,"MemoryBad"
```

FIGURE 16.4 *(Continued)*

```
          endif
          MemPtr = MemPtr+1
        endif
      endif
    else
      Action=RESPOND
      gosub DisplayAction
      Delay 200
    endif
  endif
 wend
Return
//============================================================
//============================================================
//--- Subroutine DoMovement
//--- Inputs  : Movement,MemPtr,BadPtr,CurCell,Needs[],Rx,Ry
//--- Outputs : Needs[],LastCell,Action,Status,CurCell
//--- Calls To: DispStatusMeter,DisplayAction,CheckStatus
//---
//--- Moves the robot based on the variable Movement the
//--- robot retreats from painful events, which are reported
//--- through the variable Status
//------------------------------------------------------------
DoMovement:
  SetColor LightGreen
  xyString 150,15,"# Good Mem ",MemPtr
  xyString 300,15,"# Bad Mem ",BadPtr
  Status=0
  LastCell = CurCell // save where we came from
  // Increase Needs
  Needs[HUNGRY] = Needs[HUNGRY]+1
  if random (50)>25 then Needs[SLEEPY] = Needs[SLEEPY] + 1
  Needs[BORED] = Needs[BORED] + 1
  For DM_Which = 0 to 2
    gosub DispStatusMeter
  Next
  while rCompass() <> 90*Movement
    rTurn 90
  wend
  for i=1 to 120
    rForward 1
    if rBumper()
      Action = RETREAT
      Gosub DisplayAction
      rForward -i
      Status = PAIN
      return
    endif
  next
  dRx = 0 \ dRy = 0
  if Movement = EAST  then dRx = 1
  if Movement = WEST  then dRx = -1
  if Movement = NORTH then dRy = -1
  if Movement = SOUTH then dRy = 1
  Rx = Rx+dRx \ Ry = Ry+dRy
  CurCell = 5*Rx+Ry
  if CurCell=PAIN_AREA_1 or CurCell=PAIN_AREA_2 or CurCell=PAIN_AREA_3
```

FIGURE 16.4 *(Continued)*

```
    Status = PAIN
    Action = RETREAT
    Gosub DisplayAction
    rForward -120
    Rx = Rx-dRx \ Ry = Ry-dRy
  endif
  gosub CheckStatus
return
//==============================================================
//==============================================================
//--- Subroutine DisplayAction
//--- Inputs  : Action,ACTIONS[]
//--- Outputs : none
//--- Calls To: none
//---
//--- Displays the current action in the actions list
//--------------------------------------------------------------
DisplayAction:
  SetColor Black
  xyString 735,440,"ACTION"
  line 733,457,790,457
  SetColor Yellow
  For DA_I = 1 to 7
     xystring 728,450+15*DA_I,ACTIONS[DA_I]
  next
  if Action <> NOACTION
    Setcolor Blue
    xyString 728,450+15*Action,ACTIONS[Action]
  endif
return
//==============================================================
//==============================================================
//--- Subroutine DispStatusMeter
//--- Inputs  : DM_Which,Needs[]
//--- Outputs : none
//--- Calls To: none
//---
//--- Updates the meters display
//--------------------------------------------------------------
DispStatusMeter:
  DM_Incr = DM_Which*140
  if Needs[DM_Which] > 100 then Needs[DM_Which] = 100
  DM_Value = 100-Needs[DM_Which]
  rectangle 732,50+DM_Incr,793,149+DM_Incr,white,white
  rectangle 732,50+DM_Value+DM_Incr,793,149+DM_Incr,Cyan,Cyan
  line 732,65+DM_Incr,793,65+DM_Incr,3,Red
return
//==============================================================
//==============================================================
//--- Subroutine CheckStatus
//--- Inputs  : Status,CurCell,Memory[],MemPtr,CurNeed,Needs[]
//--- Outputs : Status
//--- Calls To: DispStatusMeter,DisplayAction
//---
//--- Updates the status of the robot and displays the action
//--- being taken
//--------------------------------------------------------------
```

FIGURE 16.4 *(Continued)*

```
CheckStatus:
  if Status = PAIN then return
  // Sets Status if Pleasure
  Status = 0
  DM_Which = -1
  if (CurCell=EAT_AREA) and (CurNeed=EAT)
    Action = EAT
    gosub DisplayAction
    Needs[SLEEPY] = Needs[SLEEPY] + 15
    Needs[BORED]  = Needs[BORED]  + 10
    DM_Which = HUNGRY
  elseif (CurCell=SLEEP_AREA_1 or CurCell=SLEEP_AREA_2) and
CurNeed=SLEEP
    Action = SLEEP
    gosub DisplayAction
    Needs[HUNGRY] = Needs[HUNGRY]+5
    Needs[BORED]  = Needs[BORED]+20
    DM_Which = SLEEPY
  elseif (CurCell=PLAY_AREA) and (CurNeed=PLAY)
    Action = PLAY
    gosub DisplayAction
    Needs[HUNGRY] = Needs[HUNGRY]+20
    Needs[SLEEPY] = Needs[SLEEPY] + 20
    DM_Which = BORED
  elseif MemPtr>0
    for i=0 to MemPtr-1
      if (Memory[i,0]=CurNeed) and (Memory[i,1]=CurCell)
        // this cell is on the good list
        Status = PLEASURE
        return
      endif
    next
  endif
  if DM_Which < 0 then return
  for CS_I=Needs[DM_Which] to 0 step -1
    Needs[DM_Which] = CS_I
    gosub DispStatusMeter
    delay 50
  next
  Status = PLEASURE
  For DM_Which = 0 to 2
     gosub DispStatusMeter
  Next
return
//==============================================================
//==============================================================
//--- Subroutine CheckMemory
//--- Inputs  : Memory[],MemPtr,Needs[],CurCell
//--- Outputs : CurNeed,HaveResponse
//--- Calls To: none
//---
//--- Uses the biological needs to cause reflexive actions
//--- such as eating and sleeping. Other needs are affected by
//--- these actions.
//--------------------------------------------------------------
CheckMemory:
  // First Establish Current Need
```

FIGURE 16.4 *(Continued)*

```
    CurNeed = 0
    Maximum = Needs[HUNGRY]
    for CM_I = SLEEPY to BORED
       if Needs[CM_I] > Maximum then Maximum=Needs[CM_I]
    next
    if Maximum > NEEDS_THRESHOLD
       if Maximum = Needs[BORED]  then CurNeed = PLAY
       if Maximum = Needs[SLEEPY] then CurNeed = SLEEP
       if Maximum = Needs[HUNGRY] then CurNeed = EAT
       //Hunger is higher order priority than sleep and
       //sleep is higher order than play
    endif
    // sets variable HaveResponse to position of match
    // in memory or zero (false) if no match is found
    HaveResponse = 0
    if MemPtr=0 then return
    for i=0 to MemPtr-1
      if (CurNeed=Memory[i,0]) and (CurCell=Memory[i,1])
        HaveResponse=i // shows where found
        break
      endif
    next
return
//================================================================
//================================================================
//--- Subroutine CheckBadList
//--- Inputs  : BadPtr,BadList[],CurCell,Movement
//--- Outputs : Bad
//--- Calls To: none
//---
//--- Checks to see if a proposed move from a cell is in the
//--- bad move memory
//----------------------------------------------------------------
CheckBadList:
    // sets variable BAD
    Bad=false
    if BadPtr=0 then return
    for i=0 to BadPtr-1
      if (BadList[i,0]=CurCell) and (BadList[i,1]=Movement)
        Bad = True
        break
      endif
    next
return
//================================================================
//================================================================
//--- Subroutine DisplayInstructions
//--- Inputs  : none
//--- Outputs : Key
//--- Calls To: none
//---
//--- Displays instruction and waits for a key press then
//--- returns the character pressed in the variable Key
//----------------------------------------------------------------
DisplayInstructions:
  data MI;"Figure16.04.Bas"
  data MI;"This program transforms the robot into a 'living' creature."
```

FIGURE 16.4 *(Continued)*

```
  data MI;"It has needs as hunger, boredom, and even the need to sleep."
  data MI;"These needs have priorities and only the largest current need"
  data MI;"is a motivator.  Like a biological creature, there are inborn"
  data MI;"characteristics such as curiosity and a goal to satisfy the "
  data MI;"internal needs.",""
  data MI;"When a creature is born, all of the needs are present. The"
  data MI;"creature will explore and if a need is satisfied (by accident)"
  data MI;"the meters will reflect the fact.  The creature will also"
  data MI;"remember things that give it pain or pleasure and act differently"
  data MI;"in the future because of its memory."
  data MI;"In the beginning the creature seems to roam aimlessly. However,"
  data MI;"after an hour or so, the creature will have explored enough to "
  data MI;"have a good amount of memory.  At that time you will see that "
  data MI;"the creature has learned how to satisfy needs that motivate it."
  data MI;"At that time, it will only roam when all the needs are low as"
  data MI;"indicated by the graphs.  See the text for more information.",""
  data MI;"You can save and retrieve the memory if you wish. Currently, "
  data MI;"the program will automatically save its memory (in the files "
  data MI;"MEMORYGOOD and MEMORYBAD) when the memory size is 25, 50, and 75.",""
  data MI;"If memory has been saved, you can now load it by pressing the"
  data MI;"'Cancel' button or Esc key. Pressing 'OK' or Enter key will start"
  data MI;"the program with no initial memory and will build a new memory."
  Key = MsgBox(MI)
return
//================================================================
//================================================================
//--- Subroutine InitializeSimulation
//--- Inputs  : none
//--- Outputs : Needs[],BadPtr,MemPtr,CurNeed,
//---           CurCell,LastCell,Rx,Ry
//--- Calls To: none
//---
//--- Draws the environment and initializes all variables
//--- and arrays.
//----------------------------------------------------------------
InitializeSimulation:
  //--- Initialize system variables
  Data ACTIONS;"NO ACTION","  EAT"," SLEEP"," PLAY","EXPLORE"
  Data ACTIONS;" SAVE","RETREAT","RESPOND"
  Dim Needs[3]
  MConstant Needs,100 //all needs are maximum
  Dim Memory[100,3] // Memory saves CurNeed,CurCell,Movement
  MemPtr=0
  Dim BadList[100,2]  // BadList does not include needs
  Action = NOACTION \ Status = HUNGRY\ Movement = NORTH
  BadPtr=0 \ CurNeed=0 \ CurCell=0 \ LastCell=0

  //---Draw Environment
  //----outside border
  SetColor Black
  LineWidth 3
  rectangle 0,0,720,595
  //----Sleep area
  LineWidth 5
  circle 500,-220,940,220,LightGreen,LightGreen
  circle 580,-140,860,140,LightGreen
  SetColor Black,LightGreen
  xyString 520,2, "SLEEP"
  xyString 525,16,"AREA"
  LineWidth 3
  Rectangle 720,0,800,595,black,white
```

FIGURE 16.4 *(Continued)*

```
    SetColor Black
    line 0,0,720,0
    //---- Activity Area
    SetColor Black,Yellow
    rectangle 3,3,120,120,Yellow,Yellow
    xyString 3,2,"ACTIVITY AREA"
    //---- Fire area
    LineWidth 5
    SetColor Red
    data FirePlace;-600,0, 720,0, 720,120, 690,30, 600,0, 670,-3
    MPolygon FirePlace,Red
    SetColor Black,Red
    xyString 670,3,"FIRE"
    //----Walls
    SetColor Black,White
    data FoodArea; -240,360, 240,480, 120,480, 120,240, 360,240
    data FoodArea;  360,360, 480,360, 480,480, -360,480, 360,600
    MPolygon FoodArea
    xyString 127,460," FOOD AREA"
    //----Briar areas
    SetColor Black,LightGreen
    rectangle 240,120,360,236,LightGreen,LightGreen
    xyString 262,218," BRIARS "
    rectangle 600,480,716,591,LightGreen,LightGreen
    xyString 622,573," BRIARS "
    //----meters
    LineWidth 2
    SetColor Black,White
    xyString 740,4,"NEEDS"
    line 735,22,786,22
    rectangle 730,30,795,150
    xyString 735,32,"Hungry"
    rectangle 730,170,795,290
    xyString 735,172,"Sleepy"
    rectangle 730,310,795,430
    xyString 740,315,"Bored"
    rectangle 732,50,793,149,Cyan,Cyan // Hungry
    rectangle 732,190,793,289,Cyan,Cyan // Sleepy
    rectangle 732,330,793,429,Cyan,Cyan // Bored
    //---Place the Robot
    rLocate 420,300
    Rx = 3 \ Ry = 2 \ CurCell = 17
    rInvisible LightGreen,Yellow
return
//===============================================================
```

FIGURE 16.4 *(Continued)*

16.4.1 DEVELOPING A PERSONALITY

The first time you run the program the robot has to begin its development as if it were a baby. It can take an hour or more for the robot to store enough associations in its *good* memory to make it start behaving intelligently. When this happens, the robot begins to take on distinct characteristics that can be considered its personality. Some robots will select a single favorite path to food while other robots will have two favorite paths. In such cases the robot will choose the path that is closest to where

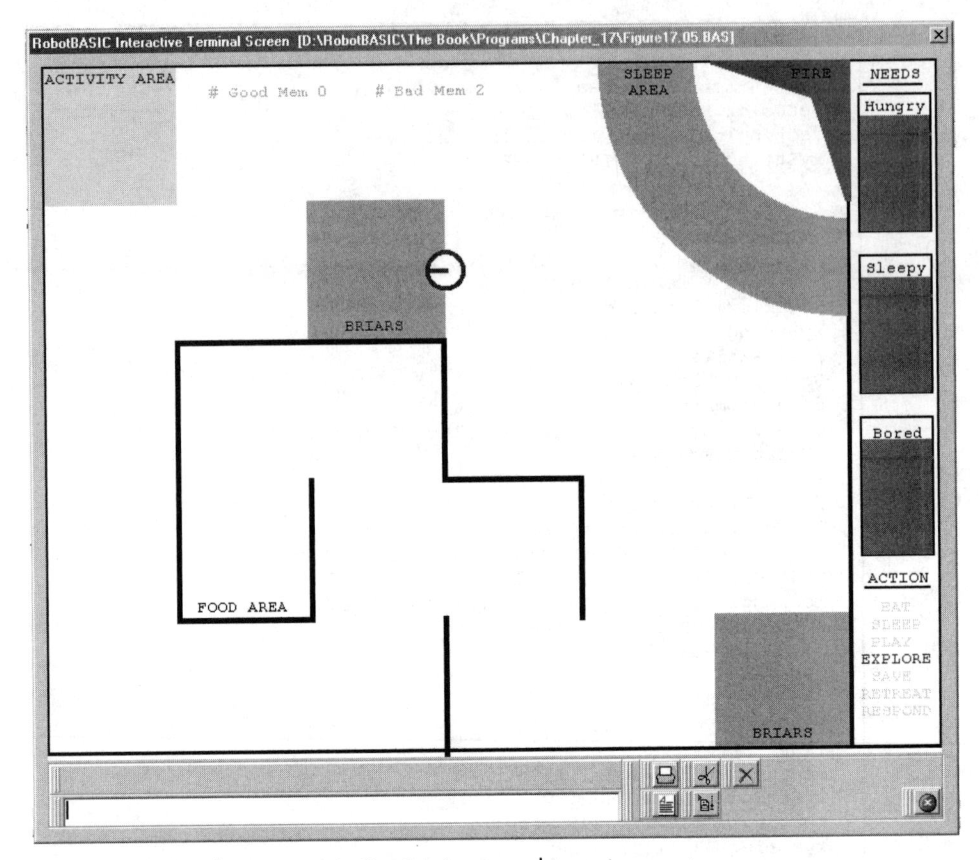

FIGURE 16.5 The program in Fig. 16.4 creates this environment.

it is when hunger becomes a motivating factor. Sometimes a robot develops the habit of pacing back and forth a few times before it actually eats (I am still unsure how this behavior develops).

The important point here is that the robot does learn, and the more it learns the faster it learns. It takes a relatively long time before the robot starts to look remotely intelligent. After the *good* memory has expanded to 40 or 50 items, though, you will see an amazing difference.

As long as no needs have reached their threshold (the red line), the robot will be content to explore its environment. However, when a need becomes a motivator, the mature robot shows its intelligence and quickly makes its way to the area in the environment that can satisfy that need.

16.4.2 DISPLAYING THE ROBOT'S ACTIONS

The area on the lower-right side of the screen shows what basic action the robot is executing as summarized in Fig. 16.6.

Action	Significance
Eat	Robot eats if it finds food when hungry.
Sleep	Robot sleeps if it finds a warm spot when sleepy.
Play	Robot plays if it finds the activity area when bored.
Explore	Robot randomly explores when needs are low.
Save	The robot is saving something to its memory.
Retreat	The robot has felt pain and is backing away from the source.
Respond	The robot is searching its memory or responding to what was found.

FIGURE 16.6 The robot has seven basic actions it can take.

16.4.3 UNDERSTANDING THE CODE

The subroutines hierarchy chart is shown in Fig. 16.7. The *MainProgram* displays instructions to the user by calling the *DisplayInstructions* subroutine then sets up the environment by calling the *InitializeSimulation* routine. If the user has indicated that the memory accumulated and saved from the last time the simulation was executed is to be used (by pressing *Cancel* or *Esc* instead of *OK* or *Enter*), the subroutine *RestoreMemory* is called to reload the robot's memory. Finally, *ComeToLife* is executed to initiate the simulation.

Above the *MainProgram* label there is a set of definitions for constants and variables that are used throughout the program.

16.4.3.1 *DisplayInstructions* This subroutine is similar to what you have seen in previous chapters. The only point to note is the fact that the variable *Key* is assigned the return value from the function `MsgBox()`. The value will be zero if Cancel or *Esc* was pressed and one if OK or *Enter* was pressed. *Key* is checked in the *MainProgram* to decide whether or not to load the memory from previous runs (it is loaded if Cancel or *Esc* is pressed).

16.4.3.2 *InitializeSimulation* This subroutine draws the environment and sets up the necessary arrays and other variables. The only point of note in this subroutine is the use of the multiple assignment construct. Notice the use of the character \ to allow for placing

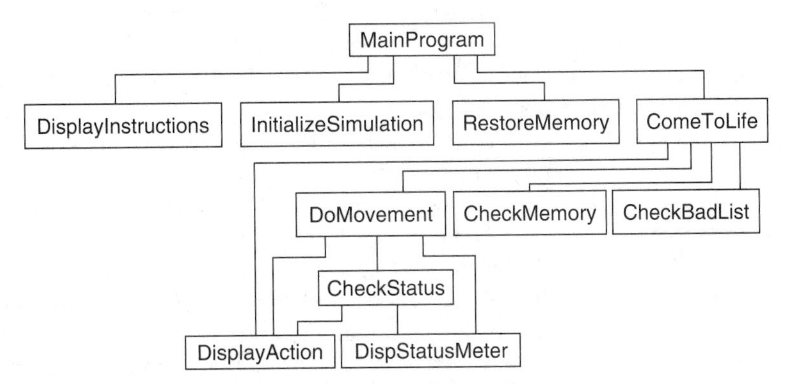

FIGURE 16.7 Subroutines hierarchy chart.

multiple assignment statements on the same line. This makes for more compact code. There is no other advantage (see Sec. B.3 for more details).

16.4.3.3 *ComeToLife* This routine calls *CheckMemory* to compare the robot's current need and position against its *good* memory. If a match is found, the variable *HaveResponse* points to the matching memory position. If *HaveResponse* is non-zero, the display will show that the robot is responding to memory and the desired movement is retrieved from memory and stored in the variable *Movement*. The subroutine *DoMovement* is then called to actually perform the movement. *DoMovement* will be discussed later.

If there are no good memories associated with the robot's current state, the software chooses a random movement (0 to 3 for north to west). The `if`-statement associated with this random choice is only performed 60 percent of the time. This means that 40 percent of the time the robot repeats its last movement creating a tendency to continue in the same direction.

The software checks to see if the random movement chosen is in the *bad* memory. If it is not, *DoMovement* is called to actually perform the movement and set the variable *Status* to PAIN or PLEASURE if that situation occurred because of the movement.

If the robot experiences pain, the current state and movement are saved to the bad memory. If the robot experiences pleasure, the current state and movement are saved to the good memory if they have not been saved sometime in the past. Notice that no check is made to see if the painful event is already in memory because the robot will never repeat painful movements (contrast this fact with real human behavior, also see Exercise 4).

The format and content of the robot's memory entries can be helpful when trying to understand the overall operation of this program. A bad-list entry contains only two items, the current cell number (specifying a region of the screen) and the action taken. For example, lets assume the entry in the bad memory was 6, 0. This would mean that when the robot was in cell six and made movement zero (north) it felt pain. If the robot is ever in cell six again and randomly chooses to move north then it will choose some other action.

The entries for the good memory contain the current cell number, the action taken, and the robot's emotional state at the time the memory was recorded. This emotional state consists of a single number specifying if the robot was being motivated by hunger, sleep, or boredom.

Let's look at an example memory entry of 1, 3, 2. This would mean that at some point in the past, the robot was in cell three and hungry (1 = eat). Furthermore we know that under these circumstances, when the robot moved south (2), the robot felt pleasure. Consequently, anytime the robot is hungry and in cell three it will know that moving south will eventually lead to pleasure. The robot may not find food immediately when it moves south because the memory entry may have been formed not because it found food but because its action (in this example, moving south) caused it to encounter a previously memorized situation that was associated with the pleasures of food. Let's see how this complex sounding situation controls the robot's behavior.

The first time the hungry robot randomly finds food, the action it performed to get there is recorded on the good list. Remember, this memory entry contains information about the robot's position (cell number) and its motivational need. Since this entry is on the good list, if the robot is hungry in the future and randomly chooses an action that moves it into this state, then the new action will also be interpreted as a pleasurable experience. Over time, the robot will learn how make its way through the environment to satisfy its need. The robot's behavior will make you think that a path has been stored in its memory, but this is not the case. The movement along the apparent path takes place based on individual associations not on a single memory of the entire path.

This algorithm is not unlike the way memory associations are formed in biological creatures including humans. Our algorithm is much simpler than a human's (it does not include value judgments) but the finished program demonstrates that the algorithm does in fact produce intelligent behavior. The use of cell numbers is just an easy way for the robot to recognize where it is. Living creatures use senses such as vision and smell to link their memories to situations and places. Also we did not implement the opportunistic behavior discussed earlier (also see Exercise 5).

16.4.3.4 *DoMovement* This module starts by displaying the current memory sizes on the screen and then saving the current position (*CurCell*) into the variable *LastCell* so that it knows where it was before the movement is made. The internal needs (hunger, sleep, and boredom) are increased appropriately and displayed on the screen meters.

The compass is used to turn the robot in the proper direction before the robot tries to move to the next area of the screen (the robot's environment is actually divided into 30 regions or cells. Each cell is given a numeric value and the current cell's number is calculated from a formula and stored in the variables *Rx* and *Ry*, which keep track of the robot's cell position). If the robot bumps into something while moving, it sets *Status* to PAIN so that it will learn not to do that movement again while in this cell. The robot automatically moves back to its original position when it encounters a painful situation. Similar actions occur when the robot moves into a briar patch or too close to the fire.

The last thing that happens in *DoMovement* is a call to *CheckStatus*, which causes the robot to react based on its status. Essentially, the robot will eat if it is hungry and in an area of food, sleep if it is sleepy and in the sleep area, and play if it is bored and in the play area. During these processes, other needs are increased reasonably. If you eat, for example, it tends to make you sleepy. If you modify the parameters in this routine you can affect the *personality* your robot develops. A major contributor to the robot's personality is the situations it encounters as it matures (just like living creatures).

16.5 Summary

In this chapter you have:

- Seen how adaptive behavior can be more effective in tackling a wider variety of unpredictably changing environments.
- Been exposed to the basic principles for creating adaptive control algorithms.
- Seen an example program that attempts to use the robot to model the behavior of biological organisms.
- Been shown that even an *artificially* intelligent robot can develop a distinct personality and unique habits based on how it grows and develops in its environment.

Now, try to do the exercises in the next section.

16.6 Exercises

1. Run the program in this chapter to see how it performs. Examine the code and try to change some of the *biological* characteristics to see how they affect the robots eventual personality and habits.
2. Modify the program so the robot will try to move toward the mouse pointer when it is exploring its environment. Help the robot learn faster by using the mouse to entice the robot to move toward food, activity, and sleep areas when a need is active. Compare this activity with teaching young children new things.
3. Modify some of the earlier programs in this book (such as, following a line or following a wall) so that the algorithm includes some form of adaptive behavior. Compare the performance of your programs with the original ones.
4. In the algorithm of this chapter, once an item was added to memory it stayed there permanently. The robot cannot forget. To make the robot behave more like a biological creature, it has to be made to forget. Implement a forgetting procedure to occasionally delete the last entry (or perhaps a random entry) in the memory arrays on a random basis. A more complicated, but more realistic simulation would be to have a long-term memory and a short-term memory and have the robot transfer some (not all) the short-term memory into long-term memory while sleeping. Also you may want some mechanism for enforcing short-term memory the more times a situation is encountered, and give preference to these reinforced items when transferring to long-term memory.
5. In the algorithm discussed in this chapter, the robot was only motivated by a single need based on biological priorities and thresholds. In other words, the robot does not develop complex value judgments. This can be observed when the robot fails to play when it enters the play area if hunger is the current motivator. Imagine how this behavior could be altered so as to be more opportunistic. The robot could decide to take advantage of being in the area and play now even though it is hungry. Some internal mechanism (perhaps opportunity thresholds) must control this behavior. The robot should not play if there is not a reasonable need to do so. Likewise, the robot should forego play and head for food if the hunger need is very large. Modify the algorithm in this chapter to implement such an adaptation.

CHAPTER 17

RELATING SIMULATIONS TO THE REAL WORLD

Building a robot is an enjoyable challenge and programming it is a rewarding task. Traditionally, you could not uncouple these two activities. In order to program a robot you had to build one. Nowadays, many people are able to easily build a very capable robot thanks to the plethora of kits and powerful components available at affordable prices. However, despite being easier, building a robot can still be an obstacle for many due to various factors.

If you are interested in programming a robot but not building one, or you want to defer building one until after you acquire more skills but want to experiment with programming a robot, RobotBASIC and this book have shown you many projects and algorithms that demonstrate the power and utility of a simulator. You may find that the simulator satisfies your curiosity about controlling a robot enough that you do not need to control a real robot. However, if you do decide to control a real robot, the role of RobotBASIC is not over. In fact, RobotBASIC can be a powerful tool in controlling a real robot, and all the knowledge and experience you have gained from reading this book, is very much transferable to the real-world. RobotBASIC can help you in moving from the simulated realm to the real one.

RobotBASIC's simulated robot is a great *prototyping* tool for doing *research and development* for robotic algorithms and ideas. You can use the simulations to demonstrate the principles of an idea without the expense of actually creating the hardware to test out the idea. Once the prototyping life cycle has run its course you will be ready to try out the ideas that have been developed on an actual robot. The characteristics of the simulated robot in RobotBASIC have been carefully designed to be achievable in a robot that can be affordably built using kits or from readily available components.

This chapter will show you how to construct a robot from an affordable kit (with some additional modifications) that emulates the characteristics of our simulated robot. However, just building the robot is not the end. You will need to port the algorithms to the robot's microcontroller. This chapter will show you *four ways* of achieving this.

17.1 A Historical Perspective

The field of hobby robotics has come a long way. A change in attitude toward the hobby is occurring. People now are not satisfied with just building a robot; rather they want to make it do useful things. The focus of the hobbyist is shifting from building a robot to programming a robot.

17.1.1 EARLY HOBBY ROBOTICS

In the 1980s when hobby robotics was just starting, building a real robot was very difficult. If you wanted circuitry to drive your motors or needed any type of sensors you had to build them yourself. This meant that you had to have a reasonable knowledge of electronics if you wanted to build a robot with even minimal capabilities.

While building everything from scratch was certainly an enjoyable challenge for the skilled few, building a robot remained more of a dream than a reality for the majority of people interested in the hobby. Most creations consisted of modified toys or crude platforms powered by windshield wiper motors salvaged from junk cars. Many robots had only one sensor—a front bumper mounted on a leaf switch. Of course, the *skilled* hobbyist dabbled with infrared light-emitting diodes (LEDs), phototransistors, and other such devices, to build obstacle-detection circuitry that enabled the robot to deal with its environment without actual collisions—but such *state-of-the-art* sensors were often unreliable.

The makeshift infrared sensors used in those days were easily *blinded* if the robot turned toward a window. A robot that worked well at home often exhibited erratic behavior when it was demonstrated at a club meeting where fluorescent bulbs illuminated the room. In the years that followed, hobbyists learned to modulate the infrared emitters to solve these problems but those without electronics experience often found it difficult to adjust their oscillators and filters properly.

More sophisticated sensors such as ultrasonic rangers and electronic compasses were not even attempted until decades later and even then they had to be individually designed and constructed. There were many how-to books that promised to guide you through the hard stuff, but the average reader often had a difficult time duplicating the authors' works.

People who managed to overcome the above difficulties faced a further challenge. A robot without a central processor (a brain) cannot make effective use of external sensors. In the early days there were no single chip microcontrollers and software development environments were limited.

People could have used some of the (expensive) personal computers (PC) of the day, but the size of these machines and the methods for powering them severely affected the mobility and size of the robot.

17.1.2 HOBBY ROBOTICS TODAY

In recent years hobby robotics has changed dramatically. Around the world, members of robot clubs are building ever more sophisticated robots as they challenge each other to expand their creativity and problem-solving skills. Certainly the low-cost and availability of powerful microcontrollers has added to the interest in robotics. Schools are discovering that adding robotics to their curriculum increases student interest, not only in robotics, but in mathematics and science as well. Regardless of the reasons for the growth, the results are a great number of people interested in hobby robotics and a booming industry catering to a tremendous demand.

Robot kits are now available from many companies. You can buy robots with wheels and robots with legs. You can get robots that balance and robots that climb. Even toy companies like LEGO offer robot kits with amazing power and versatility.

17.1.3 THE PARADIGM SHIFT

The wide variety of ready-to-use robot components is changing the nature of hobby robotics. In the past hobbyists needed to *build* a robot. Today they can buy the parts and *assemble* one with minimal effort. Because of this, the hobbyist today can concentrate on *programming* the robot. This means that now they can create a robot that can actually do something interesting and maybe even useful (a feat almost never accomplished until recently).

The change that is taking place in the field of hobby robotics parallels the revolution that took place in the PC industry. If you wanted a computer in the early 1970s, you had to build it yourself, almost from scratch. Less than 10 years later computer hobbyists could buy completely assembled systems from companies like Apple, Radio Shack, and IBM. Almost overnight, the nature of the hobby turned from soldering and bread-boarding to *programming*. The new hobby programmers were able to make the "toy" computers solve real-world problems and everything changed.

Hobby robotics is poised to change the world in a similar manner. The progress to intelligent machines will be hindered so long as robot hobbyists continue to give priority to constructing a robot and neglect the importance of programming it. It was the intelligence of software that gave simple hobby computers capabilities and utility that made many corporations prefer them over the giant mainframe systems they were using. This quickly transformed personal computing from a hobby to a business that changed the world.

The hardware needed to build a robot is now readily available, and creating a capable robot is within the reach of anyone. If you visit a robot club today you will see many robots that have been assembled from a variety of standard parts and sensors and you will observe that the emphasis is shifting from *building a robot* to *programming a robot*.

Programming a robot is a crucial activity in robotics. Constructing the robot's hardware and electronics is a great deal of fun and can be very satisfying, but, the robot will not be able to achieve much without the right *artificial intelligence (AI) algorithms* that enable it to *autonomously* carry out useful tasks.

The hardware and software on a robot should work together to achieve the task. Nevertheless, the *algorithm* is what matters. The hardware required to achieve the task and the details of how it is to be made to function is of importance ultimately. However, the amount, type, and configuration of transducers (sensors), actuators (motors), and the shape, size, and type of the robot itself are details worked out as a result of the algorithms that are developed to solve the tasks the robot is meant to solve.

Programming an actual robot can be a daunting task due to the development cycle that has to be followed:

1. Build a robot and equip it with a microcontroller.
2. Mount and align the sensors.
3. Connect the microcontroller to a PC running a software development environment.
4. Download a program to the robot.
5. Unplug the robot.
6. Place the robot in the test environment.
7. Turn the robot on and watch the results of your programming.

If the robot does not function properly, you will have to repeat steps 2 to 6 until you arrive at a working program. Finally, when the robot functions you still need to test the robot in other environments to ensure that the robot will function in all environments it is likely to encounter. Devising and building test environments can be time-consuming and costly. Additionally, the robot may get damaged if it does not function properly. Furthermore, if the robot that is being used is not adequate for the task, you may have to abandon the hardware and start from square one.

You can see that this cycle is cumbersome. If problems are encountered, the debugging process can be unwieldy. Due to the awkwardness of the entire process, the programmer may tend to accept the first-working solution instead of continuing to hone the program until it performs optimally.

RobotBASIC lets everyone get right to the heart of robotics. They don't have to build anything or even assemble anything. They can start learning how to program a robot right away. There is no need to build environments for testing. There is no need to safeguard the robot from damage. There is no reason why you cannot keep improving the algorithm to the optimum point. There is no reason why you cannot test the robot exhaustively. You can experiment with different configurations and combinations of sensors. You can experiment with different algorithms and different situations. This is a *major paradigm shift* in *thinking about robotics*.

Most hobbyists who use RobotBASIC to learn how software can give real intelligence to a robot will eventually want to build an actual robot. After programming simulations, hobbyists will have a much better appreciation of the type and configuration of hardware their robot needs to achieve the target tasks. Knowing what type of hardware you need and how to put it together can be a major cost- and time-saving. The remainder of this chapter discusses a variety of ways to use the algorithms developed in RobotBASIC with real-world robots.

17.2 Constructing a Robot

Many companies offer a wide variety of hardware and kits that make building a robot easier than ever before. Many hobbyists, however, often want to build their own robots, or at the very least, significantly enhance the functionality of a basic kit.

RobotBASIC's robot has been designed to have sensors and instruments that simulate ones that could actually be built or affordably purchased by a typical hobbyist. Nowadays, there are many companies offering sensors, but many specialize in only a few items. One company that we had purchased items from over the years truly stands out. This company is Parallax, Inc. Not only do they offer a wide variety of both robot kits and sensors, they also share our educational views that students of all ages can benefit from courses and projects involving robots. You can certainly use parts from *any* source to create a robot that emulates RobotBASIC's simulated robot, but since Parallax offers virtually all of the sensors we have on our robot, we will discuss their kits and sensors as the primary basis for constructing a real-world version of our simulated robot.

Another reason we chose Parallax is that they provide a wealth of documentation (spec sheets, program listings, application notes, discussion forums, and more) on their web site to help you utilize their products effectively. This allows the discussions in this chapter to be general in nature. If you need more help, visit the Parallax web site (www.parallax.com) and also refer to the list of resources at the end of this chapter. Parallax offers an educational robot kit called Boe-Bot (see Fig. 17.1). It has two servo-powered wheels and one caster, all mounted on a rectangular body. It comes with an easily programmable microcontroller (BASIC Stamp 2 [BS2]) and enough parts to give anyone a good introduction to robotics. A detailed manual outlines a series of experiments that

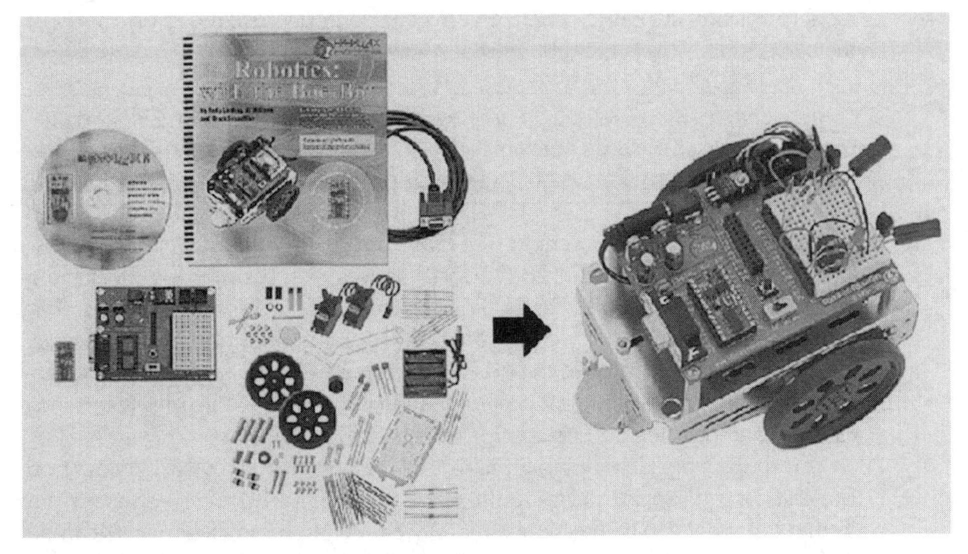

FIGURE 17.1 Boe-Bot educational robot kit.

are very helpful and informative. The Boe-Bot can easily serve as the basis for our real-world robot even though the basic kit does not have all the functions provided by our simulation. The photograph in Fig. 17.1 as well as others in this chapter that depict Parallax products are courtesy of Parallax, Inc.

The discussions that follow, offer suggestions for customizing the Boe-Bot to enhance its abilities. Remember, our simulated robot has all the sensory capability most hobbyists would ever want on a mobile robot. Use the simulator to determine what you would like the robot to do and to establish what capabilities it needs to accomplish your target project. When you are ready to build a real robot you can purchase only the sensors you need for the projects you wish to do. Furthermore, if you have utilized the simulator effectively, you can feel confident that the money you spend on the hardware project will result in an operational robot with the desired capabilities.

17.2.1 WHEEL AND BASE ASSEMBLY

RobotBASIC's simulated robot is a circular robot (no corners to hook things) that can turn around its center. This movement has the advantage of allowing the robot to turn within its footprint without having to worry about bumping objects around it. Since the robot is round and rotates around its center, if it is not in contact with an object before it turns, it will not hit an object as it turns.

The robot in the Boe-Bot kit has two side-by-side wheels and one caster to keep the robot from tipping. If both wheels turn in the same direction at the same speed the robot will move forward (or backward). If one wheel turns forward while the other turns backward, the robot will rotate around its center.

In order to generate the motions described above, each wheel has its own motor. We could have successfully used DC motors, stepper motors, or servomotors but both DC and stepper motors require external drive circuitry. Servomotors (similar to those used in model airplanes) are easy to control with digital logic because the drive-circuitry is built into the servo itself. This is the type of motor supplied with the kit.

NOTE: Standard servomotors have a fixed range of motion and are not suitable for driving the wheels of a mobile robot. Servos that can rotate continuously can be purchased from Parallax or standard servos can be modified.

The width of the pulses sent to Boe-Bot's servo drive motors control the speed of each wheel. You can control how far the robot moves or turns by controlling the number of pulses sent to each wheel. This is an *open-loop* feedback system though, that does not guarantee that the wheels have moved the desired amount. One servo, for example, might be more efficient than the other and turn slightly more even when given the same number of pulses. The Parallax documentation explains how to calibrate the servos so they can operate together, but some error is to be expected.

You should consider adding Parallax's wheel encoders (they easily mount on the Boe-Bot chassis and provide pulses to the BS2 that indicate how far the wheels have turned) so that the robot can be positioned more accurately. The stock encoders can only resolve the robot's movement to about $\pm$ in of linear wheel travel, but adding a custom reflecting disk to the wheels could improve on this limitation.

17.2.2 BUMPER SENSORS

When we built the real-world version of our simulated robot, we could have constructed everything from scratch. As mentioned earlier, we decided to use the Boe-Bot kit because it made the assembly a lot easier. The rectangular shape of the Boe-Bot did require some modification to make it circular as in the simulation.

Our simulated robot has four bumper switches mounted around the front, back, and sides of the base. There are many ways of building these bumpers for a real robot. They could be made using leaf switches, for example, with a plastic or wire bumper glued to the actuator lever. We decided to get really creative and design a custom bumper system that is integrated into our robot's body.

We used ±-in foam-board (found at many craft and office supply stores) and the inner ring of a wooden embroidery hoop to construct the robot's round body (see Fig. 17.2).

The inner ring of the embroidery hoop is glued around the edge of the foam board. This wooden edge is covered first with a thin layer of foam tape and then a layer of real copper tape (also found at craft stores). Make sure you use real copper because the surface must be conductive to electricity.

The outer embroidery hoop is cut into four pieces (130° on the front and rear and 50° on the sides) that will become the actual bumpers. Each of these pieces is also lined with copper tape along the bottom edge as shown in Fig. 17.3. Also shown in the figure is the Boe-Bot body from Parallax which was modified in several ways. First, the servos were mounted with spacers to give a wider, more stable wheelbase for the new larger body. Second, the original

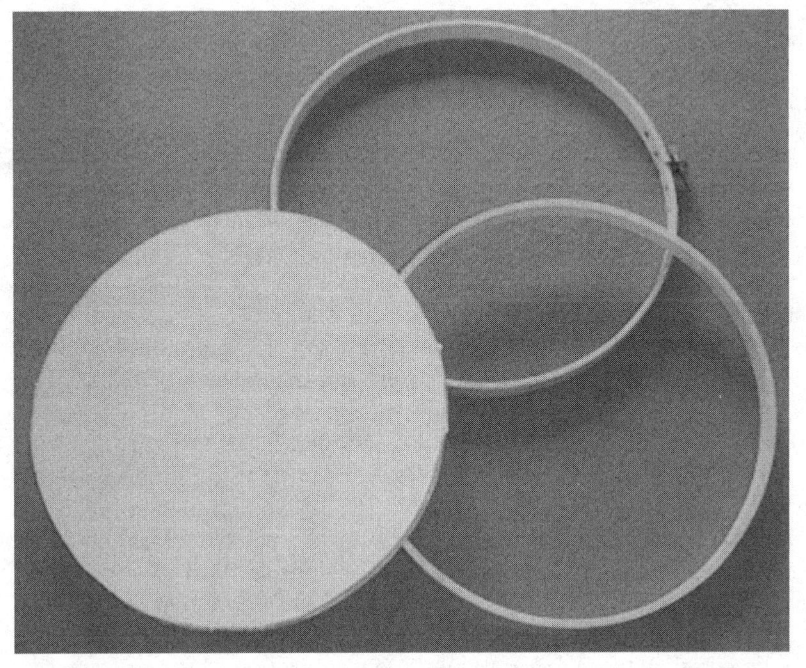

FIGURE 17.2 Foam-board can be easily cut to form the circular body of the robot.

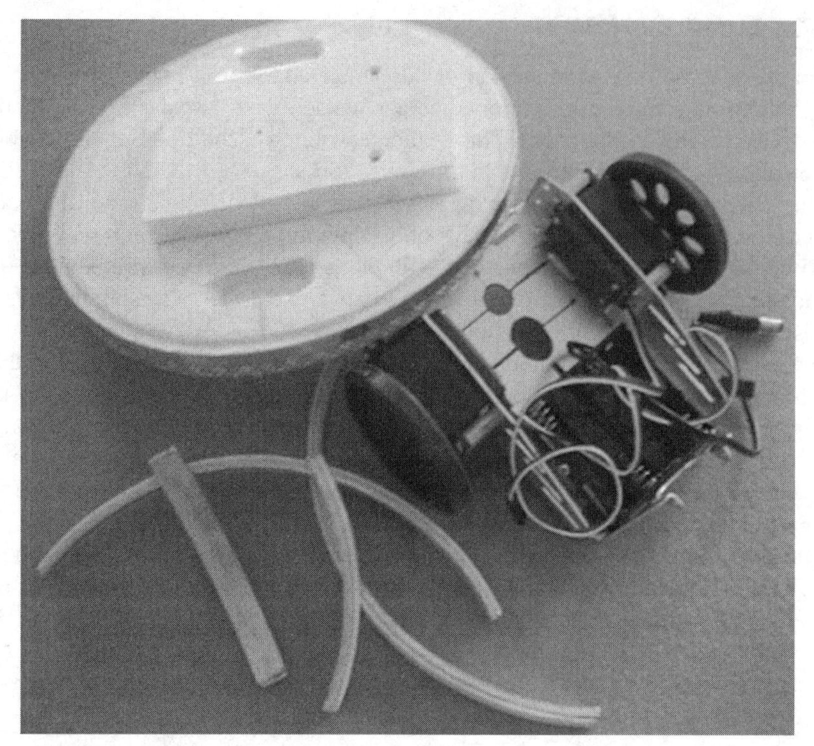

FIGURE 17.3 Copper tape creates the conductive surfaces for the bumper switches.

hole (used to pass wires from the servos to the BS2) was enlarged (more on how this will be used later). A second hole was added to accommodate the wiring. Notice also that another piece of foam-board is used to build up the area where the Boe-Bot will be mounted, and that recessed areas are used to ensure that the wheels will have sufficient clearance. This is easily done because foam-board can be cut with a razor blade or hobby knife.

A piece of thin rubber was glued to the top of the foam-board body and allowed to stick out an inch or so around the edges. The four bumpers were glued to the protruding rubber around the edge of the robot as shown in Fig. 17.4. If the rubber is too stiff you may have to make small cuts between the bumpers and the body so that the bumpers flex easily. In our case, we also had to cut the front and rear bumpers in half so that they could move more freely (the two pieces are wired in parallel so they still appear to be one bumper electrically). After the glue that holds the bumpers to the foam has dried, the excess foam should be trimmed away.

Figure 17.4 also shows how the Boe-Bot body mounts to the foam-board. Notice also the brass tube that passes through the very center of the robot (originally used by Parallax to feed wires to the top of the robot). A felt tip marker can be placed in this tube if you want the robot to draw as it moves (see Chap. 10). A pen used in this way cannot be raised and lowered as it does in our simulations, but this capability could be added by the creative hobbyist using a solenoid (or servomotor). We raised the Boe-Bot's circuit board (see later figures) to make room for such a modification should it be desired in the future.

FIGURE 17.4 The Boe-Bot mounts to the round foam-board body.

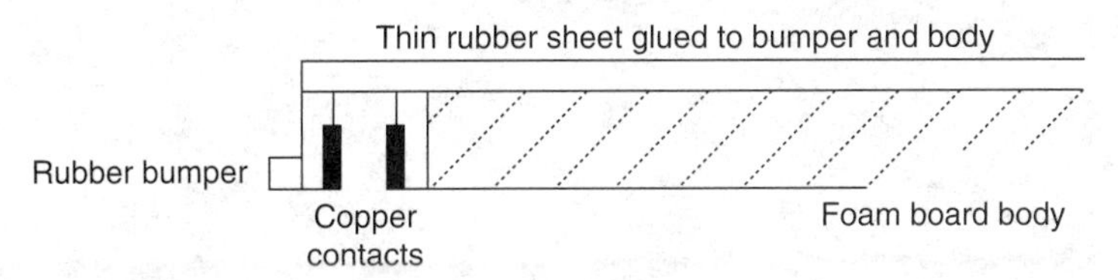

FIGURE 17.5 This diagram shows the operation of the bumper switches.

If you look closely at the bumpers in Fig. 17.4 you will see that thin strips of rubber have been glued to the lower edge of the wooden embroidery hoop. These rubber strips not only provide extra cushion during collisions, but they also help ensure good contact of the copper surfaces because they ensure that the pressure from the contact will be applied to the lower half of the bumper. When a collision occurs, the bumper bends inward to make contact as shown in the diagram in Fig. 17.5.

17.2.3 INFRARED PERIMETER SENSORS

Even though our robot will have physical bumpers to indicate when a collision occurs, we wanted our robot to detect objects in its path before a collision actually happens. One way to do this is with light. The idea is simple. When we want to determine if there is an object close to the robot we turn on a light source that will project its light away from the robot. If an object is close by, some of that light will be reflected back. If no reflection is detected with a phototransistor we can reasonably assume there is no object near that sensor.

As stated earlier, this idea is easy to understand. Unfortunately, the practical implementation of it can be a little harder. To begin with, we can't use normal light unless we want our robot to operate in total darkness. An easy solution is to use infrared light and place a luminance filter that blocks visible light, but passes infrared over the phototransistor.

The above solution works, but only if there are no other sources of infrared light in the robot's work area. Unfortunately, fluorescent lights generate a lot of infrared. Even sunlight has enough infrared to cause problems. An easy solution to this is to *modulate* the light source (i.e., have it pulse at a specified frequency). An electronic filter can then be added to the phototransistor circuit to ensure that only the modulated light will be detected.

The above sounds (and is) complicated. Fortunately nowadays, you can buy phototransistors that have an embedded electronic filter and an integrated luminance filter. The Boe-Bot kit includes two of these parts and two infrared LEDs, and the manual gives detailed experiments that show how to use them. Additional parts can be purchased separately.

RobotBASIC's simulated robot has five sensors of this type to help it detect objects in its path. Figure 17.6 shows how the real robot also has five sensors. The wiring for each

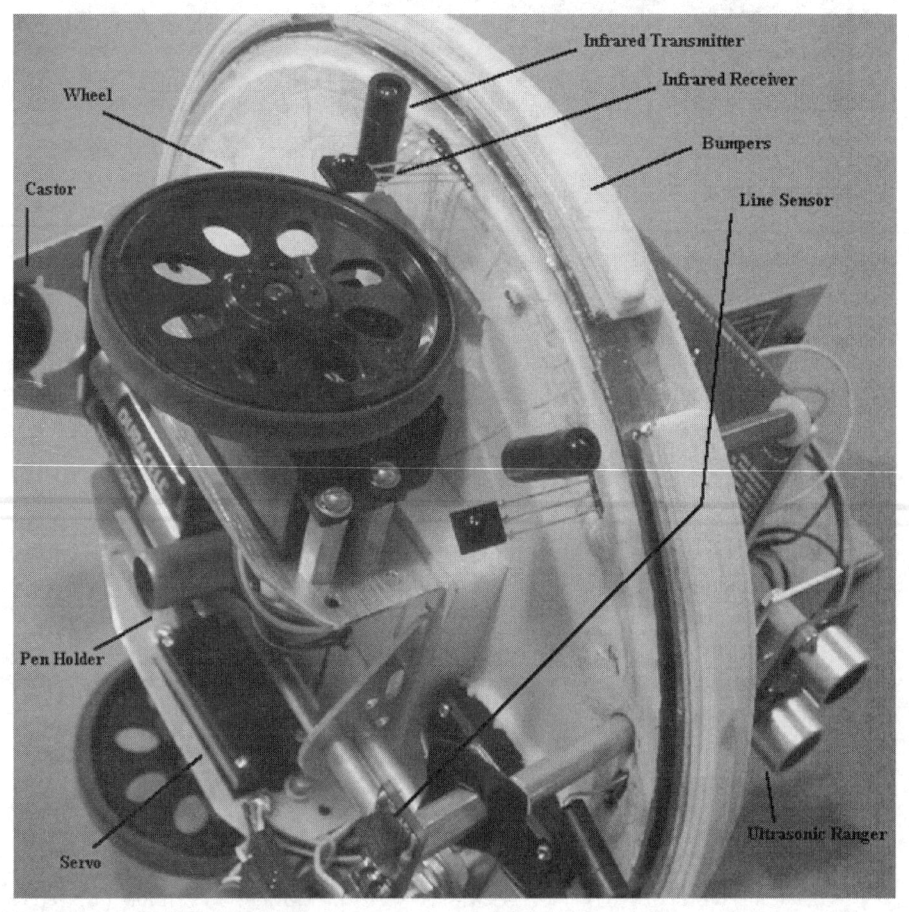

FIGURE 17.6 Infrared sensors are composed of an infrared LED (transmitter) and a photo-sensitive detector (receiver).

sensor travels through the foam-board body and to a breadboard so that the sensors can be connected appropriately to the BS2. Different hobbyists will certainly connect things differently (more on this later).

17.2.4 LINE SENSORS

The simulated robot has the ability to detect the presence of a line drawn on the floor. In the real world, this would be done in much the same manner as the infrared perimeter sensors described above. If the emitter and detector are mounted very close to the floor though, we don't really have to worry about other sources of infrared light, which means that we don't have to modulate the light source. The closeness of the line to the sensor does mean that the physical orientation of the emitter to the detector must be just right to get a reliable operation.

Parallax has a QTI sensor that integrates both the emitter and detector into one package so that the two components are always properly aligned for line-detection and drop-off detection applications. Figure 17.7 shows both a picture of the sensor and its electrical schematic. The black lead connects to ground and the white lead to +5 V.

This sensor was designed to operate in an analog mode (allows reading a gray-scale) where the microprocessor determines the time required for the capacitor to charge. For our purposes, we only need a digital output so we placed a 10 kΩ resistor between the white and red leads as instructed in the Parallax documentation. The output (red lead) will then be either a 0 or 1, depending on whether the surface is reflective (white) or not (black). Making the sensor digital not only makes it easier to read (just read the value of the input pin it is attached to), it also makes reading the sensor much faster.

We mounted three of the QTI sensors near the front of the robot very close to the floor as shown in Fig. 17.11. We also mounted two more QTI sensors to be used as drop-off detectors, which can be used in projects similar to the one in Chap. 9.

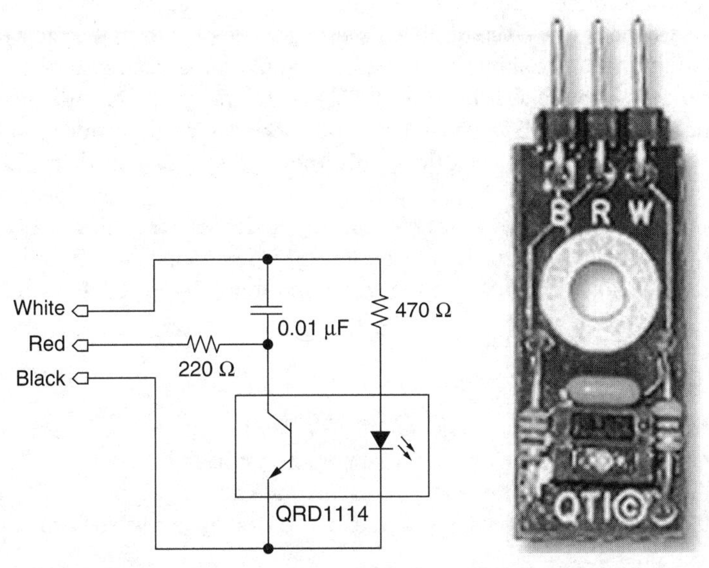

FIGURE 17.7 The QTI sensor makes detecting a line on the floor easy.

FIGURE 17.8 The Ping))) ultrasonic sensor determines the distance to objects in the robot's path.

17.2.5 RANGING SENSOR

The infrared perimeter sensors detect objects that are close to the robot (hopefully before the bumpers are engaged) but they have a very limited range—typically about 3 to 6 in. A robot often needs to be able to sense objects much further away. Ideally, the robot should be able to not only detect these objects, but determine how far away they are.

One method for implementing such a sensor is with ultrasonic transducers. When the robot wants to check for a distant object (perhaps from 1 to 6 ft) it directs an ultrasonic wave (sound above the limits of human hearing) in the desired direction. If an object is present, the sound will be reflected back to the robot and detected. The amount of time it takes for the wave to reach the robot is directly proportional to the distance from the object. Parallax offers an ultrasonic sensor called the Ping))) as shown in Fig. 17.8.

Parallax also offers a motorized turret that can rotate the sensor so the robot can look in different directions without actually rotating its body. The Ping))) sensor and its servo-controlled turret are shown mounted on the real-world robot in Fig. 17.11.

17.2.6 THE COMPASS

The simulated robot has a compass accurate to 1°. Parallax offers a low-cost electronic compass with an accuracy of about 6°. For many applications this is sufficient. Their documentation offers many ideas for dealing with sensor limitations. Figure 17.9 shows the compass module. Electronic compasses accurate to 1° or less are available from other companies, but at a higher cost.

FIGURE 17.9 The compass module from Parallax is accurate to about 6°.

17.2.7 THE GPS

The simulated robot has a global positioning system (GPS) capable of reporting the robot's location to within 1 screen pixel. Real-world GPS are often accurate to only 20 ft or so. Ways for improving on this accuracy and alternative approaches for determining a robot's location have been discussed in Chap. 15. Figure 17.10 shows a GPS module from Parallax.

17.2.8 THE CAMERA

Digital cameras are an ideal way to give vision to a robot. In our simulation, the camera is limited to detecting colors in its field of vision. This limitation makes sense for the simulation because there are no real objects of which to take pictures. Additionally, the discipline of robotic vision is a complicated field of study that requires a book in its own right. To be able to analyze and make use of visual data, extensive mathematics and calculus is required.

Parallax distributes a camera developed by the Seattle Robotics Club that can provide all of the functions needed for our robot and much more. The camera is shown mounted on our real-world robot in Fig. 17.11.

Version 2.0.1 (and later) of RobotBASIC has support for serial and Bluetooth communication so that it is possible for a skilled hobbyist to download actual camera pictures

FIGURE 17.10 The GPS module from Parallax is not nearly as accurate as the one in our simulated robot.

from a real-world robot to a RobotBASIC array. Advanced users can then use the mathematical and matrix capabilities of RobotBASIC to analyze and react to pictures in ways often only done at professional research and development laboratories. The skilled user could easily create a color image on RobotBASIC's terminal screen from the camera data, allowing for remote observation of the robot's environment.

17.2.9 BEACON DETECTION

To our knowledge no one currently offers a beacon detection system for hobby robots. As you saw in Chap. 15, an appropriate beacon system can provide excellent navigational capabilities for a home or office based robot.

Our simulated robot can detect a beacon mounted above other objects in the room, meaning that it can be seen even if there are objects on the floor between the robot and the beacon. The RobotBASIC simulation also provides the distance to the beacon if you wish to use it. In the simulation, the beacon is just an object of a specified color. Our real-world robot could actually use its camera to perform this function by looking for a specific color. The vertical position of the color on the camera's image can be used to give an approximate distance to the beacon (the beacon color should appear higher on the image as the robot gets closer).

There are many options for beacon detection. For example, instead of making the beacon a specific color, it could be a visible light flashing at a fixed frequency such as two pulses per second. The microcontroller could detect the beacon with the camera by comparing two pictures taken $^1/_4$ second apart.

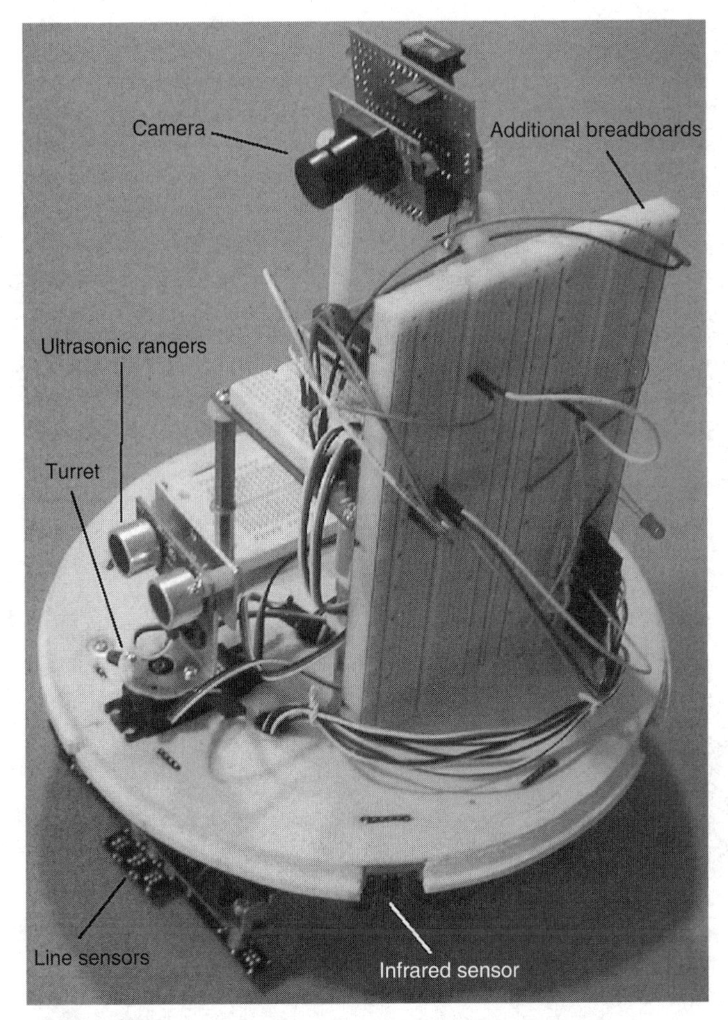

FIGURE 17.11 A real-world robot emulating RobotBASIC's simulated robot.

If you don't want to use a camera for beacon detection, you could use circuitry similar to that described for the infrared perimeter sensors described above. Both the emitter and the detector would need some form of lens to make them more directional and to increase the operating distance. You might even consider experimenting with an old television remote control as the transmitter.

As you can see, even though you might have to create your own beacon detection system there are many options. Perhaps Parallax or some other vendor will offer such a sensor in the future. An ideal kit would come with a detector and several infrared beacons that the robot can individually turn on and off by remote control. Such a system would put realistic navigation within the budgets of most hobbyists.

Figures 17.11 and 17.12 show our prototype for a real-world robot with capabilities similar to our simulation. Figure 17.11 shows a camera on top, the ultrasonic sensor and

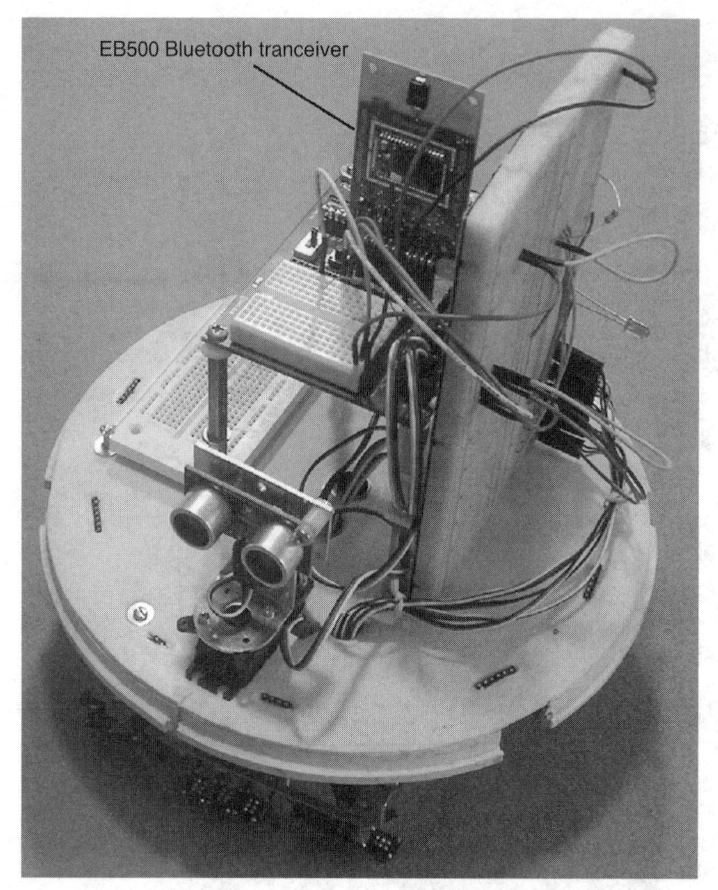

FIGURE 17.12 Another view of a real world version of our simulation.

its turret on the front, and the line sensors underneath. Notice the breadboards for prototyping circuits.

Figure 17.12 shows another view of the robot. In this case, the camera has been replaced with Parallax's EB500 Bluetooth transceiver to allow direct control from RobotBASIC (more on this later).

17.2.10 PRACTICAL CONSIDERATION

Even though parts are available to allow you to create a real-world version of our simulation there are potential problems that should be mentioned. Every motor and sensor needs one or more input/output (I/O) pins on the robot's microcontroller. Most controllers have a very limited number of I/O pins available.

You could use only a couple of sensors at a time and change them based on the experiments or applications you are pursuing. This is relatively easy to do if you utilize breadboards like those shown on our robot. If you want to keep all of your sensors connected so they can be used together, you could multiplex the I/O pins (more on this later) or even link several processors together, greatly increasing the number of I/O lines available.

Parallax's BS2 microcontroller can easily exchange data with other BS2s using their built-in serial communication capabilities. With multiple controllers available, each could be given specific tasks such as controlling the motors, gathering sensory information, managing a wireless link, and so on. Of course that complicates the programming required to construct a real robot, which emphasizes just how valuable a simulation can be for those wanting to learn how to program a robot. With RobotBASIC you have all of the sensors most people could imagine, and they are available to be used and experimented with immediately.

Another advantage of using a multiprocessor system on your robot is speed. Since each processor is dedicated to specific tasks, they could constantly read data from the sensors (instead of waiting for it to be requested by the main processor). When the distance to a distant object is needed, the processor could use the most recent data available instead of waiting for the ultrasonic sensor to perform its task. This technique would not be wise for sensitive sensors such as bumpers, infrared, and line detection, but it can be very advantageous for time-consuming sensors like the camera or ultrasonic ranging, especially since the data obtained from these sensors changes relatively slowly over time.

Another problem with the Parallax hardware we used for our real-world prototype is that both the camera and the EB-500 communication module are designed to fit into the same socket on the Boe-Bot's main BS2 carrier board (see Figs. 17.11 and 17.12). This means that you can only use one of these items at a time unless you create a custom cable or expansion board. If you do decide to utilize a multiprocessor system as described above, you could use two carrier boards, thus giving you sockets for both the camera and the EB-500. Both controllers could then communicate with each other and with other controllers, should they be required for your sensory needs.

17.3 Controlling the Real Robot

There are various ways you can give a robot a brain:

1. You can program a microcontroller with all the necessary software to interrogate transducers (sensors) and activate actuators (motors). The same microcontroller can be used to implement the algorithms to create intelligence. However, unless your AI tasks are simple, most microcontrollers lack the memory space and math engines to be able to achieve a complex AI.
2. You can mount a PC (laptop) on top of the robot and use it to control the robot through a combination of serial (USB or RS-232), parallel port, and ISA BUS based hardware. This solution gives you all the power of a PC to execute effective AI programs. Also the software development cycle becomes easier, since there is no need for cross-compilers and plugging and unplugging of hardware and wires. However, unless your robot is quite large and is capable of supplying lots of AC or DC power, this solution can be unwieldy.
3. You can use a microcontroller on the robot that has the ability of doing all the interrogation of the transducers and activation of actuators. This microcontroller would need

additional software to be able to communicate with a PC wirelessly (or wired but with less mobility). In this case, the PC would have software to carry out the task of AI. The PC receives from the microcontroller the status of the robot's sensors and applies the AI algorithms and then sends the desired actuations back to the microcontroller. The microcontroller receives the commands from the PC and acts upon them.

This method has many advantages, one of which is *distributed computing*. This way you can have one PC controlling many robots or many microcontrollers on the same robot. Additionally, you can envisage projects where the PC receives data from various widely distributed sites to be able to determine the appropriate command sequence for the robot's situation. The robot's microcontroller can be simple, yet able to achieve a lot due to the distributed computing, and sensing power that this option provides.

In the next section we will discuss how you can translate the algorithms you develop using RobotBASIC's simulated robot, to a real robot using each of the three methods above.

17.3.1 CONTROL BY A MICROCONTROLLER

One of the easiest and most straightforward ways of using the algorithms developed in this book with other robots is to translate the RobotBASIC programs into the native language of the target robot's microcontroller. We will use the PBASIC language used by Parallax's BS2 controller, but the principles shown here can be applied to any language on any microcontroller.

Let's look at a specific example to make the translation process easy to follow. If you recall, Chap. 7 developed algorithms for following a line. Figure 17.13 shows one of the algorithms from that chapter.

In order to follow a line, the code in Fig. 17.13 relies on the two RobotBASIC commands, `rForward` and `rTurn` and the function `rSense()`.

The command `rForward` moves the robot forward a specified amount. Likewise, `rTurn` rotates the robot as requested. The function `rSense()` returns a number with the lower three binary bits representing the line sensors as discussed in Chap. 7.

In order to convert the code in Fig. 17.13 to Parallax's PBASIC, we need three subroutines to duplicate the functionality supplied by RobotBASIC. We will *assume* we have these subroutines and three variables as described below. We are using PBASIC syntax here to provide a concrete example, but the concepts (and even most of the syntax) apply to many other microcontroller languages. If you are familiar with the language of the

```
while true
  rForward 1
  while rSense() & 1
    rTurn 1
  wend
  while rSense() & 4
    rTurn -1
  wend
wend
```

FIGURE 17.13 This algorithm to follow a line was developed in Chap. 7.

microcontroller you wish to use, you should be able to create programs for it that emulate most of the commands and functions in RobotBASIC.

Subroutine	Variable	Action when the subroutine is called
FORWARD	DIST	Moves the robot forward by the amount specified by the variable DIST.
TURN	ANGLE	Rotates the robot by the amount specified by the variable ANGLE.
SENSE	SENSORS	Places a value in the variable SENSORS indicating the status of the 3 line sensors.

We will develop the code for each of these subroutines in a moment, but for now just assume that we have them. They are slightly harder to use than their RobotBASIC equivalents. Instead of issuing the command `rForward` 6, for example we would have to do the following:

```
DIST = 6
GOSUB FORWARD
```

Reacting to the line sensors is almost as easy. Compare the two code fragments below. They demonstrate examples of the subroutines *TURN* and *SENSE*.

RobotBASIC Code	PBASIC Equivalent

`If rSense( ) = 6`
`   rTurn 2`
`endif`
	`GOSUB SENSE`
	`IF SENSORS = 6 THEN`
	`   ANGLE = 2`
	`   GOSUB TURN`
	`ENDIF`

As you can see, it is very easy to create code that emulates the simulator's capabilities as long as we have subroutines that provide the functionality normally handled by RobotBASIC.

Figure 17.14 shows the PBASIC version of the code in Fig. 17.13. The first three lines in the figure are comments. The first two comments are directives to the compiler, informing it which processor to generate code for (refer to Parallax's documentation for more information).

The PBASIC compiler requires that each variable be declared. The second section of code in Fig. 17.13 establishes all three variables with a size of Nib (4 bits). PBASIC also has a byte size (8 bits) and a word size (16 bits), but due to the small memory (often typical on microcontrollers), you should never use more bits for variables than your program needs.

As you proceed through the code, you reach the main portion of the program. PBASIC has several types of loops, but in this example we are using a `do`-loop that loops continuously until one of the `if`-statements forces it to exit. The PBASIC compiler can be downloaded from the Parallax web site and contains extensive help.

The program in Fig. 17.14 should be easily compared to Fig. 17.13. You will notice that the three subroutines that do all the work have been left blank (except for comments).

Each of these routines must access the controller's I/O ports in order to turn on motors or read data from the line sensors. These actions must be tailored to the processor you

```
' {$STAMP BS2}
' {$PBASIC 2.5}
'=============================================================
' variables must be declared first
DIST    VAR Nib
SENSORS VAR Nib
ANGLE   VAR Nib
'=============================================================
DO
    DIST = 1
    GOSUB FORWARD
    DO
        GOSUB SENSE
        IF NOT(SENSORS & 1) THEN EXIT
        ANGLE = 1
        GOSUB TURN
    LOOP
    DO
        GOSUB SENSE
        IF NOT(SENSORS & 4) THEN EXIT
        ANGLE = -1
        GOSUB TURN
    LOOP
LOOP
'=============================================================
FORWARD:
    ' place code here to make the robot
    ' move forward DIST units
RETURN
'=============================================================
TURN:
    ' place code here to make the robot
    ' rotate ANGLE units
RETURN
'=============================================================
SENSE:
    ' place code here to read the line
    ' sensors and give SENSORS the
    ' appropriate value
RETURN
```

FIGURE 17.14 This code is the PBASIC equivalent to Fig. 17.13.

are using and to the specific I/O ports to which you have attached your motors and sensors.

Parallax's BS2 processor has 16 I/O pins. We will connect the two servo motors to I/O pins 12 and 13. The connections for the three QTI line sensors (modified to operate in a digital mode as described in Sec. 17.2) will use pins 7, 8, 9, and 4 as described later.

The servomotors are controlled by pulsing them with the PBASIC statement `PULSOUT 12,750`. This statement sends a pulse of 750 units (each unit on the BS2 is approximately 2 microseconds) to the device attached to I/O pin 12. Sending a series of pulses with a duration of approximately 750 units will hold a continuous rotation servo motor (when properly calibrated as described in the Boe-Bot manual) in its current position, as if brakes are being applied. If the pulse is larger than 750 units then the motor will turn (lets say forward, but the direction is relative). The longer the pulse, the faster the motor will turn,

```
FORWARD:
    ' move the number of units specified
    ' by the variable DIST
    DIR = 1
    IF DIST&128 THEN                           'is it negative
       DIR = 0
       DIST = (DIST^255)+1                     ' two's complement
    ENDIF
    FOR t = 1 TO DIST
        ' This code moves the robot about
        ' 1/4 inch on our prototype
        IF DIR = 1 THEN
            PULSOUT 12,775
            PULSOUT 13,725
        ELSE
            PULSOUT 12,725
            PULSOUT 13,775
        ENDIF
        PAUSE 10
    NEXT
RETURN
```

FIGURE 17.15 This code moves the robot forward or backward as specified by the variable *DIST*.

and it will keep turning as long as the pulses are periodically applied. If the pulse is less than 750 units, then the motor will turn in reverse. The smaller the pulse, the faster the motor will move.

The *FORWARD* subroutine in Fig. 17.15 is an application of these principles. If you have limited experience with how to control a servomotor refer to the Boe-Bot's manual which has a lot of explanation on all this and many example programs for controlling the servo motors.

The code in Fig. 17.15 is complicated by the fact that Parallax's PBASIC does not truly handle negative numbers. Mathematical operations create the proper negative results using two's complement binary numbers. Unfortunately, the numbers are always treated as positive so it is up to the programmer to interpret them properly. Many microcontrollers often have many limitations such as this.

The beginning of the subroutine sets the variable *DIR* equal to one to indicate movement forward. The code then checks the sign bit of *DIST* and if it is negative it sets *DIR* to zero (indicating reverse) then converts *DIST* to its two's complement (effectively creating the absolute value of the original negative value).

A `for`-loop is used to move the robot the specified number of units. The code inside the `for`-loop does all the work. Depending on whether the robot is to move forward or backward, the servomotors are pulsed to make the robot move about $^1/_4$ in. Note that since the wheels are on opposite sides of the robot, one servo moves clockwise while the other moves counter-clockwise.

This subroutine could have been made more accurate (and more complicated) if the code monitored wheel counters (see Sec. 17.2) instead of just timing the robot's movement. Accurate movements are not really needed for this application though, because the robot will constantly correct its movements based on the data from the line sensors. These corrections counteract any errors made by the movement routines.

```
TURN:
    ' This subroutine will turn
    ' the number of units specified
    ' by the variable ANGLE
    DIR = 30
    IF ANGLE&128 THEN                     'is it negative
        DIR = 0
        ANGLE = (ANGLE^255)+1             ' two's complement
    ENDIF
    FOR T = 1 TO  ANGLE
        ' The following code turns the
        ' robot about 1 degrees
        PULSOUT 12,735+DIR
        PULSOUT 13,735+DIR
        PAUSE 15
    NEXT
    PAUSE 10
RETURN
```

FIGURE 17.16 This code rotates the robot in units specified by the variable *ANGLE*.

The subroutine *TURN* is nearly the same as *FORWARD*. The major difference is that the two wheels move in opposite directions to cause the robot to rotate. Figure 17.16 shows the code for *TURN*. The numbers were chosen experimentally to make the robot move a very small angle for each unit specified by the variable *ANGLE*. Wheel counters (or even an accurate electronic compass) could have been used to increase the accuracy of this routine too.

In order to get the proper pulse duration, the subroutine *TURN* sets the variable *DIR* to 30 for a right turn and to 0 for a left turn. The value of *ANGLE* is converted to its two's complement if the original value is negative. A `for`-loop then executes the code that turns the robot the requested number of times.

The code inside the `for`-loop uses the value of *DIR* to control the pulses sent to the servos making the robot turn in the proper direction. The robot moves about 1° each time the `for`-loop is executed, but the actual amount is very inconsistent because the motors are attempting to move in such tiny increments. Again, this presents no problem for this example because the program is constantly adjusting positions based on the line sensors. As mentioned earlier, if more accurate movement is needed, you might consider using wheel counters or an electronic compass to monitor how far the robot actually turns (see Sec. 17.2). You can certainly make a real robot as accurate as our simulation, but the more accuracy you require, the more complicated and expensive your robot will be. This reminds us again of the advantage of using a simulator. You can learn to program a very powerful and expensive robot without the cost or hassles of dealing with hardware.

Figure 17.17 shows the code to implement the *SENSE* subroutine. The power connection (the white wire) for each of the QTI sensors connects to I/O pins 7, 8 and 9 on the BS2. The output lines from all three sensors are connected to I/O pin 4. This may seem a bit unusual until you see how these sensors are operated.

In order to get reliable operation it is important that the light from one sensor does not affect the detector for another. The code in Fig. 17.7 solves this problem. The power to each sensor is applied individually by forcing the appropriate pin high (5 V). After a short pause, the data from that sensor is read on pin 4 and placed into its proper bit position

```
SENSE:
   ' This subroutine reads information
   ' from the line sensors and places
   ' the values into the variable SENSORS
   HIGH 7: PAUSE 1: SENSORS.BIT0 = IN4: INPUT 7
   HIGH 8: PAUSE 1: SENSORS.BIT1 = IN4: INPUT 8
   HIGH 9: PAUSE 1: SENSORS.BIT2 = IN4: INPUT 9
RETURN
```

FIGURE 17.17 The line sensor data is stored in the variable *SENSORS.*

in the variable *SENSORS*. All the sensors can be read on the same pin because only one device is active at a time. After reading the data from a sensor, the power is removed by making the controlling pin an input.

When the routines in Figs. 17.4 through 17.7 are combined and downloaded to the BS2, the Boe-Bot is ready to follow a line. Figure 17.18 shows an environment created with black electrical tape on white poster board. The real-world robot successfully followed the line.

The movements generated by the *FORWARD* and *TURN* subroutines presented here work fine, but there is a slight jerky action often associated with servos. The Boe-Bot

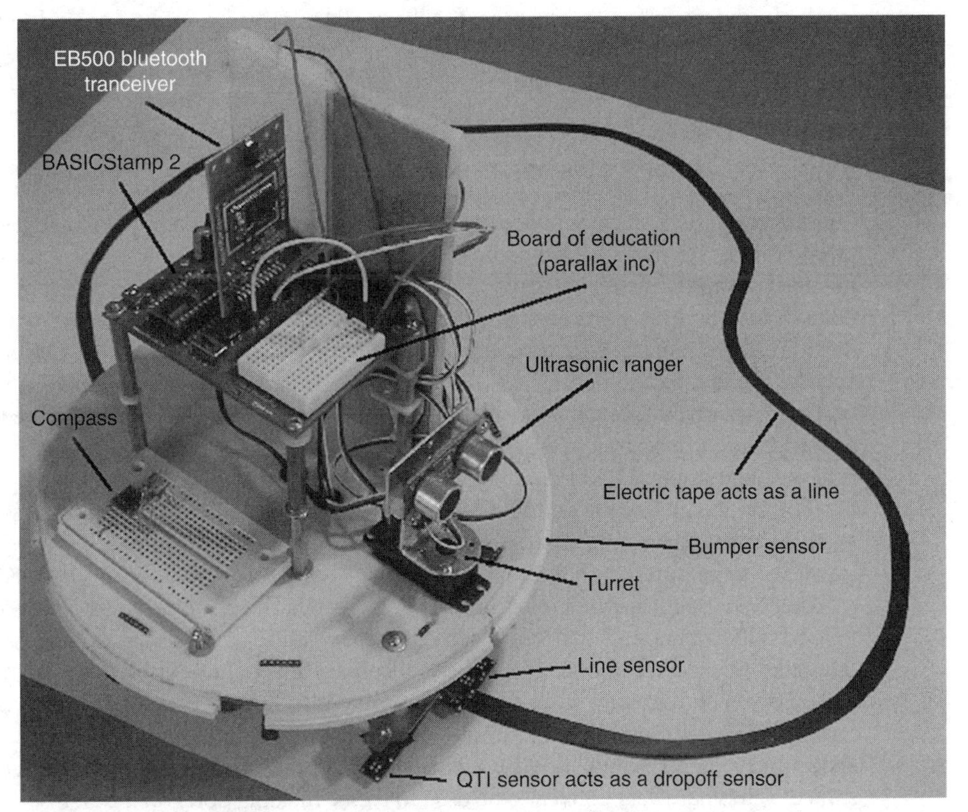

FIGURE 17.18 Black electrical tape is an easy way to make a line to follow.

documentation shows ways of *ramping* the servos up to speed to limit the jerks. Perfect routines that make a robot move forward and turn will depend on what microcontroller is used and the type of motors used (you might choose to use stepper or DC motors). Also you should consider adding additional sensors such as a compass and wheel counters to make your controlling subroutines as accurate as possible. Since this book is really not about hardware, the details of these functions are left as an exercise for the reader.

The above examples will get you started if you desire to use RobotBASIC algorithms on a real robot. The details are cursory explanations for how to make a real microcontroller read sensors and activate motors. For more power and effectiveness you have to be familiar with your microcontroller of choice and you have to be familiar with many electronics principles too.

Parallax's BS2 controller is relatively easy to use and programming it is much easier than programming many other microcontrollers. Also the PBASIC language is a very good language that enables you to utilize the BS2 effectively and with relative ease. However, you still need knowledge and skill in building and connecting electronic components.

Introducing people with minimal skills to robotics using a simulator is an effective and motivating choice. The beginner gets to be skilled at devising robotic algorithms without having the opportunity to destroy expensive components. Also, people can start playing with a robot immediately without the frustration of having to build one before they truly understand it.

Many readers will enjoy using the simulator and appreciate not having to deal with all the quirks of a physical robot. For others, we hope the information provided in this section helps you reach your goals.

17.3.2 CONTROL BY AN ONBOARD PC

The approach used in the previous section works well for translating RobotBASIC algorithms for use on many of the embedded controllers used in robotics. If you need more memory or more processing power though, you might consider mounting a laptop computer directly on your robot. Doing so can give you all the computational power you should need for any project. It also means you won't have to write your code on one machine and download it to another.

There are some disadvantages with using the laptop approach though. Since laptops are much larger than an embedded controller, the robot itself must be larger. This translates to bigger motors and larger batteries. In addition, you will need some way to interface the computer to your motors and sensors. Version 2.0.1 (and later) of RobotBASIC has built-in support for both serial and parallel ports. USB and Bluetooth are supported via virtual serial ports.

You could use these ports to access sensors directly. The standard printer port on most PCs is bidirectional and can be multiplexed easily into four 8-bit input and four 8-bit output ports. In fact, RobotBASIC has special commands to handle the multiplexing hardware for you so that it appears to the programmer to actually have eight ports (see the RobotBASIC help files). These virtual-port commands assume you have the hardware shown in Fig. 17.19 appropriately connected to a printer port. The techniques demonstrated by this hardware could also be used to create multiple ports on a BS2 as mentioned earlier in Section 17.2.

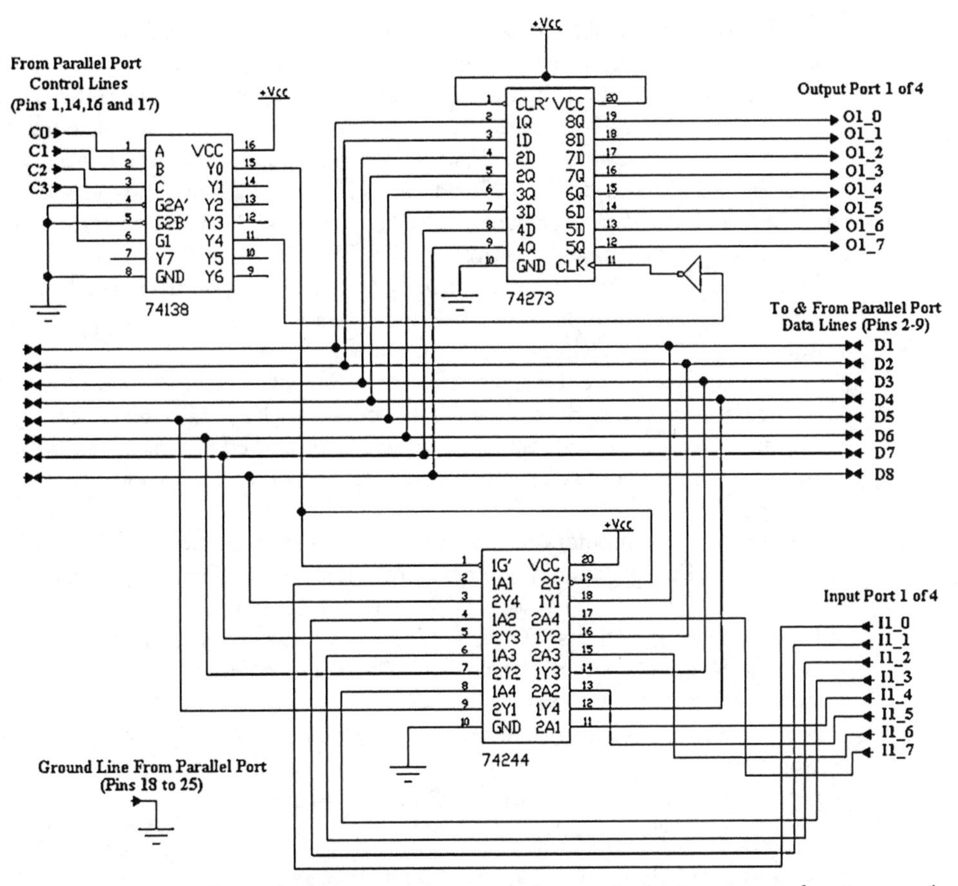

FIGURE 17.19 This multiplexing circuit expands the standard printer port to four input and four output ports.

The circuit diagram shown in Fig. 17.19 is a schematic. The details of the necessary connections are shown. However, the actual circuit will have to have filtering capacitors across all integrated circuits (ICs) and the power supply. This is necessary since this circuit may have to switch at high clock speeds and without the appropriate capacitor values to filter out spikes the circuit may not function properly.

The diagram shows the details for one input port and one output port. The other six ports (three input and three output) will be connected in a similar manner. The Unconnected pins Y1 to Y3 on the 74138 IC will be used to HighZ/Activate the three additional 74244 ICs, and pins Y5 to Y7 will be used to clock/latch the three additional 74273 ICs. The data lines on the additional 74273 and 74244 ICs have to be connected as with those shown in the schematic. The inverter between the output of the 74138 and the clock inputs to the 74273 chips (4 total) can be from a 7404 IC or any other equivalent chip. The control lines C0, C1, and C4 from the PC parallel port are normally inverted, but no inverters are necessary in hardware since RobotBASIC already takes care of the inversion in software.

The intent of the circuit is to latch the data bus from the parallel port onto one of the 74273 registers for the output operation. For the input operation the circuit will activate the correct buffer chip to allow the input data from the port to pass to the parallel port data bus. When the reading operation is over the buffer will be deactivated (HiZ).

The control lines C0 to C2 will indicate which chip is to be clocked or activated. The control line C3 will activate the 74138 to decode the address and thus activate/clock the right chip, once the data is present on the bus, or the software is ready to read the bus. Once the read/write operation is over the 74138 chip is deactivated to make all buffers HiZ. All the timing and control is handled by RobotBASIC's virtual parallel port protocol commands (see Sec. D.4). It is extremely important to ensure that no input buffer is ever activated when the parallel port is in output mode.

You could write your entire programs using RobotBASIC and create subroutines to do the actual work as described above. If you have a background in electronics, you could build your own motor drivers and sensor interfaces and communicate with them using the standard printer port as described above, or one of the many port systems available for the PC.

Many modern computers no longer have a parallel printer port, so we provided other options. For example, if you are comfortable with interfacing to Parallax's BS2, you could attach it to your PC via a serial port or USB port and let the BS2 collect data and control motors with its I/O pins. Hobbyists' magazines such as *Nuts and Volts, Servo,* and *Robot* have many articles and advertisements that can help you find appropriate solutions for your needs.

It is also worth mentioning that the port capabilities of RobotBASIC along with its rigorous mathematical functions make it an ideal tool for hobbyists experimenting in nearly any area of electronics and control applications.

17.3.3 CONTROL BY A REMOTE PC WIRELESSLY

The above two options for controlling a robot can be very functional. However, there is yet a better option. Writing complex AI programs to control a robot requires the power of a PC because it provides lots of memory, floating-point math, matrix operations, and much more. These capabilities are crucial for writing software that can control a robot to do complex tasks. This means that most microcontrollers would not be very suitable and controlling a robot directly with an onboard PC can be prohibitive too.

Imagine combining the advantages of both the options above and removing the disadvantages. This can be accomplished by having a microcontroller that controls the hardware of the robot. The microcontroller reads sensors and activates motors. It will not be required to think or devise strategies. This task is left up to a program running on the PC. The PC sends commands to the microcontroller and receives status data back. All this is done via a wireless link. This way the robot can be compact and not require lots of power and the controller can be simple and cheap. Furthermore, as you will soon see, the robot's microcontroller only has to be programmed once.

The PC decides on what the robot has to do in terms of moving, turning, and reading sensors. The PC program does not know the details of how the robot turns on motors or how it interfaces with its sensors. The PC program tells the robot to move forward 2 units and expects that the robot will perform the task properly. The PC program will decide why and when the robot has to move, but it delegates the details of actually making the move to the robot itself.

Think of a manager and her secretary. The manager tells the secretary to type a letter, but does not care how the task is accomplished. She does not care which word-processor program the secretary uses nor does she know the details of *how* the secretary prints the letter. All the manager cares about is that the command of printing a letter is achieved.

The secretary on the other hand does not care *why* the manager needed the letter printed. The secretary however, has to be able to type the letter and print it using the available equipment. Together the manager and secretary can achieve a lot more than either could alone.

A similar cooperation as above can be achieved by combining the PC and microcontroller. The task of the hardware-level control is performed by a microcontroller mounted on the physical robot. The microcontroller's program recognizes the commands given to it by the manager (the PC) and decides on what motors to activate and so on, to carry out the requested task. The actions taken by the controller will depend on the types of motors being used and how they are connected to the controller. The motors could simply be turned on for some period of time or the duration might depend on the data obtained from wheel counters or an electronic compass. These low-level tasks are the responsibility of the robot's controller and are concealed from the managing PC.

Remember though, that these tasks are all simple ones. The controller program has no decision-making ability as to *why* it should move. This way the controller program can be simple. If the number of sensors and devices to be handled exceeds the capacity of the controller you can use multiple controllers, as mentioned earlier. There are many inexpensive microcontrollers available, so using this option is not prohibitive, and in some cases can be more effective than using individual electronic components. The PC takes the role of the high-level manager that decides why a move is needed, but never has to worry about the details of *how* it is made.

RobotBASIC (version 2.0.1 and up) provides commands that allow you to use wireless communication devices such as a Bluetooth transceiver connected to a USB port. The commands allow you to send and receive data between RobotBASIC and any device that can communicate serially (wireless or wired).

Using the serial commands you can control a microcontroller by sending command sequences that specify motor movements or request data from a specified sensor. RobotBASIC can provide all the communications required by the PC end of the link with its `SerIn` and `SerOut` commands. Refer to App. D and the help files for more information.

The robot's microcontroller must be programmed with subroutines that enable control of actuators (motors) and interrogation of transducers (sensors). The main body of the microcontroller program sits in a loop listening for commands from the PC via a wireless serial communications medium. When the PC sends a valid command the controller's main program deciphers that command and calls the appropriate subroutine to carry out the actual work. Some commands may require that the controller send some data back to the PC. The controller program sends this data and then repeats the entire process of waiting for the next command and so on (see next section and Fig. 17.21 for a specific example).

The details of how the data is sent and how many bytes to send and receive can be determined by you, if you wish, but you will need to devise a protocol to achieve this.

Of course, you will have to create RobotBASIC subroutines that replace all the commands and functions of the simulator in a manner similar to how we translated code for the BS2 earlier in this chapter. But, before you consider doing this, there is a better option still.

17.3.4 CONTROL BY A REMOTE PC WIRELESSLY USING AN INBUILT PROTOCOL

You have seen how easy it is to control our simulated robot using the commands and functions of RobotBASIC. Wouldn't it be great if you could use these very same commands to control a real robot? You can do this with the built-in protocol provided in RobotBASIC (version 2.0.1 and up). The protocol provided follows the same principles discussed in the previous section, but instead of having to use raw serial communication commands such as `SerIn` and `SerOut`, you can use the very same commands the simulator uses. You can say `rForward` 10 and the command will generate the necessary communications to make a properly prepared real robot execute the command as discussed above. You do need to make the simulator know when to use the simulated robot and when to use the real robot. This is easily achieved using a special command.

When you issue the command `rCommPort` *portNumber,* the RobotBASIC programming environment enters the nonsimulated mode and starts operating via the serial communications medium to control a real robot. Once in the nonsimulated mode, the commands `rForward` and `rTurn` no longer animate the on-screen simulation. Instead, they *automatically* send data to the serial port (either real or virtual) identified by *portNumber* (see the RobotBASIC help file or Sec. D.6 for more information).

The data sent from the PC consists of an ID (identification) code indicating the action to be taken and a second byte in the form of an integer, indicating the amount to move or the amount to turn.

You can connect the serial port specified to any radio transceiver with serial I/O or you can use a Bluetooth adapter. Most Bluetooth USB adapters provide a virtual serial port option so they can be an easy solution, especially since Parallax offers a Bluetooth Boe-Bot that handles the wireless communication using their EB500 module.

The important point here is that this option is very flexible. You can create any type of communication link using nearly any transceivers you wish. At the robot end, you will need another compatible transceiver, of course, that is able to communicate with the robot's processor.

When the robot's processor receives the data from RobotBASIC it needs to call appropriate subroutines based on the ID code. It is your responsibility to write these subroutines so that they perform the requested functions. If the data received, for example, is the ID for *forward* and the parameter was five, then your subroutine must command your robot's motors so that your robot moves forward 5 units.

It is important to realize that *you* must write these subroutines because they must be custom designed for *your* hardware. Only you know what type of motors you have and how they are interfaced to your robot's controller. Of course, if you are a member of a club or classroom and everyone is building the same robot, then you can share your routines. Typically though, the hardware-level subroutines for controlling your robot must be designed specifically for your hardware.

The subroutines you write for *TURN* and *FORWARD* have another responsibility beyond that of moving the robot. After they have commanded the motors to create the requested movement, they *must* read the line sensors, the infrared sensors, and the bumpers. This data (3 bytes) must be sent back to RobotBASIC via the serial port or radio link (again, see the help files for more information).

Actually the amount of data sent back to RobotBASIC is always 5 bytes in order to provide consistency. The additional 2 bytes are required in some commands [e.g., `rLook()`] where data has to be returned in addition to the status of the above three sensors. When these 2 bytes are not required, they must be set to 0 and sent anyway. The reason for this is to make the number of bytes received and sent back the same for all the commands. This makes for a simpler and faster communications protocol on both the microcontroller and in RobotBASIC.

The protocol requires that the robot using this protocol always expects to get 2 bytes from RobotBASIC and always send back 5 bytes. If your robot does not have bumpers or infrared or line sensors, it must send 0 as values for the bytes.

RobotBASIC will store the sensor values received in a buffer and supply them to your programs when requested by the functions, `rSense()`, `rFeel()`, and `rBumper()`. This gives you tremendous flexibility. Let's look at an example. Assume for a moment that a robot you are building does not have line sensors but does have some form of custom sensor. You can implement your sensor and have your robot's controller send the appropriate data back for that sensor in the proper position (the third byte) in the 5 bytes always returned to RobotBASIC. Anytime you use the `rSense()` function it will return the value for your custom sensor. Even though the simulated `rSense()` provides only 3 bits of data, the real-world version provides all of the bits sent back from the robot when the function is used. This is also true for `rFeel()` and `rBumper()`. This lets you provide many custom features on your real-world robot. You could, for example, easily let your robot have eight bumper sensors since the data returned is an 8-bit byte.

Notice that each time the robot is commanded to move, it sends back sensory data. This back-and-forth handshaking is very important because the transceivers must know when to transmit and when to listen or they can get out of sync. This methodology also speeds the communication process because no additional transmissions are necessary to request the standard sensory data (line, infrared, and bumper).

There are several more RobotBASIC statements that communicate over the serial link when the `rCommPort` mode has been invoked. In order to make the communication efficient, each of these commands will perform as requested and then return 5 bytes. If all of the bytes are not needed, then 0s are sent in the unused bytes. Unless otherwise specified, the unused bytes are at the end of the transmission. These bytes cannot be omitted even though they are unused because RobotBASIC will be expecting them. See Sec. D.6 for additional details on the protocol.

The table in Fig. 17.20 shows the two commands discussed earlier as well as the remaining sensory functions and summarizes their functionality. To create consistency, all of these commands will send an ID and a second byte even if no parameters are required. Some simulator commands such as `rTurn` are sent to the robot as two versions (turn left and turn right). This allows a full range of 360° to be handled with a single byte of data. If you use the command `rTurn` 350 in the simulator, RobotBASIC would automatically order the robot to turn left 10°, and if you say `rTurn` –350 a 10° right turn is commanded.

Command	Bytes Sent	Bytes Received
rLocate	3, 0	bumper, infrared, line, 0, 0
rForward	6, units to move	bumper, infrared, line, 0, 0
(backwards)	7, units to move	bumper, infrared, line, 0, 0
rTurn (right)	12, degrees to turn	bumper, infrared, line, 0, 0
(left)	13, degrees to turn	bumper, infrared, line, 0, 0
rCompass	24, 0	bumper, infrared, line, degrees
rLook (right)	48, 0 or angle	bumper, infrared, line, color
(left)	49, 0 or angle	bumper, infrared, line, color
rBeacon	96, Color number	bumper, infrared, line, distance
rRange (right)	192, 0	bumper, infrared, line, distance
(left)	193, 0	bumper, infrared, line, distance
rPen	129, 0, 1 for up, down	bumper, infrared, line, 0, 0
rSpeed	36, speed	bumper, infrared, line, 0, 0
rGPS	66, 0	x, y, 0

Note: Distance, degrees, color, x, and y are 2-byte integers (MSB first)
Refer to Appendix D for more details.

FIGURE 17.20 Data sent and returned during communication.

Figure 17.21 shows a sample skeleton control program. It is written for the BS2 in PBASIC. Notice that the sending and receiving of the serial data is expected on pins 0 and 1 of the BS2. How this communication data arrives at these pins is immaterial. You can use Parallax's EB500 Bluetooth module to perform the communication or any other wireless transceiver or even a direct wire, if you are doing some testing and don't want to commit to a specific transceiver. Even if you are using another controller or other hardware, this structure should help you develop an appropriate control program for your robot. Typical code has been given for several routines. All the others have been left as an exercise for the reader. Remember, the example code here is only valid for the modified Boe-Bot robot built in Sec. 17.2. Use this code as a guide to preparing a program for your system.

The capability described above puts RobotBASIC in a class of its own. Once you have created a physical robot, and programmed it with control software that can receive and execute the 2-byte commands and return the required 5-bytes sensory data, you have a fantastic robot development system.

After you develop your algorithm using the simulator as we have done throughout this text, you simply add the following line:

rCommPort *portnum*

to the program immediately before the `rLocate` statement. Then, when you run the program it will *automatically* connect with your real-world robot and control it with the program just used for the simulation.

In general, the real-world robot should respond to objects in its environment in the same manner as the simulated robot, but this of course depends on the quality of your algorithm. Programs that perform open-loop movements (e.g., having the robot draw a triangle) may have trouble emulating the simulator's behavior exactly. If your algorithm uses sensors such as wheel encoders and a compass to control the movement commands, the robot should respond properly. Programs that utilize sensory data to control all movement should

```
' {$STAMP BS2}
' {$PBASIC 2.5}
'~~~~~~~~~~~~~~~~~~~~~~~~~~~~~~~~~~~~~~~~~~~~~~~~~~~~~~~~~~~~~~~~~~~~
'========== assign pins
  ReceivePin PIN 0
  SendPin PIN 1

'=================Declare necessary variable
  SENSORS     VAR Nib
  Direction   VAR Byte
  T           VAR Byte
  bData VAR Byte(2)     'Received data buffer from RobotBASIC
                        'byte 0 is the command code and byte 1
                        'is a parameter for the command.
  bBuffer VAR Byte(5)   'Buffer of data to be sent to RobotBASIC
                        'byte 0 is the bumper sensor. Byte 1 is the
                  'infrared sensors. Byte 2 is the line sensors.
                  'Byte 3 and 4 is a 16 bit integer for the data
                  'to be returnd as the result of some commands.
                  'Byte 3 is the MSB, byte 4 is the LSB.
                  'See RobotBASIC help for more details.
'================================================================
'================================================================
Main:
  PAUSE 1000   'WAIT FOR the eb500 radio TO be ready
  DO
    SERIN ReceivePin,84,[STR bData\2] 'receive command and its
                                      'parameter from RobotBASIC
    IF bData(0) = 3  THEN
       GOSUB rLocate
    ELSEIF bData(0) = 6 THEN
       Direction = 1
       GOSUB rForward
    ELSEIF bData(0) = 7 THEN
       Direction = 0
       GOSUB rForward
    ELSEIF bData(0) = 12 THEN
       Direction = 1
       GOSUB rTurn
    ELSEIF bData(0) = 13 THEN
       Direction = 0
       GOSUB rTurn
    ELSEIF bData(0) = 24 THEN
       GOSUB rCompass
    ELSEIF bData(0) = 48 THEN
       Direction = 1
       GOSUB rLook
    ELSEIF bData(0) = 49 THEN
       Direction = 0
       GOSUB rLook
    ELSEIF bData(0) = 96 THEN
       GOSUB rBeacon
    ELSEIF bData(0) = 192 THEN
       Direction = 1
       GOSUB rRange
    ELSEIF bData(0) = 193 THEN
       Direction = 0
       GOSUB rRange
    ELSEIF bData(0) = 129 THEN
       GOSUB rPen
```

FIGURE 17.21 This skeleton program shows the principles required for writing a microcontroller-based program to act as a server for the RobotBASIC protocol.

```
    ELSEIF bData(0) = 36 THEN
       GOSUB rSpeed
    ELSEIF bData(0) = 66 THEN
       GOSUB rGPS
    ELSE
       GOSUB NotRecognized
    ENDIF
    SEROUT SendPin,84,[STR bBuffer\5] 'Send back 5 bytes
                                       'to RobotBASIC
  LOOP
END
'==============================================================
'==============================================================
rLocate:
     ' this routine initializes the sensor buffer so that
     ' the functions rBumper, rFeel, and rSense will
     ' work properly even if no movement has been made
     GOSUB ReadSensors
RETURN
'==============================================================
rForward:
    ' DEBUG "rForward ",DEC bData(1),CR
    ' move the number of units specified
    ' by the variable bData(1)
    FOR T = 1 TO bData(1)
        ' This code moves the robot about
        ' 1/4 inch
        IF Direction = 1 THEN
            PULSOUT 12,775
            PULSOUT 13,725
        ELSE
            PULSOUT 12,725
            PULSOUT 13,775
        ENDIF
        PAUSE 10
    NEXT
    GOSUB ReadSensors
RETURN
'==============================================================
rTurn:
    ' DEBUG "rTurn ",DEC bData(1),CR
    ' This subroutine will turn
    ' the number of units specified
    ' by the variable bData(1)
    IF Direction = 1 THEN Direction = 30
    FOR T = 1 TO  bData(1)
        ' The following code turns the
        ' robot about 1 degrees
        PULSOUT 12,735+Direction
        PULSOUT 13,735+Direction
        PAUSE 10
    NEXT
    GOSUB ReadSensors
RETURN
'==============================================================
rCompass:
     ' place code here to read the compass sensor
     ' and place the appropriate data in bytes 3 & 4
     ' of the buffer (see below)
```

FIGURE 17.21 *(Continued)*

```
     GOSUB ReadSensors
     bBuffer(3) = 0
     bBuffer(4) = 0
RETURN
'==============================================================
rLook:
     ' place code here to read the camera sensor
     ' and place the appropriate data in bytes 3 & 4
     ' of the buffer (see below)
     GOSUB ReadSensors
     bBuffer(3) = 0
     bBuffer(4) = 0
RETURN
'==============================================================
rBeacon:
     ' place code here to read the beacon sensor
     ' and place the appropriate data in bytes 3 & 4
     ' of the buffer (see below)
     GOSUB ReadSensors
     bBuffer(3) = 0
     bBuffer(4) = 0
RETURN
'==============================================================
rRange:
     ' place code here to read the ultrasonic sensor
     ' and place the appropriate data in bytes 3 & 4
     ' of the buffer (see below)
     GOSUB ReadSensors
     bBuffer(3) = 0
     bBuffer(4) = 0
RETURN
'==============================================================
rPen:
     ' place code here to lift or lower the pen based on
     ' the value of bData(1)
     GOSUB ReadSensors
RETURN
'==============================================================
rGPS:
     ' place code here to read the GPS sensor and place
     ' data in the buffer (not the zero's shown below)
     bBuffer(0) = 0
     bBuffer(1) = 0
     bBuffer(2) = 0
     bBuffer(3) = 0
     bBuffer(4) = 0
RETURN
'==============================================================
rSpeed:
     ' place code here to control the speed of robot
     ' for example, a value to control pulse width
     ' bData(1) holds the parameter from RobotBASIC
     GOSUB ReadSensors
RETURN
'==============================================================
NotRecognized:
    DEBUG "Not A Command ",DEC bData(0),"  ",DEC bData(1),CR
    bBuffer(0) = 255   'set all returned values to -1
    bBuffer(1) = 255
    bBuffer(2) = 255
```

FIGURE 17.21 *(Continued)*

```
    bBuffer(3) = 255
    bBuffer(4) = 255
RETURN
'=============================================================
ReadSensors:
   ' place code here to read bumpers and infrared sensors
   ' example code is given for Parallax line sensors

   'read bumper sensors
   'none for now
   bBuffer(0) = 0

   'read infrared sensors
   'none for now
   bBuffer(1) = 0

   'read line sensors
   Right:  HIGH 7: PAUSE 1: SENSORS.BIT0 = IN4: INPUT 7
   Center: HIGH 8: PAUSE 1: SENSORS.BIT1 = IN4: INPUT 8
   Left:   HIGH 9: PAUSE 1: SENSORS.BIT2 = IN4: INPUT 9
   bBuffer(2) = SENSORS

   'set the last two bytes to zero
   bBuffer(3) = 0
   bBuffer(4) = 0
RETURN
'=============================================================
```

FIGURE 17.21 *(Continued)*

respond very well, even if your real robot is not capable of moving as accurately as the simulation.

Imagine the power! You don't have to translate any programs. You don't have to download anything to the physical robot. You don't even need a connection between the PC running RobotBASIC and your robot. You just change one line and your robot *immediately* responds to your algorithm. And if you find a situation that needs a more advanced algorithm, just comment out the `rCommPort` statement (or set the port number to 0) and use the simulator and all its debugging features to develop your improvements—then try it again for real. This power and ease of use gives hobbyists capabilities never even dreamt of only a short time ago.

We have tried to provide support for all the commands we expect most hobbyists would want on a real-world robot. If your project needs additional commands, you can use the `SerIn` and `SerOut` statements discussed earlier to meet your needs.

17.4 Resources

This is not a book about hardware, and it would be impossible to support all of the possible options and answer the many questions people are sure to have about their particular configuration. If you want to build a real robot, you are just going to have to do your homework and prepare yourself for the challenge.

There are many books and web resources that provide almost everything you need. We have a few suggestions to enable you to locate the help you may require. Of course, the vendor where you purchased your hardware should be first on your list of resources. In fact, it is a good idea to check out the online help and technical support available from any company before you decide to make a huge purchase.

Some of the resources we use are listed below:

- *Robot Builder's Bonanza*, a hardware-oriented robot book for hobbyists published by McGraw-Hill and available in bookstores everywhere.
- Parallax Inc (www.parallax.com), a supplier of microcontrollers and hobby robotics items. They also carry numerous books that discuss and teach the use of the BASIC Stamp.
- *Servo Magazine* (www.ServoMagazine.com), a magazine for robot enthusiasts with many articles to assist you and many ads for companies that cater to the robot hobbyist.
- *Robot* (www.BotMag.com), another magazine for robot enthusiasts.
- *Nuts and Volts* (www.NutsVolts.com), a magazine that caters not only to robot enthusiasts but to all areas of electronics and computer technology.
- www.RobotBASIC.com for
  - The latest version of RobotBASIC.exe.
  - Updated listings of all the programs in the book.
  - Solutions for some of the exercises in the book.
  - Any corrections to errors that may have slipped into the book.
  - Other information and news.

17.5 Summary

In this chapter you have:

- ❑ Reviewed a brief history of hobby robotics.
- ❑ Seen that today's hobbyists can build sophisticated robots with off-the-shelf technologies including motors, sensors, and complete kits.
- ❑ Seen several methods for using the algorithms developed with RobotBASIC on a real-world robot including specific examples aimed at a modified Boe-Bot from Parallax Inc.
- ❑ Seen that RobotBASIC has a wide variety of I/O and communication options that can be used in robotics, as well as other electronics and control applications.
- ❑ Seen how a special communication protocol can allow you to control a real-world robot directly from RobotBASIC, using the same programs developed to control the simulated robot without any change.

CHAPTER 18

CONTESTS WITH ROBOTBASIC

Competition is a great motivator for people to strive toward excellence. Contests can be a great forum for people to compete against each other with a spirit of cooperation and friendly rivalry, while sharing their innovations with others.

The robotics field is full of innovation, cutting-edge ideas, and inventions. Contests have been the traditional way for hobbyists and even serious researchers to test their ideas and inventions against the standards set by their peers.

Contests create a positive feedback mechanism for an increasingly improving standard. People compete and find out that others have better ideas, so they go back and improve theirs further. The cycle continues spiraling upward toward progressively more challenging improvements that benefit every one in the field.

18.1 RobotBASIC-Based Contests

A robotic contest often requires a tremendous amount of planning and effort on behalf of the organizers of the event. Typically, a significant amount of open space is required. Often mazes or other environmental items have to be constructed, stored, and transported.

A contest imposes many demands on the participants as well. Considerable time and resources are needed to turn an idea into a physical robot suitable for a contest entry. Often, motors, sensors, and other parts have to be ordered. After delivery, the robot must be constructed, programmed, physically modified to account for unanticipated problems, and then programmed again and again before it meets the standards for an acceptable entry into the competition. This time-consuming process limits the frequency of events.

RobotBASIC alleviates all the above obstacles. In a RobotBASIC-based contest participants can begin programming immediately—which is really the core of robot design. A contest can easily be held at every club meeting, and since you can hold more contests, you can have a greater variety of types and difficulty levels. This can encourage greater attendance at meetings and as members learn how to program simulations they will gain experience that will make them more likely to spend the time, effort, and expense needed to build more sophisticated physical robots.

Competitions can even be held over a web site where the contest's rules and parameters are posted and the participants can post their entries. Members can participate without ever attending a physical venue, which opens the contest to more participants. The audience and judges can watch the various entries performing by downloading the programs and running them on their version of RobotBASIC.

When it is time to hold real-world contests, RobotBASIC can help because it makes designing and programming a real robot easier and faster. After developing ideas with the simulation software, you will have better insight into what sensors a real robot needs and how these sensors should be situated and configured. As you have seen in preceding chapters, the type, quantity, and placement of sensors greatly influence the algorithms needed to develop robots that successfully solve the imposed challenges. Also, as you have seen, the simulator's debugging tools allow algorithms to be developed much faster than with a real robot.

18.2 Types of Contests

It is often important to hold contests with a wide range of difficulty levels. Challenges can be straightforward, such as avoiding obstacles or drop offs, or more complex such as line- or contour-following. If there are advanced participants you may consider more sophisticated contests such as locating hidden objects, solving mazes, or navigating from point-to-point around a cluttered environment.

Variations on a theme can make a previously assigned contest a whole new challenge. Consider line-following as an example. If the robot is only allowed to use one line sensor instead of three, the contest will have a totally new flavor. Another variation would be to make the line very wide, like a road, and require that the robot stay on the line simulating a car. Further variety may be achieved by varying the width of the line along the track, requiring the robot to change the algorithm it uses for staying on route. You could also make the line intermittent (made up of dots and dashes) forcing the robot to have to reacquire the line as it moves along the path.

The number of variations is limited only by the imagination. The goal is to induce the participants to innovate and be enjoyably challenged. The level of difficulty of a contest should be geared to the range of abilities of the expected participants. Contests within the RobotBASIC environment can be as diverse in difficulty and variety as any real physical robotic contest, and the experiences gained in meeting the challenges assigned in a simulated robotic contest are applicable on numerous levels to nonvirtual contests.

18.3 Scoring a Contest

Determining the criteria for winning a contest is a crucial part of the planning process. Using the line-following example again, the winner could be the robot that completes the task in the least amount of time or maybe the one that exits the line the least number of times before completing the course. The winner could also be the robot that performs the task with the most efficiency. RobotBASIC provides a variety of methods for comparing the performance of participating robots.

18.3.1 SCORING WITH THE POINTS SYSTEM

RobotBASIC has an internal point system that can provide a way to score a contest. Each time the robot moves forward or turns, an internal counter is incremented by two points and each time the robot interrogates any sensors the counter is incremented by one point (see Sec. C.9 for more information). The winner of a contest could be the robot that accomplishes the task using the least amount of points.

The efficiency of an algorithm is dependent upon a variety of design considerations. Choosing the type and combination of sensors can be just as important as the frequency and method of interrogating these sensors. Also the number of times the robot makes the wheels turn and stop can be an indication of the effectiveness of the robot. In general, it is better to analyze sensory data using mathematics and logic in order to minimize superfluous movements since movement generates more points than reading the sensors.

Algorithms can be made more powerful and more efficient with movement and sensory acquisitions in many ways. Proper planning with mathematics and logic can minimize the trial and error approach to tackling challenges. This is a really important concept when the robot is used as a teaching tool (see Chap. 19) because it helps students see real-life relevance and application for many of the topics they study.

Another way of decreasing the points used by an algorithm is to utilize more efficient code. Example 1 in Fig. 18.1 uses `rFeel()` to avoid a collision. If the value of the sensors happens to be two then the sensors will be read three times. Example 2 makes the robot do exactly the same thing, but much more efficiently. The value of the sensors is read only once and is stored in a variable. The value of that variable is used in the `if`-statements to make the required decisions, thus avoiding reading the sensors more than one time.

Example 1

```
if rFeel()&1
  rTurn -1
elseif rFeel()&4
   rTurn 1
elseif rFeel()&2
   rForward -5
endif
```

Example 2

```
a = rFeel()
if a&1
   rTurn -1
elseif a&4
   rTurn 1
elseif a&2
   rForward -5
endif
```

FIGURE 18.1 The program fragment on the right generates fewer points.

18.3.2 SCORING WITH THE BATTERY

If you recall from Chap. 13, RobotBASIC has a simulated battery. The robot has the ability to determine the amount of charge left on the battery and to recharge the battery. If the battery option is enabled by issuing an `rIgnoreCharge` *false* statement the robot will cease to operate if the battery becomes depleted. This provides another method for scoring contests. The winner could be the robot that accomplishes the most of whatever the robot is supposed to achieve before the battery runs out.

Just like in car racing you could allow the contestants' robots to pull into a pit to recharge their batteries by docking with recharging stations. This would cause the robot to abandon its tasks to seek a station, but that just adds more variety to the contest. Robots that can find the charging station quicker have an advantage, and efficient algorithms are rewarded because they allow the robot to work longer before having to charge the battery.

18.3.3 SCORING WITH THE QUALITY OF CODE

Another way of judging the entries would be on the programmers' coding style and efficiency. With physical robots, contestants use a variety of microprocessors and programming languages like C, BASIC, assembly, and so on. This makes it hard to have a standard for judging the code since there is no homogeneity between programming languages.

In RobotBASIC-based contests all the contestants are using the same programming language. Since the code to achieve the simulations can be made readily available to judges, the style and readability of code can be an element in the scoring standards. This would encourage contestants to code more professionally. This method of comparing entries can be more subjective rather than objective, but with a proper set of standards the subjectivity can be minimized. However, the field of robotics and programming is as much an art as it is science and, like any art contest, the contestants' creativity and style can be a valid subject for judgment.

18.4 Constructing Contest Environments

Considerable time, effort, and expense are typically required by the contest organizers to create a challenging and suitable contest environment (mazes for instance). Also the contestants themselves need to construct similar environments in the process of creating and testing their robots.

Organizers often have to post a complex description of the environment that will be used in the contest. These specifications have to be very clear, precise, and understandable. On the day of the contest, if contestants have misinterpreted these specifications, their robots and/or code may require a considerable amount of modification to adapt to the unanticipated environment.

RobotBASIC allows many methods for contest organizers to distribute very precise specifications of their intended contest environment:

1. Subroutines can be distributed so as to be incorporated into the contestant's code to create a similar (or exactly the same) environment as the target environment.

2. The `WriteScr` command can be used to create a bitmap file of the environment, which can then be used by a contestant with the command `ReadScr` to create an environment similar to (or the same as) the intended environment.
3. Files created with the `MWrite` command can be given to contestants who then use the `MRead` command to create an array that will be used to recreate (using `mPolygon`) the contest environment.

Contestants write their code to use the distributed files and judges can run these programs using whatever file they desire to test and judge the entries. Using this method, judges may wish to test the entries on various files that simulate environments of varying difficulty or complexity, and thus can judge winners on various levels.

18.5 Summary

In this chapter you have seen that:

- RobotBASIC contests can be motivational and how they can be used to increase participation in a club or classroom.
- Contests can be designed to appeal to contestants of various skills and abilities.
- The types of contests and scoring methods are limited only by the imagination.
- Contests can emphasize the utility of mathematics and logical thinking.
- Contest organizers and contestants can save time, effort, and expense when participating in robotic competitions using RobotBASIC.

18.6 Suggested Activities

1. Design several contests that are appropriate for participants with different skill levels. Revise your contests to give renewed enthusiasm to the participants.
2. Devise ways that RobotBASIC can be used to hold mini-contests on a regular basis at club meetings or even over the internet.
3. Design an internet-based contest that encourages sharing of principles and ideas among the participants. Explore the methods discussed in Sec. 18.4 to distribute contest environments to the participants.
4. Design a contest where the rules of judging are based on several criteria. Discuss how you chose the weights to give each criterion.

CHAPTER 19

ROBOTBASIC IN THE CLASSROOM

RobotBASIC provides a platform that helps teachers create a building block approach to teaching many subjects:

- Robotics and artificial intelligence (AI)
- Applied mathematics
- Computer programming
- Problem solving
- Logical thinking
- Engineering principles

In this chapter we propose that RobotBASIC can serve as a tool for teachers to provide an enjoyable and effective learning experience for students.

19.1 RobotBASIC within the Learning Process

When RobotBASIC is used as a teaching tool the learning process will be filled with tangible, memorable, and rewarding experiences with immediate feedback on what has been learned in the lesson. This can result in desirable behavioral changes that are part of the learning process.

When students use RobotBASIC they can realize an immediate and appreciable purpose and utility in what they learn. This makes learning easier and the student is able to retain the material longer. RobotBASIC gives students vivid sensory feedback to all the actions they make within the environment. Students actively participate in the knowledge acquisition process using many of their senses, allowing them to relate better to the knowledge gained.

RobotBASIC can be a valuable aide to teachers because it provides a means for combining many facets of learning into one convenient medium. Assignments in RobotBASIC are based on solving interesting and relevant problems. When student groups analyze problems and develop solutions, they improve their conceptual and perceptual abilities, all the while developing problem-solving skills.

RobotBASIC promotes incidental learning where students learn about dry intangible subjects while engaging in exciting and relevant activities. For example, as students learn to make the robot's pen draw shapes or maneuver within the screen environment, they are exposed to many principles of algebra, geometry, and trigonometry.

RobotBASIC provides many anchor points for relating multiple subjects to a meaningful, tangible, and positive learning experience where students can appreciate personal gain and rewards (material or social) as a return for their efforts.

A student's first experience with learning a new subject can affect his or her future interest in the subject. RobotBASIC makes the initial learning process an interesting and enjoyable experience so students will like and have positive feelings about the subject and will be willing to pursue further studies.

RobotBASIC provides a medium that enhances perception. It encourages students to perceive precisely and accurately and to group their perceptions into a meaningful whole that leads to insight into the subjects learned.

RobotBASIC makes learning an active process within an interesting environment where simple specific exercises can evolve into complex open-ended projects. Each time a challenge is achieved a variation can be added to create renewed interests and challenges. This makes students more disposed to learn because they can recognize clear well-defined objectives for the material they are required to learn.

As students increase the robot's abilities by using modules they have previously built, they learn to correlate what they have learned and to apply that knowledge in new and innovative ways.

A teacher can use RobotBASIC to demonstrate a skill using clear step-by-step examples, which then lead to the students performing the skill themselves. Students can use their personal copies of RobotBASIC to practice until the skill is learned. Students can have their own copy at school and at home giving them the opportunity to practice as often as they wish. Also, due to the fact that RobotBASIC provides an enjoyable experience, students are more likely to practice. RobotBASIC provides the teacher with a means to devise criteria for evaluating whether the students have learned the skill.

19.2 RobotBASIC as a Motivator

RobotBASIC provides students with the satisfaction of displaying their abilities to their peers which leads to self-fulfillment and satisfying a basic desire for esteem and status among one's peers.

Assignments beyond a student's ability hinder motivation. RobotBASIC allows assignments to be created for a wide variety of skill levels. This customization helps students feel that they are valued individuals.

Using RobotBASIC, a teacher can create a forum where students are able to display their abilities to the rest of the class during informal contests. The natural human desire to compete drives many students to work hard in order to excel. Other students will be inspired by the projects they see and strive to make theirs better. The teacher may reward students by letting them assist groups that are having trouble. This fosters a spirit of cooperation and leadership.

19.3 RobotBASIC within the Teaching Process

RobotBASIC provides concrete and clear communication between the teacher and student with a shared experience, avoiding confusing abstractions and ambiguous information content.

RobotBASIC is an aide to teachers in every step of the lesson-plan preparation. Objectives, for example, are easily defined. Supplies, materials, and equipment are minimized since RobotBASIC provides a presentation platform that subsequently becomes the application platform where the students can practice the required skills. When it is time for evaluation and critique, RobotBASIC provides an objective, flexible, comprehensive, constructive, and specific mechanism. Whether the teacher opts to use the instructor-student, student led, group-guided, written, or self-guided methods for evaluation, RobotBASIC can play a role.

RobotBASIC can be utilized in any of the teaching methods a teacher wishes to use as the means for conveying the desired knowledge. Whether in lectures, guided discussions, or demonstration-performance sessions, RobotBASIC can be a suitable platform for keeping the material organized.

Running a completed RobotBASIC program to introduce an assignment can provide an effective attention grabber and provides clear motivations and objectives. The nature of RobotBASIC allows you to capitalize on the students' interest and develop their knowledge from past to present, simple to complex, known to unknown, frequently used to less frequently used, thus enhancing and enforcing the teaching process.

19.4 RobotBASIC at Every Level of Education

RobotBASIC can be utilized at all levels of education with quantifiable and satisfactory results.

19.4.1 GRADE SCHOOL

When teaching grade school students you may start by teaching the students how to move the robot around an empty room. The goal might be to move the robot several times and then return to its original position. After the students master this ability, they could be taught how to draw some simple shapes on the screen to simulate obstacles and then tackle the original objective within a cluttered room (at this level they should not be told about the sensors). There is no need to introduce any form of loops or `if`-statements. It could be motivational though, to increase the interest in graphics by showing the students how to change line widths and colors. Young students can find basic graphics interesting and challenging. They also can learn about distance measurement, coordinate systems, and angles by moving the robot around the screen.

If the robot's pen system is explained to students they can be asked to draw rectangles or triangles on the screen. The idea of variables can be introduced on a very rudimentary level (such as to specify how far the robot moves or how much it turns). After the students master these types of problems, they should be able to understand simple subroutine concepts.

19.4.2 MIDDLE SCHOOL

Middle school students can be introduced to RobotBASIC with many of the same examples used for younger students. They will master those tasks very quickly and will enjoy tackling problems that require simple loops and decisions. In order to have problems that are interesting, the student should be introduced to sensors, but in typical classrooms, it is probably wise to only tell the student about the bumper sensors.

Because students at this level are generally familiar with algebra, you will be able to use variables in more exciting ways. For example, subroutines can be created that draw shapes at positions specified by the variables x and y. These subroutines can then be combined to draw shapes throughout a room. At that point students are often ready to learn about random numbers.

19.4.3 HIGH SCHOOL

High school students can typically master the previous concepts (which can serve as an introduction to the language) very quickly. At that point they can tackle problems such as following a line or wall. Advanced students might even tackle problems like goal-seeking.

At this level you should consider using RobotBASIC to demonstrate the use of mathematical concepts such as the pythagorean theorem and other principles of geometry, algebra, mathematics, and trigonometry.

Many high schools offer courses that use real educational robots. Often students have to work in large groups or wait in line to try the programs they write because the school can only afford to buy a few robots. In such cases RobotBASIC provides the perfect solution. Every student in the class can have their own free copy of RobotBASIC. Individual students and groups can develop their own solutions to assignments and test their ideas with the simulator. When a group completes an assignment, they can move their program to a real robot to see their programs operate in the real world (see Chap. 17 for more details).

19.4.4 COLLEGE LEVEL

At the college level, students should be able to deal with any subject in this book, which can be used as a course text. RobotBASIC is an excellent first language for engineering students. In keeping with the building block approach to teaching, the RobotBASIC language provides a perfect stepping stone for engineering students to progress to more complex languages such as C/C++.

RobotBASIC is a powerful yet easily learnable language that serves to introduce students to programming without overwhelming them with extraneous knowledge that only serves to confuse the beginner.

Once students become adept at handling programming constructs such as looping, conditional execution, modular design, and more complex principles such as arrays, they will be better able to proceed to programming with more complicated languages.

RobotBASIC is a wonderful language for early engineering classes because it is an interpreter replete with powerful commands and functions including matrix operations such as inverting, sorting, transposing, and regression analysis (to mention only a few). Using RobotBASIC, students can easily write programs to help with their assignments. RobotBASIC can serve as a versatile and powerful programmable scientific graphics calculator that provides numerous set of functions and commands to complete many engineering and mathematics projects.

RobotBASIC can also be used in hardware courses where the objective is to build an actual robot. Simulations can be used to decide on the type, quantity, and placement of sensors, and of course, the algorithm for achieving the specified goal. Additionally, RobotBASIC's port I/O commands and serial I/O commands help in creating programs that use the PC as a hardware interface platform.

19.5 Summary

In this chapter you have seen that RobotBASIC:

- Helps make the learning process more effective, meaningful, and enjoyable.
- Can provide the means for motivating students to excel.
- Helps the teacher during the planning, delivery, practice, and evaluation of lessons and skills.
- Can aid in teaching mathematics, engineering, problem-solving, logical-thinking, programming, and robotics.
- Can be used as a tool for teaching students at any age level.

The following section contains a set of suggested teaching tasks at various levels of education.

19.6 Suggested Teaching Tasks

19.6.1 GRADE SCHOOL

1. After the students have been shown how to move the robot around the screen and how to draw with the pen, they should be asked to make the robot draw a triangle or rectangle on the screen.

NOTE: If the students are too young to understand coordinates, they can create the shape using trial and error.

2. Students should organize earlier assignments (such as drawing a rectangle and drawing a triangle) into subroutines and then create a program that calls the subroutines in order to create several shapes on the screen.
3. Students should be asked to *predict* what shape would be drawn if the robot were to move forward then turn, move forward and turn, over and over again. They should try writing such a program (without loops) to see how the robot responds. At that point they should be ready for a simple introduction to the `for`-loop.

19.6.2 MIDDLE SCHOOL

1. Students should make the robot roam randomly around an empty screen.
2. Students can be asked to create a subroutine that will draw a rectangle at the coordinates specified by the variables x and y. Advanced students can be asked to make the rectangle a random size. This can be done with the robot or with graphic statements.
3. Students should use a loop to draw 10 shapes at random positions on the screen.
4. Students should modify the roaming program above so that the robot avoids obstacles placed randomly on the screen.

19.6.3 HIGH SCHOOL

1. Students should create a robot that follows a line.
2. Students should modify the above program so that the robot will roam randomly until it finds a line. Once the line is found the robot should follow the line until the end and then resume roaming. Use only line and bumper sensors.
3. Modify the line-following program so the robot reacts appropriately when it encounters an object that blocks its path along the line.
4. Complete the above assignments for wall-following using the infrared sensors.
5. Write a program that will cause the robot to locate a beacon in an empty room and find its way to the beacon. Advanced students can add objects to the room.

19.6.4 COLLEGE STUDENTS

College students can complete all the exercises at the end of each chapter in this text.

PART 5

APPENDICES

The four appendices act as a user manual and reference for the RobotBASIC IDE and programming language:

- Appendix A is a user manual for the IDE (integrated development environment).
- Appendix B is an overview of the programming language structures and concepts.
- Appendix C is a comprehensive listing of all the functions and commands in the language, as well as the functions and commands for the robot simulator.
- Appendix D lists commands that enable input and output to the parallel and serial ports. Additionally, it details a protocol that facilitates control of a real robot using the simulator commands and functions via a serial wireless communication medium like Bluetooth.

The details of all the features in the language are subject to change as alterations and upgrades are implemented. Check our web site www.RobotBASIC.com for the latest version of the software. The help files accessible from the latest IDE will have the most valid up-to-date descriptions of all the functionalities of the language. Make sure to always download the latest version and to consult the help files for any new and modified features.

APPENDIX A

THE ROBOTBASIC IDE

The RobotBASIC IDE consists of an Editor Screen, a Terminal Screen, a Help Screen, and a Debugger Screen. Each screen has many buttons and menus that facilitate the various actions required in each one. This appendix will discuss each screen and what actions can be achieved in each.

A.1 The Editor Screen

The Editor Screen (Fig. A.1) has various buttons and menu items that facilitate the creation, editing, and running of programs. If you place the mouse cursor on a button and wait for a second, a hint will pop-up showing the button's intended function (Fig. A.2). In addition, each button has an icon that helps in remembering the button's action.

It is also possible to perform any button's action by using drop-down menus. Additionally, each menu action has a keyboard combination (shortcut) that executes the action by pressing a function key or a combination of *Ctrl* or *Alt* and a key.

As you can see from Fig. A.3 the edit menu can be dropped down by clicking the *Edit* option on the main menu, or by pressing *Alt+E*. Each sub-action in the menu can be chosen by clicking that action or by pressing the correct *Alt* combination.

In many cases, as you can see to the right of each option, there is a shortcut key that can invoke the menu action without accessing the menu. For example, to invoke the *Find* option, you can press *Ctrl+F*.

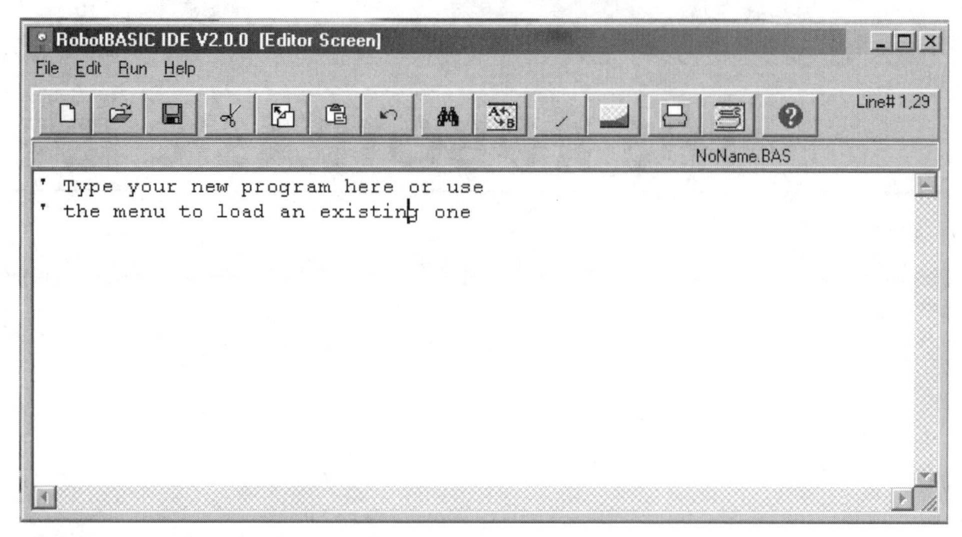

FIGURE A.1 The Editor Screen.

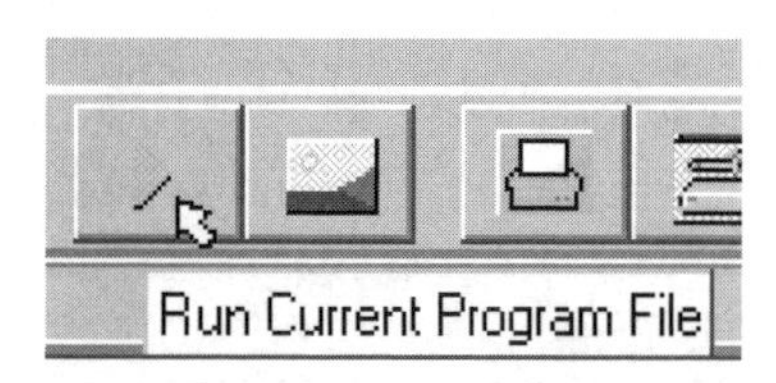

FIGURE A.2 Button hints.

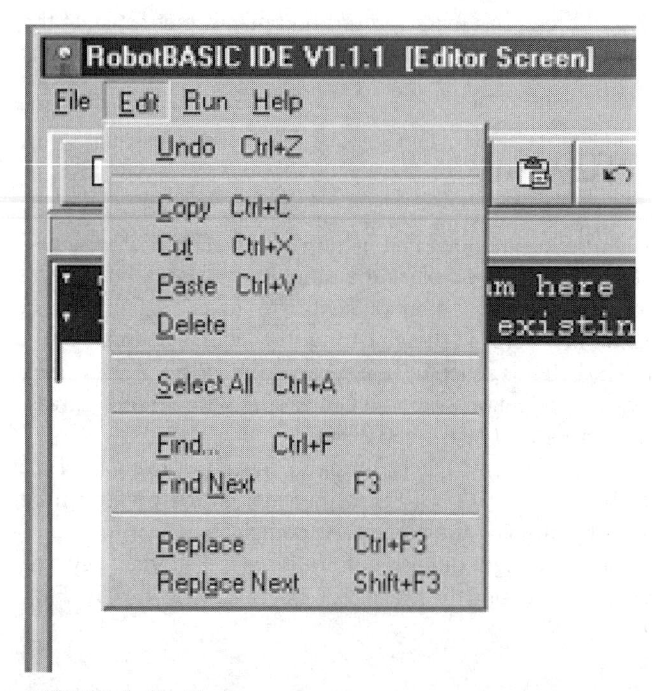

FIGURE A.3 Menus.

Another feature of the Editor Screen is the panel under the buttons. This panel contains the name of the file currently being edited. If the file is a new one that has not been previously saved and named, it will hold the name NoName.BAS, otherwise it will have the file's full name.

On the far right of the buttons panel, you will see a line and column number field. This indicates the line number where the cursor is currently positioned in the editor and the number after the comma is the character number in the line.

To run the program currently loaded in the editor either, click the *Run* menu and the *Run Program* submenu, or click the button, or use the *Ctrl+R* shortcut on the keyboard. While a program is running this button and its corresponding menus are disabled. It will be enabled when you halt the program or the program ends, or an error occurs during the program run.

Running a program will bring up the Terminal Screen and display any program interaction on this screen (see Sec. A.2).

The editor in RobotBASIC is very similar to Windows' Notepad program. The menu and speed buttons allow for many helpful features. You can save and open files. You can search and "search and replace". You can *Cut, Paste, Copy, Undo,* and *Print*. Additionally, you can view a help file that contains information on all the features of the language.

A.2 The Terminal Screen

The Terminal Screen (Fig. A.4) is where program input and output takes place. This screen has various buttons, a prompt panel, an input box, and a display screen.

The display screen is where all the output from the program is displayed. Commands such as `Print` and `Circle`, display their results on the screen. The display screen is 800 × 600 pixels (width × height). If you are using a Windows system where the monitor limits are less than 1024 × 768 resolution, the Terminal Screen will resize to a smaller size. You can find the limits of the *x, y* coordinates by using the `ScrLimits` command (see Sec. C.7).

If you get a runtime error or the program is ended the Terminal Screen will remain on top of the editor. You can close it or move it to the background (see below). There is a shortcut key (*Ctrl+T*) while in the editor, to reopen the Terminal Screen, or you can use the menu option or the speed button. The display from the previously run program will not be cleared until the current program is executed again or a new program is run.

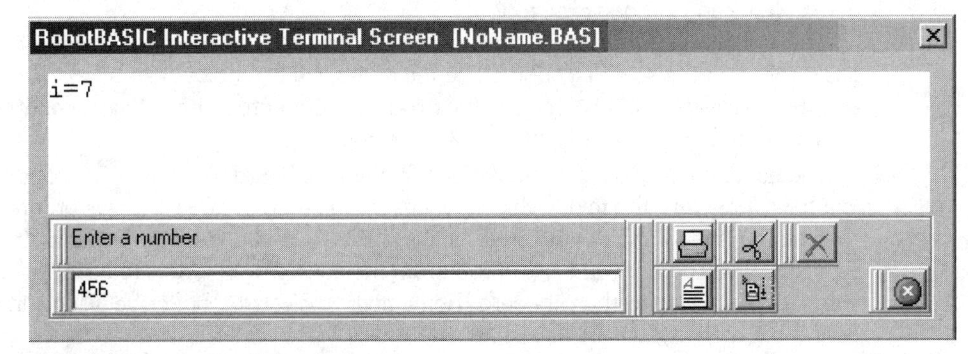

FIGURE A.4 The Terminal Screen.

```
Print "i=",7  //display a string and number
Input "Enter a number",j //prompt and wait for an input
End
```

FIGURE A.5 Sample input/output program.

The bottom of the screen is where all input is done using the commands `Input`, `Waitkey`, `GetKey` and `GetkeyE` (see Sec. C.7). The edit box will only be activated when an `Input` command is issued. Any entry in the box must be terminated with an *Enter*.

The input box has a buffer that holds any previous inputs. To access a previous input, press *Ctrl+P* or *Ctrl+N* to step through prior inputs. This facilitates easy repetition of inputs.

When you issue an `Input` *Expr, Var* statement, the *Expr* will be printed just above the edit box as a prompt. If *Expr* results in a zero length string then character > will be printed as a prompt instead of the null string (see Sec. C.7). Input data is treated as a number if it is a legal number otherwise it will be returned as a string. You can use functions to manipulate inputs as needed (see Sec. C.8). The display shown in Fig. A.4 results from the simple program in Fig. A.5.

Notice that the result i = 7 is shown in the screen area. The words "Enter a number" are displayed in the prompting panel and the user input 456 is currently being shown in the input box.

The buttons on the bottom right allow certain actions to occur. Each button, just as in the Editor Screen, shows a hint when you place the mouse cursor over it.

The button allows you to abort the currently running program and return to the Editor Screen. You can also abort a program by closing the Terminal Screen window. The button allows you to view the Editor Screen. The Editor Screen has a button to display the Terminal Screen again. The program will continue to run when you swap between the Terminal and the Editor Screen; it will not stop.

The button causes the screen area to be printed on the default printer. The screen area is printed as a graphic. Only what is visible on the screen will be printed, any output that has scrolled off the screen will not be printed, but see below for how to retrieve any text that has scrolled off the screen.

The button will turn on the debugging feature. See the `Debug` command in Sec. C.7. The buttons and facilitate the use of a special feature of the Terminal Screen that can be a powerful tool.

Text printed with the `Print` command is stored in a buffer, regardless of whether it has scrolled off the screen or not. Once the buffer is full (it holds 8 kb of text) it will clear and start to fill again. If this occurs in the middle of your output the buffer will only be holding part of the output when you use the button.

The buffer content can be sent to the Windows clipboard using the button. Once you have done this, the text in the clipboard can be used in any Windows program. The button is used to clear the buffer and the Windows clipboard at the same time. This is needed if you do not want any output from previous program runs to be included when you copy the buffer to the clipboard. Remember to clear the buffer between runs if you desire, but before you start the run.

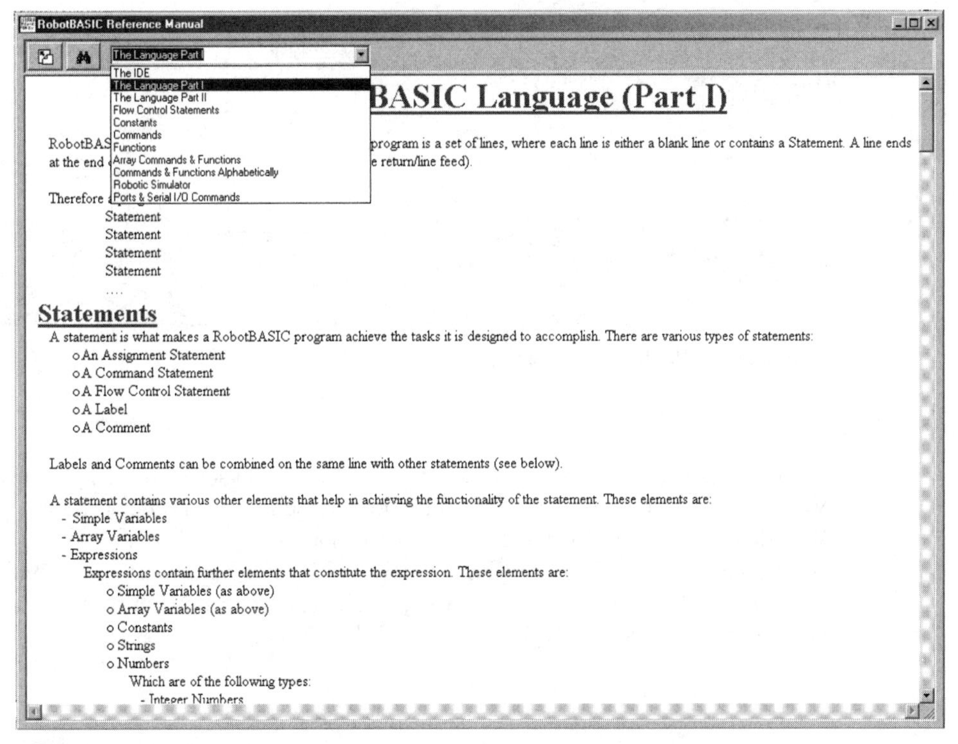

FIGURE A.6 The Help Screen.

A.3 The Help Screen

The Help Screen (Fig. A.6) provides explanations about the use of RobotBASIC's commands, functions, and various other aspects of the entire system. The screen has a drop-down combo-box that contains a title heading for all the help pages from which you can choose. These pages cover every aspect of the system and closely parallel Apps. A, B, C, and D. The help text can be selected and copied to the Windows clipboard using the button or *Ctrl+C*. The button (or *Ctrl+F*) allows you to search the text in the currently displayed section.

A.4 The Debugger Screen

The debugging feature of RobotBASIC is discussed in detail in Sec. C.7. The buttons are self-explanatory. Refer to the `Debug` command in Sec. C.7 for more detailed information.

The Debugger Screen (Fig. A.7) buffer will become full if there is too much output. To prevent this from causing an error use the *Clear* button occasionally to clear the buffer.

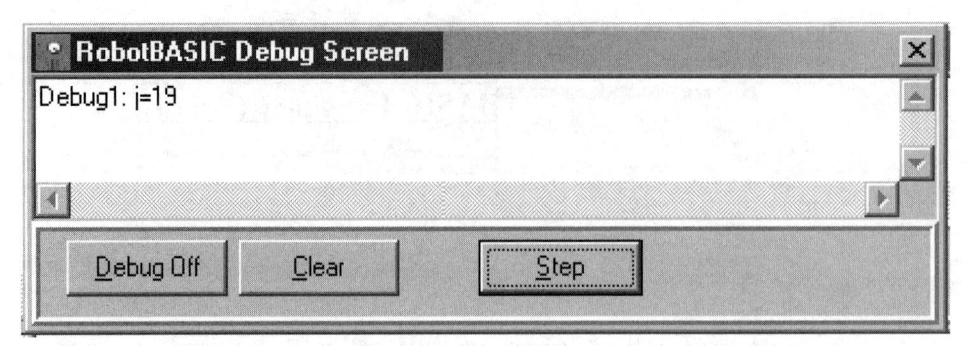

FIGURE A.7 The Debug Screen.

You can resize and move the Debug Screen, and you can swap between it, the Terminal Screen, and the Editor Screen. Be careful not to lose it behind other windows. You can always drag it to the side while viewing other windows.

If you happen to lose the debugger window under other windows, you can bring it back to the top by using the menu option *Bring Up Debug* from the *Run* menu on the Editor Screen, or you can use *Ctrl+D* while in the Editor Screen.

APPENDIX B

THE ROBOTBASIC LANGUAGE

RobotBASIC is a line-oriented programming language. A program is a set of lines, where each line is either a blank line or a program statement. A line ends at the end of the line. In more technical jargon, a line ends with a CR+LF (Carriage Return + Line Feed).

So a program is a collection of statements as shown in Fig. B.1.

NOTE: Always refer to the help pages within the RobotBASIC IDE for the most up-to-date information regarding material mentioned in this appendix.

B.1 Statements

A statement is what makes a RobotBASIC program achieve the tasks it is designed to accomplish. There are various types of statements:

- Assignment statements
- Command statements
- Flow-control statements

```
Statement
Statement

Statement

Statement
Statement
...
```

FIGURE B.1 A program is a collection of statements.

- Labels
- Comments

Labels and comments can be combined on the same line with other statements (see below). A statement contains various other elements that help in achieving the functionality of the statement. These elements are:

- Simple variables
- Array variables
- Expressions, which contain further elements that constitute the expression. These elements are:
  - Simple variables
  - Array variables
  - Constants
  - Strings
  - Numbers, which are of the following types:
    - Integer numbers
    - Floating-point numbers
  - Operators, which are of the following types:
    - Parenthesis
    - Mathematical operators
    - Comparison (relational) operators
    - Logical operators
    - Bitwise operators
  - Functions, which can contain expressions

B.2 Comments

Comments are an important part of any programming language. Without good comments a program can be hard to understand. Without good comments, even the person who wrote the program may not be able to follow the program's logic when reading the program at a later time.

Comments are not executable statements. They are used to annotate a program and format the text of the program to look nice and make the logic of the program understandable by someone trying to read the program. Comments can also divide the program into sections, making it easy to see where the various operations begin and end.

Comments can be in a line of their own, or they can be inline with a statement. Any text in a line after and including the comment indicator will be ignored and is not considered part of the statement.

RobotBASIC allows for two methods of indicating a comment. The character combination // is one and the character ' is another. For example:

```
//this is a comment on a line by itself
'this too is another comment
d = 9      //this is a comment inline with a statement
b = d*2    'so is this comment
```

B.3 Assignment Statements

An assignment statement is where you would assign a value to a variable. *Variables* are storage spaces for values that will be used in the program logic. Variables can be simple variables or array elements. See the sections on variables (Sec. B.7.3) and arrays (Sec. B.7.4) for more detail on what a variable or array is and how to use it. In the discussion below *Var* means a simple variable and *VarA*[...] means an array element. See the section on expressions for further details of what an expression is. Assignment statements are like this:

```
 Var = Expression
or
 VarA[...] = Expression
```

For example:

```
Distance = 5.4  //—assigns the variable Distance the
            //—value 5.4
TimeTaken = 2.1
Speed = Distance/TimeTaken //—Speed will be assigned
    //—the result of the
    //—expression
Dim MailBoxes[20,10] //—see later for what this means
MailBoxes[5,3] = 9   //—assign an array element
```

NOTE: You can place multiple assignment statements on the same line by using the \ character. So instead of saying:

```
a = 1
b = 2
c = 3
```

you can say:

```
a = 1 \ b = 2 \ c = 3
```

This helps make more compact code.

B.4 Command Statements

RobotBASIC has an extensive set of commands that achieve a variety of tasks like drawing lines on the screen, accepting input from the user, printing results on the screen, and so on. These commands are the heart of RobotBASIC and are what makes the language achieve most of its actions.

Some commands have parameters and some do not. Some commands require a specific number of parameters and some have optional parameters. Parameters are how values are passed to the command and results obtained from it. A comma must separate all parameters. Commands look like this

```
CommandName parmater1, parameter2...
```

If the parameter is a value *passed* to the command then it must be an **expression** that results in the value (for details on expressions see Sec. B.7). Some parameters may have to be numeric (**float or integer**) and some may have to be **strings**. In either case the expression must result in the appropriate data type.

If the parameter is a value to be *obtained* from the command, then the parameter must be a simple **variable** (see Sec. B.7.3).

The command will use the values (expressions) passed to it, perform its action and then assign a value to the variables passed to it (when appropriate).

Some commands act on an entire array, in that case the parameter must be an array name (see Sec. B.7.4 for details on **arrays**). For a list of commands and a description of their action see Sec. C.7.

Commands *are not* case sensitive, so `Print` and `print` are the same. Here are some examples of commands in action:

```
rLocate 100,100,90 //initializes the robot and puts it
                   //at location 100,100 with heading 90
rForward 20 //make the robot go forward 20 pixels
rTrurn 40   //make the robot turn 40 degrees
```

B.5 Labels

A label is a marker to a certain location in the program. The marker is given a name and can be used within certain statements to refer to that location. You will see how this is done in the flow-control statements in Sec. B.6.

A label can be in a line by itself, or it can be inline with another statement, but it must be the first thing in the line. A label must end with the colon (:) and must start with a letter but after the first letter any combination of numbers and letters can be used (see Sec. C.1 for other label styles).

Labels *are* case sensitive, so *Label1:* and *label1:* are not the same. Here are some examples of the use of labels:

```
ProgramStart:  //this is a label on its own
 Input "Enter a number",n
 Print "You have entered ",n
```

```
 If n < 5 then goto ProgramStart //goes to the position
                                  //marked by the label
                                 //if input is less than 5
 If n > 30 then goto Pos1 //so does this one if the
                          //input is greater than 30
 K = n*2
 Print K
End   // stops the program here
Pos1: Print "Too high a number" // this shows a label in
                           // line with another statement
```

B.6 Flow-Control Statements

Flow-control statements make RobotBASIC:

- Take actions depending on a certain condition.
- Repeat actions a number of times.
- Repeat actions until a certain condition is fulfilled.
- Repeat actions while a certain condition is fulfilled.
- Go to a sub-part of the program, return from there, and continue execution from where the program branched off.

Flow-control statements *are not* case sensitive, so `While`, `WHILE`, and `while` are the same.

Here is a simple program that shows three different flow-control structures:

```
for I = 1 to 5  //forward the robot 5 pixels
    rForward 1
next
Input "Enter a number",N
if N < 5 then print "too small"
if N > 30 then print "too large"
while N > 0 // make the robot turn N times
    rTurn 1
    N = N-1
wend
End
```

Refer to Sec. C.6 for detailed use of all the flow-control structures provided in RobotBASIC.

Subroutines in RobotBASIC are created by surrounding a group of statements with a label and the command `Return` like this:

```
SomeRoutineName:
   Statement
   Statement
 ...
Return
```

You can then, from anywhere in your program, branch off to the subroutine. The subroutine's statements will be executed and when `Return` is executed the program flow will go back to the line right after the line that called the subroutine. Here is an example:

```
Statement
Statement
   ...
gosub SomeRoutineName
Statement
Statement
End
SomeRoutineName:
     Statement
     Statement
    ...
Return
```

See Sec. C.6 for more details on the `Gosub` statement and all the other flow-control statements.

B.7 Expressions

If you have used a scientific calculator you know what expressions are. An expression is a formula that manipulates numbers, strings, or other expressions and returns a result.

You can use numbers, strings, variables, constants, and/or functions in an expression (see Secs. B.7.1, B.7.2, and B.7.3 for details on each of these elements).

An expression results in any of the three basic elements of RobotBASIC, an **integer number, a floating-point number, or a string** (see Sec. B.7.1 for details).

In many statements, parameters are needed for the statement to function. Anywhere a parameter is needed, an expression can be used that would result in the correct data type required by the statement.

Operators are used in an expression to manipulate the sub-parts of an expression. These operators are parenthesis, comparison operators, math operators, logical operators, and bitwise operators (for details see Sec. B.7.5).

There are many functions in RobotBASIC that accept parameters and return a value. These functions can be used within expressions, are themselves expressions, and are given expressions as parameters (for details see Sec. B.7.7).

B.7.1 NUMBERS

Numbers are used in all computer programs. There will hardly ever be a program that does not use numbers. In RobotBASIC there are two types of numbers:

Integer Numbers These are whole numbers. The range for integer numbers is from −2,147,483,648 to +2,147,483,647

For example:

```
Age = 46    //assigns the variable Age the
            //integer value 46
```

Floating-Point Numbers These are numbers with decimal fractions. The range for floating-point numbers is from $\pm 2.23 \times 10^{-308}$ to $\pm 1.79 \times 10^{308}$

You can use the letter "e" or "E" with a + or − to indicate powers of 10 as shown in the example below.

For example:

```
//assigns the variable Distance the floating
// point value 5.78
Distance = 5.78
C = 3.1e+8        //speed of light is 310000000 m/s
Mass = 0.1E-6     // Mass is 1/10000000 grams
```

RobotBASIC has many functions and commands that require and/or return numbers. There are functions to convert numbers and manipulate them. When an integer number is required if you pass a float instead, the system will truncate the number into an integer by removing the decimal fraction.

When RobotBASIC evaluates mathematical operations with a mixture of floats and integers the result will be a float. However, if all the numbers involved in an expression are integers then the result will be an integer. This is especially important when performing division. For example:

```
A = 5 \ B = 6 \ C = 4.2
Print A/B      //prints 0
Print C/B      //prints 0.7
Print 1.0*A/B  //prints 0.83333
```

B.7.2 STRINGS

Strings are text and must be enclosed in quotes. In many programs strings are used to communicate with the user. Any time you need to display results to the user in a friendly fashion, or prompt the user regarding what to do next or how to use your program, strings are required.

For example:

```
Print "Hello World" // Hello world is a string
```

RobotBASIC has many functions and commands that require and/or return strings. There are functions to convert strings and manipulate them. Strings can be as long as required, there is no limitation on the string size.

You define a string within your program by using the quote (") character. A string is surrounded by a pair of quote characters. If you need to use the quote character within the string as part of the string you must make it a double quote ("").

For example:

```
Print "Hello ""World""" // will print Hello "World"
```

B.7.3 SIMPLE VARIABLES

A variable is a named storage area to keep numbers or strings. Variable names must start with a letter but then any combination of letters and numbers can be used. **Variable names *are* case sensitive, so *Distance* and *distance* are different.**

The length of a variable name can be up to 255 characters, so you can use names like *DistanceToBeacon_1*. It is important to use meaningful names in your programs because this makes your programs easier to read and understand.

A variable can be used anywhere a number or string is used. As long as the variable has been assigned a value before using it, it can substitute for its value. A variable is assigned a value in an assignment statement (see Sec. B.3) or by passing it as a parameter to a command that will assign it a value.

Unlike standard BASIC and many other programming languages, variables in RobotBASIC do not have to be predefined as being of a specific type. Any variable can store any value type (float, integer, or string). Additionally, you can reassign a value of a different type to a variable that already holds a certain type. If a variable exists when it is being assigned a value, it does not matter what the old value is or of what type it is, the new value will replace the old one. If the original value is an integer and the new value is a string the variable type will be changed to a string, and vice versa. The same is true for floats and integers and floats and strings.

In certain operations RobotBASIC will convert between variable types as necessary if it can. For example, if you add an integer to a string the integer will be converted to its string representation and concatenated to the string. If the operation cannot be performed the system will return an error.

For example:

```
Message="Hello World" //assigns a variable a value
Today=Date(1) //assigns a variable the value of a function
Print Message // will print Hello World
Print "Today is ",Today //will print Today is 2007/03/01
rGps X_Pos,Y_Pos  //will assign X_Pos the robot's x
                  //position and Y_Pos the y position
Distance=PolarR(X_Pos,Y_Pos) //calculates using a
                             //function using variables
Message = 8+4  //change Message to a number
```

B.7.4 ARRAYS

An array is a collection of variables given a single name. Each variable is an element in the array. Each element in the array is referenced by its position in the array. **Array names *are* case sensitive, so *Dist*[ ] and *dist*[ ] are different.**

Each element in an array can be of any data type (float, integer, or string). Additionally, you can reassign a value of a different type to an element that already holds a certain type. If an element has previously been assigned and it is being reassigned, it does not matter what the current value is or of what type it is, the new value will replace the old one. If the original value is an integer and the new value is a string the variable type will be changed to a string, and vice versa. The same is true for floats and integers and floats and strings.

Remember the elements of the array do not all have to be of the same type. Each element can be of any type.

Think of an array of mailboxes. They are collectively called *MailBoxes*. To access the third element from the left on the second row we would say *MailBoxes*[1,2]. The reason we have 1, 2 and not 2, 3 is because the first element is 0, the second is 1, and so on.

RobotBASIC allows you to have arrays of any dimension with as many elements in each dimension as you want. The dimension of the array is the number of indexes it has. For example *Array_1*[5,7,8] is a three-dimensional array while *Array_2*[4,6,7,8] is four-dimensional.

The dimensional constraints of an array are the extent of each dimension. For example *Array_3*[5,7] is a two-dimensional array with constraints of five rows and seven elements in each row (or 7 columns). Remember that even though we have five rows, the fifth row is row number four since the first row is row number zero.

Once an array is defined, each element in the array can be used as if it were a simple variable. Anywhere a simple variable can be used an array element can be used. And just like a simple variable, the array element needs to be assigned a value before it is used.

NOTE: In commands that require a variable to be passed, so it can be assigned a value, you cannot substitute an array element; you must use a simple variable. The simple variable can be later stored in the array element if needed.

Some commands and functions in RobotBASIC act on an entire array at once. These commands require the array name as a parameter.

An array is created by using the `Dim` command or the `Data` command. See Sec. C.7 for details on these commands. Here are examples of array manipulations:

```
Dim Array1[3,4]
Array1[1,1] = "testing"
Array1[0,0] = 9
Array1[0,1] = 8.4
Print Array1[1,1];Array1[0,0]
Array1[2,2] = Array1[0,0]+Array1[0,1]*4
```

B.7.5 OPERATORS

Expressions are formulas that manipulate expressions. Expressions are manipulated using operators. Some operators have precedence over other operators. For example multiplication has precedence over addition. So if you write 4+5*3 the result is 19 not 27. That is because the numbers 5 and 3 are multiplied first then the 4 is added. If you wanted to do the addition first you must write (4+5)*3 which will result in 27. There are five types of operators (*listed in order of precedence*):

NOTE: The following list of operators is in the order of their precedence

Parenthesis "( )" Any expression surrounded by parenthesis will be evaluated before it is passed on for further evaluation outside the parenthesis. Thus, parenthesis are used to trump any **operator precedence rules** (see above). You can think of any combination of expressions within the parenthesis as an expression in itself just as if it were a single number or string.

Use parenthesis around operations when you are not sure how they would evaluate due to operator precedence, or to make the intent of the formula clear, or to override operator precedence.

Examples:

```
Print 3*4+5   //prints 17
Print 3*(4+5) //prints 27
```

Math Operator (listed in order of precedence)

Unary Negate (–) Put in front of a numeric expression makes the result a negative if it is positive and positive if it is negative. For example:

```
a = 5
print -a  //print -5
print  -(a-7)  //prints 2
```

Raise to the Power (Exponentiation, ^) Raises an expression to the power by another (the exponent).

For example:

```
Print 4^2+1         //prints 17
Print 4^(2+1)       //prints 64
Print (3+1)^(2+1) //prints 64
Print 4^-2.0        //prints 0.0625
Print 4^-2          //prints 0
Print 4.0^-2        //prints 0.0625
```

Divide (/), Multiply (*), Percentage (%), Modulus (#)

/ divides a number expression by another.

* multiplies a number expression by another.

% calculates the percentage of a number expression.

calculates the integer remainder after dividing an integer expression by another.

In all the above operations, if either expression is not a number an error will occur. All these operators have the same precedence and will be evaluated from left to right if they are in sequence.

When you divide two integers you will get the result of an integer division not a floating-point division so 2/3 is 0 not 0.6666. If you want to make sure the result is a floating-point number make sure that at least one of the expressions is a floating-point number: 2.0/3 results in 0.666

Modulus is an integer operation only. If any of the expressions is a float it will be truncated to an integer. The percentage operation will yield a float result always.

Examples:

```
Print 5/6       //prints 0
Print 4/3       //prints 1
Print 4.0/3     //prints 1.333333
```

```
Print 4*3         //prints 12
Print 400%3       //prints 1.2 notice the float result
Print 5* 8 # 5    //prints 0
Print 5* (8 # 5)  //prints 15
Print 8.3 # 5.3   //prints 3
Print "ttt"/3     //ERROR
```

Add (+), Subtract (–)

+ adds an expression to another. If the two expressions are numbers then addition will occur. If the two expressions are strings then concatenation will occur. If the first expression is a string and the second is a number then the second expression will be converted to a string and concatenated to the first expression. If the first expression is a number and the second is a string an error will occur.

– subtracts an expression from another. If either value is not a number an error will occur.

Both operators have equal precedence and are evaluated from left to right if they are in sequence.

Examples:

```
Print "Test"+"ing"    //print Testing
Print "Test"+5        //prints Test5
Print 5+"test"        //ERROR
Print 7+5             //prints 12
Print 7-5             //prints 2
Print "Test" - ""ing"  //ERROR
```

Comparison (Relational) Operators Comparison operators compare one expression to another and return the value 0 if the result is false and 1 if the result is true. For example 5 > 4 results in 1 (true) but 5 > 10 results in 0 (false).

The result of this type of operation is often used in flow-control statements to determine whether to take action or not. But the result of a comparison operation can be used anywhere an expression can be used, just as if it where a number (which it is, the number 0 or 1).

Comparison operators have equal precedence and are evaluated from left to right. But it is very confusing if you combine comparison operators in an expression. If you need to do so, use logical operators (see below) for better clarity and use parenthesis to clarify the meaning.

Many operators have multiple forms that perform the same operation. This way you can use any style you might be familiar with from other programming languages.

All operators operate on string or number expressions but both expressions must be of the same type. Except, for the "$" operator where the expressions can only be strings.

If you compare two string expressions beware of lower and upper case letters. Letters are compared in order of their ASCII codes. Upper case letters have less value than lower case letters. Comparing strings may give a result you do not expect depending on the length of strings and the letters in the strings. For example "Test" is not equal to "test"; it is less.

In the following list true = 1 and false = 0 (see Sec. B.7.6).

Operator	Description
$	To see if the left string expression is *contained* within the right string expression. Both expressions must be strings or an error will result. For example, `"st" $ "testing"` returns true.
>	To see if the left expression is *greater than* the right expression. For example, 5 > 4 returns true.
<	To see if the left expression is *less than* the right expression. For example, 4 < 3 returns false.
= or ==	To see if the left expression *equals* the right expression. For example, 4 = 7 returns false.
>= or =>	To see if the left expression is greater than *or* equal to the right one. For example, 5 >= 7 returns false.
<= or =<	To see if the left expression is less than *or* equal to the right one. For example, 5 <= 7 returns true.
<> or >< or !=	To see if the left expression is *not* equal to the right one. For example, 5 != 4 returns true.

Logical Operators Logical operators are usually used to combine results from comparison operators. A logical operator will consider the expressions it operates upon as false if the expression results in a zero, or true if the expression results in other than zero (negative or positive). You can perform logical operations on any numerical expressions. If any of the expressions results in a string an error occurs.

All logical operators have equal precedence and will be evaluated from left to right if they occur in sequence. Use parenthesis if you are not sure how the combination will perform.

There are two forms for each operator. This way you can use the style you are familiar with. **The letter formats *are not* case sensitive. So AND, and, AnD are the same.**

Logical AND (AND, &&)
For example

(5 > 4) AND (4 < 3)	results in false
(5 > 4) && (4 < 3)	results in false
(5 > 4) AND (4 >= 3)	results in true
(5 > 4) && (4 >= 3)	results in true

Logical OR (OR, ||)
For example

(5 < 4) OR (4 < 3)	results in false
(5 < 4) || (4 < 3)	results in false
(5 > 4) OR (4 < 3)	results in true
(5 > 4) || (4 < 3)	results in true

Logical XOR (XOR, @@)

For example

(5 < 4) XOR (4 < 3)	results in false
(5 < 4) @@ (4 < 3)	results in false
(5 > 4) XOR (4 < 3)	results in true
(5 > 4) @@ (4 < 3)	results in true

Logical NOT (NOT, !)

For example

!(5 < 4)	results in true
NOT(5 < 4)	results in true
!(5 > 4)	results in false
NOT(5 > 4)	results in false

Here is a table of how the various operators will do the logic:

Operators	Left expression	Right expression	Result
AND	0	0	0
	0	1	0
	1	0	0
	1	1	1
OR	0	0	0
	0	1	1
	1	0	1
	1	1	1
XOR	0	0	0
	0	1	1
	1	0	1
	1	1	0
NOT		0	1
		1	0

Bitwise Operators Bitwise operators only work with numeric expressions. If either the right or left expression results in a string an error will be issued.

Bitwise operators perform the equivalent logical operation on each bit of the numbers that result from the expressions. So for example if we do 9 bAND 4 then since 9 is 01001 and 4 is 00100 the result will be 00000 which is 0. If we do 9 bOR 4 the result will be 01101 which is 13. Just remember that bAND, for instance is an AND but performed on a bit-by-bit basis.

All bitwise operators have the same precedence and are evaluated from left to right if they are in sequence. There are two formats for each operator. This is so you can use the format you are familiar with. **The letter formats *are not* case sensitive so bAnd, band, BAND are the same.**

```
Bitwise AND (bAND,&)
    7  &  2                     equals 2
    7 bAND 2                    equals 2
Bitwise OR (bOR, |)
    6  |  1                     equals 7
    6 bOR 1                     equals 7
Bitwise XOR (bXOR, @)
    6 @ 2                       equals 4
    6 bXOR 2                    equals 4
Bitwise NOT (bNOT, ~)
    ~ 1                         equals -2
    bNOT 1                      equals -2
    ~ 0                         equals -1
    bNOT 0                      equals -1
    ~ 5                         equals -6
    ~ (-6)                      equals 5
Shift Right (bShiftR, >>)
    514 >> 1                    equals 257
    514 bShiftR 1               equals 257
Shift Left (bShiftL, <<)
    5 << 4                      equals 80
    5 bShiftL 4                 equals 80
Rotate Right (bRotR)*
    514 bRotR 1                 equals 1
```

This bitwise operator operates only on a byte. If the number is greater than 255 the lowest byte alone is used and all other bytes are zeroed.

```
Rotate Left:bRotL
    5 bRotL 4                   equals 80
```

This bitwise operator operates only on a byte. If the number is greater than 255 the lowest byte alone is used and all other bytes are zeroed.

B.7.6 CONSTANTS

Constants are numerical values defined within RobotBASIC, but instead of using the number you can use a name for the number. This is the same as using a simple variable. The names have been defined within RobotBASIC and you can either use the number if you can remember what it is or use the name anywhere the number is needed.

Constants can be used anywhere a numerical expression is required. Constant names *are not* case sensitive so Red and red are the same.

There are constants that define colors, there are constants to define things like true and false, and many more.

NOTE: Refer to the constants help page within the IDE for a list of all constant values

B.7.7 FUNCTIONS

Functions are expressions that use expressions as parameters and return a number or string. A function can be used anywhere a number or string resulting expression can be used. The form of a function is as follows:

```
FunctionName(parameter,parmeter,...)
```

The parameters are expressions and can also be other functions, since a function is an expression. If the function does not take parameters then you only have the parenthesis and nothing in between them, for example `rFeel()`.

RobotBASIC has many functions that return strings or numbers and accept strings or numbers as parameters. There are functions to get the sine of an angle, to convert a string to uppercase, to convert a float to an integer, to convert a number to string, and much more.

Some functions operate on an entire array. In those functions you pass the array name as a parameter to the function.

For example:

```
Print sin(40*pi()/180)
B = ToUpper("test") + Spaces(30)
A = Left(B,20)+ "__"
If !(rFeel() & 2) then rTurn 3
```

APPENDIX C

COMMANDS, FUNCTIONS, AND OTHER DETAILS

Throughout this appendix there will be reference to the following items:

- *ExprN* implies that a numeric resulting expression is required.
- *ExprS* implies that a string resulting expression is required.
- *Expr* implies that either a numeric or string expression is acceptable.
- *{Expr}* or *{Var}* implies that it is optional and *{Expr...}* means many can be given.
- *Var* implies that a variable name must be given.
- *VarA* implies that an array name must be given. In some cases the array must exist and must have been previously dimensioned using the `Dim` command or created using the `Data` command.
- *VarA*[...] implies that an array element is required. The element specification [...] must be valid within the array's dimensional constraints.
- If a *Var* is expected in any of the commands, then if *Var* exists it will be assigned the result otherwise it will be created and assigned the result.
- If *VarA* is expected then it must be a previously dimensioned array, but in some cases where *VarA* is created by the command it does not have to be previously dimensioned.

NOTE: Always refer to the help pages within the RobotBASIC IDE for the most up-to-date information regarding material mentioned in this appendix.

C.1 Labels

See Sec. B.5 for a detailed discussion of how labels fit within RobotBASIC's language. Here we give a few additional details. There are three styles for labels:

C.1.1 ALPHA-NUMERICAL STYLE 1

This is the clearest style of all the other styles and is the easiest to pick out while scanning the source code of a program. In this style of labels you must begin the label with a letter followed by any combination of digits and letters and finally ended with a colon (:). You must end this label style with the colon (:) and commence with a letter.

For example:

```
//this is a label on its own
Label_1:
     print 20
//a label inline with another statement
Label_23: print 40
     goto Label_1
//when you refer to the label do not use the :
```

NOTE:

- When referencing a label in a statement like Gosub or Goto do not include the colon (:) character. The colon (:) is put as part of the label only at the label position to define the label and is not used in statements referring to the label.
- Labels *are* case sensitive. So *Loop:*, *loop:*, and *LOOp:* are all different.
- Labels can be up to 255 characters long, so do *use* meaningful names for subroutines and looping labels.

C.1.2 ALPHA-NUMERICAL STYLE 2

This style is similar to programming languages other than BASIC and is supported by RobotBASIC for people who like this style. A label of this style must begin with the colon (:) followed by any combination of digits and letters. For example:

```
//this is a label on its own
:Label_1
        print 20
//a label inline with another statement
:2_Label  print 40
//when you refer to the label do not use the :
        goto Label_1
```

C.1.3 NUMERICAL STYLE

This style supports the standard BASIC line numbers and is not a recommended style. You can use any combination of digits that are the first thing in a line or the only thing in the line. A variation to this style is supported where after the first digit you can use a combination of letters and digits, but you must start the label with a number. For example:

```
//10 is the numerical label inline with another statement
10   print 4
20   print 9
//4A2C is the other style inline with another statement
4A2C print 100
//200XYZ1 is the other style in a line on its own
200XYZ1
     print "Hi there"
//30 is the numerical label in a line on its own
30
      goto 10
```

C.2 Assignment Statement

See Sec. B.3 for detailed information about the assignment statement. A few more details are provided here.

In RobotBASIC you do **not** have to specify the data type of a variable. The type of the variable is determined by the data that is stored in it. Also the type of a variable can change if you reassign a different type value to it.

Variable names cannot include "%," "$," or "#" as in other BASIC implementations. You *cannot* use these characters as part of variable names. These characters are used as operators (see Sec. B.7.5).

RobotBASIC has a powerful array structure compared to other programming languages. In addition to a versatile `Data` command, you can have almost limitless (only limited by memory) dimensional arrays and limitless elements in each dimension. Additionally, **each** element in the array can be any of the three types in RobotBASIC (strings, integers, and floats). Any of the elements can be assigned and reassigned any of the types. You can manipulate and perform operations on arrays that include inversion, multiplication, and so on. There is also a set of statistical operations like variance, regression, and more.

The assignment statement is used to set the value of a variable or an array element like this:

```
Var = Expr  (a variable name followed by an equal sign (=) then an
expression)
```

or

```
VarA[ExprN{,ExprN....] = Expr  (an array name[element specification]
followed by an equal sign [=] then an expression)
```

A variable does not have to be specified prior to assignment, but if you use this format $a = a+4$ then the variable must be previously specified because otherwise the interpreter will not know what the original value was and an error will occur.

If the resulting expression is a string, integer, or float the variable will be created and assigned that value and type. If the variable already exists but is of a different type and/or value it will be erased and assigned the new type and value.

The array must have been created using the `Dim` or `Data` command. The element specification must be legal for the array dimension specified in the `Dim` command (see array commands in Sec. C.7). The array element is treated as if it is a variable name and the above information regarding variables is applicable.

C.3 Expressions

See Sec. B.7 for details on expressions and what constitutes an expression. A few additional details are given here.

Expressions can contain previously assigned variables and/or array elements. If an array element is specified then it must have been assigned and be within the range of the array's dimensional constraints as specified in the `Dim` command. Expressions can also contain functions (see Sec. C.8). Expressions can also contain bitwise, logical, comparison, and math operators (see Sec. B.7.5)

```
print sin(4+5/3.0)*(3|4)+ ((a>b)&(c < 5))
n = "this{"+b+"} = "+c
```

NOTE: If all the numbers involved in an expression are integers the result will be an integer. If any of the numbers is a float then the result will be float.

```
B = sin(pi(2*Theta^2))/(3+4*Acceleration)
```

Even though all the numbers are integer the result is still float because `Sin()` returns a float.

C.4 Strings

See Sec. B.7.2 for details about strings. A few more details are given here.

You can add two strings (concatenation)

```
temp = "test" + "ing"    //temp will be "testing"
```

You can add a string and a variable that is string

```
temp = "test"
print "I'm "+temp+"ing"      //will print I'm testing
```

You can add a string and a number as long as the string resulting expression comes before the number. The number will be converted to a string. Use the `ToString()` function to do the same thing regardless of order.

```
a = 12.3
b = "Result = "
c = b+a
print "The "+c      //prints The Result = 12.3
```

```
print "The ",b,a  //prints The Result = 12.3
print "The ",b+a  //prints The Result = 12.3
print "The ",b+ToString(a) //prints The Result = 12.3
print a+b         //Gives An Error
```

You can compare two strings for equality. But be careful about > and < since upper and lower case letters will affect the result. You can also use the $ operator

```
if  "tin"  $ "testing" then print "yes"   //prints yes
//ExprS1 $ ExprS2 will test to see if ExprS1
//occurs within ExprS2
```

You can use the `InString()` function to do a similar action

```
if InString("Testing","tin") > 0 then print "yes"
//will print yes
```

C.5 Variables

See Sec. B.7.3 for a detailed discussion of variables. A few additional points are given here:

- All variables *are* global. This means that once a variable has been defined and assigned a value it is accessible from any statement in the program from that time onward. So if you use a counter inside a subroutine in a `for/next` statement like `for I = 1 to 6` and *I* is being used in a counter in the calling section, unpredictable logical errors may occur. So be careful in variable assignments and especially within subroutines.
- Variables are case sensitive. So *Theta, theTa,* and *theta* are all different. This can cause logical errors if you mistype the names. Take care.
- Variable names can be up to 255 characters long so do *use* meaningful names to avoid variable clashes. For example, for a counter in a subroutine called `Delay` use *Del_1, Del_2,* etc.
- Do *not* use variables with the same name as commands, functions, constants, or labels.

C.6 Flow-Control Statements

See Sec. B.6 for details on how flow-control statements fit in the syntax of the RobotBASIC language. Here we will list the various flow-control constructs. **Flow-control statements are *not* case sensitive, so `While`, `WHILE`, and `while` are the same.**

C.6.1 `If-Then` STATEMENT

```
if ExprN then statement
```

The **in-line if-style**. If *ExprN* results in a number not equal to zero (*true*), then the statement after the `then` will be executed. Otherwise (*false*) the program flow will go to the next line ignoring the statement after the then. *ExprN* can be any expression that results in a number. If the result is zero then the condition is redeemed to be *false*. If the result is other than zero (positive or negative) then the condition will be redeemed as *true*. See Sec. B.7.5 for comparison and logical operators.

```
a = 3 \ b = 4 \ c = 5 \ d = 1 \ e = 0
if a then print "true"  // prints true
if d then print "true"  // prints true
if e then print "true"  //will not print
if a > b then print "true"      // will not print
if !(a > b) then print "true"  // prints true
if  !(a > b) && (c < 9) then print "true" //prints true
if "es" $ "test" then print "true"  // prints true
if InString("test","es") then print "true"  // prints true
```

C.6.2 `If-ElseIf` STATEMENT

```
if ExprN1
   statement
   statement
   ...
{elseif ExprNn}
   statement
   statement
   ...
   .
   .
   .
{else}
   statement
   statement
   ...
endif
```

This is the **structured-`if` style**. Notice that there is no `then` after the `if`. This is how it should be. The interpreter will distinguish between the previous style and this style by this one difference. The conditions are evaluated as above.

This structured-`if` style is made up of blocks of lines. There is the `if`-block, the `elseif`-block and the `else`-block. You can have multiple `elseif`-blocks or none. You can have only one `else`-block or none and it must be the last block.

If *ExprN1* evaluates to true then the `if`-block lines will be executed. All the other blocks will be ignored.

If *ExprN1* evaluates to false then the interpreter will evaluate the `elseif` conditions (*ExprNn*) in sequence until the first one that evaluates to true. If none evaluate to true then the `else`-block lines will be executed (if used).

The lines of the first `elseif`-block that evaluate to true (if *ExprN1* is false) will be executed and all other blocks (including the `else`-block) will be ignored.

This `if`-style can be nested in any combination. In-line-`if` does not count as a nested one. You can nest other `if/.../endif` structures within any block of the structure.

C.6.3 `For-Next` LOOP

```
for Var = ExprN1 To ExprN2 {Step ExprN3}
  statement
  statement
  ...
next
```

The interpreter will repeat the statements between the `for/next` a number of times equaling `Abs(`*ExprN2-ExprN1)/ExprN3*+1. The interpreter will put *ExprN1* result into *Var*. The statements between the `for/next` structure will be executed. The interpreter will then decrement *Var* by *ExprN3* if given or 1 if not. The interpreter will then check if the result is greater than *ExprN2*. If it is, program flow will continue with the statement after the `next` (i.e., exit the `for/next` loop). If not, flow will go back to the statement right after the `for` and repeat the whole process.

You can optionally give a `Step` size. *Var* must be a variable but it does need not be previously defined. *EprN1, EprN2,* and *EprN3* must result in integers. *ExprN3* must be greater than 0. If it is 0 then it will be made 1. If it is less than 0 then it will be made positive.

If *ExprN1* is greater than *ExprN2* the interpreter will decrement *ExprN1* until it is less than *ExprN2*. You can nest `for`-loops. An error will occur if you try to nest too deep.

NOTE: You can modify the *Var* counter within a `for/next` loop, just as any other variable. But *beware* of your logic. Use `Break/Continue` along with `if/else/endif` or `if/then` to force an early abort or reloop of the loop (i.e., before reaching the end-count).

Example:

```
For I = 0 to 10 step 2
    Print I
Next    //will print 0 2 4 6 8 10
For I=10 to 0 step 2
    Print I
Next    //will print 10 8 6 4 2 0
```

C.6.4 `Repeat-Until` LOOP

```
repeat
    Statement
    Statement
    ...
until ExprN
```

Will execute the statements between the `repeat-until` as long as *ExprN* evaluates to zero. The lines within the loop will be executed at least one time, since *ExprN* will not be evaluated until the loop has been executed the first time.

You can use `Break` within the loop, which will force the loop to terminate immediately, ignoring any lines after the `Break`. You can use `Continue` within the loop, which will force the loop to reloop immediately, ignoring any lines after the `Continue`.

Example:

```
I = 0
repeat
     Print I
     I = I+1
     If I # 2 <> 0
        I = I+1
        If I > 10 then break
        continue
     Endif
until I > 20//will print 0 2 4 6 8 10
```

C.6.5 `While-Wend` LOOP

```
while ExprN
    Statement
    Statement
    ...
wend
```

Will execute the statements between the `while-wend` as long as *ExprN* evaluates to other than zero. The lines within the loop may never execute if *ExprN* evaluates to zero upon entry into the loop.

You can use `Break` within the loop, which will force the loop to terminate immediately ignoring any lines after the `Break`. You can use `Continue` within the loop, which will force the loop to reloop immediately ignoring any lines after the `Continue`.

Example:

```
I = 0
while I < 20
    Print I
    I = I+1
    If I # 2 <> 0
        I = I+1
        If I > 10 then break
        continue
    Endif
wend      //will print 0 2 4 6 8 10
```

C.6.6 `Break` STATEMENT

Used inside a `for/next`, `repeat/until`, and `while/wend` loops to break the loop immediately. The lines beyond the `Break` statement will not execute and the current level of the `for/repeat/while` loop will be abandoned to the line following `next/repeat/wend` line associated with the current level. This is useful if a loop needs to be abandoned before reaching the end of the loop.

C.6.7 `Continue` STATEMENT

Used inside a `for/next`, `repeat/until`, and `while/wend` loop to reloop the loop immediately. The lines beyond the `Continue` statement will not execute. The program flow will go to the end of the loop (`next/until/wend`). Execution will continue on that statement. The loop will be executed as per normal. This is useful if a loop needs to be repeated before reaching the end of the loop.

C.6.8 `Case` CONSTRUCT

The case construct is not implemented in the language. However, you can *emulate* a case construct using the structured-`if`.

```
If ExprN //Case_1 condition here
    //Do stuff
      ...
```

```
ElseIf ExprN //Case_2 condition here
   //Do stuff
     ...
ElseIf ExprN //Case_3 condition here
   //Do stuff
     ...
     .
     .
     .
Else
   //Do Stuff that will be done if no case is true
     ...
Endif
```

C.6.9 GoSub STATEMENT

```
Gosub Label
Gosub Expr
Return
```

Flow will go to the statement immediately following the *Label* and will continue until a `Return` statement is encountered after which the flow will return to the line following the line that called the subroutine. A subroutine is marked by a *Label* and the `Return` statement as the last line. You can use other `Return` statements within the logic of the subroutine, but always have the last line of the subroutine as a `Return`, just in case your logic gets there. If you do not do so any program lines following the subroutine will be executed as part of the subroutine.

You can call a subroutine from within another subroutine. An error will be issued if you try to nest too deep. Be careful when you do this that you do not inadvertently create a circular endless loop. Recursion can also cause trouble if you do not have the correct windup conditions.

If you use the format `Gosub` *Expr*, then *Expr* must result in a number or string that is a name of a valid label. If *Expr* results in a number then it will be converted to a string. So saying `Gosub` 1000, `Gosub` 100*10, or `Gosub` "1000" are all equivalent and would cause the program flow to branch to the label 1000.

The second format is useful if you want to branch to a subroutine name that will be calculated depending on some logic. For example, you may have an array of labels and you can make the program go to one of the labels depending on some number that is used to index in the array.

Example:

```
Data Labels; "L1",100,"1000"
while true
  Input "Enter a number 0 to 2",I
  If !Within(I,0,2) then continue
  Gosub Labels[I]
wend
End
L1:
  print "I am In L1"
  Gosub Substring("I'm going to Test_Sub, Bye",14,8)
  //notice the above line calculates the subroutine
```

```
  //name using functions
Return
100
   print "I am In 100"
   Gosub Labels[I]*10
   //notice the above line how it goes to a subroutine
   // name that is mathematically calculated
Return
1000
   print "I am In 1000"
   a = "Test"+"_Sub"
   Gosub a
   //notice above goes to a subroutine name defined
   //by a variable
Return
Test_Sub:
   print "I am in Test_Sub"
Return
```

NOTE: Return statements within if-structures inside subroutines must be carefully considered; be extra careful when using Return statements out of subroutines within if-structures. *Check your logic.*

C.6.10 `OnErroR` STATEMENT

```
OnError Label
OnError Expr
OnError
```

This is a special format of a `GoSub`. If an error occurs during the run of a program the system will direct the flow to the subroutine defined by *Label* or the label defined in the variable *Expr* (just as in the `GoSub` statement above). No error message will be issued and the program will continue to run as if you have just issued a `GoSub` statement on the line that caused the error. In the subroutine you can use the command `GetError` *VarN1{, VarS{, VarN2{, VarN3}}}* to find out the error number, message, line number, and character number. It is up to you to handle the error and how to redirect the program flow accordingly. You can issue this statement with different labels at different times to override the routine that will be used if an error occurs. You can also issue the statement without a label (or *Expr*) to turn off the feature and have the interpreter handle errors as normal

C.6.11 `End` COMMAND

Will cause the program flow to terminate and return to the editor. The Terminal Screen will remain on top but you can close it or use the button on the bottom right corner to switch to the editor window. Or use Windows methods to do so. You can review the terminal screen if you close it, by using the menu or speed button on the editor window.

C.6.12 `Goto` STATEMENT

```
Goto Label
```

Program flow will branch to the statement immediately following the *Label*. The label can be on any line in the program.

Be very careful using this flow control. Most of the time you can avoid using `Goto` statements by using good structured programming techniques with the help of the structured flow-control constructs detailed above.

The main use for a `Goto` statement is to create loops, but it is more advisable to use `repeat/until`, `while/wend` loops and `for/next` loops along with `Break`, Continue and the appropriate `if/else/endif` combinations.

Certain precautions must be taken if you use `Goto` within `if/else/endif`, `repeat/until`, `while/wend`, or subroutines.

- When a `Goto` statement is executed, the interpreter will **reset** any `if` nesting and any `for/next`, `repeat/until`, or `while/wend` nesting. It would be as if there are no pending `if/else/endif` any more and likewise no more `for/next`, `repeat/until`, or `while/wend` loops. So if you use a `Goto` within an `if` to go beyond the entire nested `if` structure you would be ok, but if you `Goto` within the next `if/else` or `else/endif`, etc. you will get an error. Likewise if you are within a `for/next`, `repeat/until`, or `while/wend` use a `Goto` to get out of the entire nested structure, otherwise you will get an unbalanced `next/wend/until` error.
- It is not a good idea to use `Goto` within *looping-structures*. If you must do so, make sure your program logic accounts for it.
- In a subroutine, try to use `Goto` only to branch within the subroutine. Make sure that somewhere within the subroutine a `Return` is eventually issued. Using `Goto` to get entirely out of the subroutine is legal, but beware of the logic of your program.
- You can use `Goto` to go beyond an End statement, but check your logic.
- Try to *avoid using* `Goto`. Use of `Goto` makes the program flow hard to understand, hard to follow, hard to debug, and can cause logical and semantic errors that are hard to locate. However, `Goto` can be useful in handling errors to redirect the program flow to one place in the code to handle all errors (but also see the OnError statement).

C.7 Command Statements

See Sec. B.4 for a discussion on how commands fit within the RobotBASIC language. The commands are listed here in order of functionality. An alphabetic order can be found in Sec. C.12. **Commands *are not* case sensitive. `ClearScr`, `clearscr`, and `clearSCR` are all the same.**

C.7.1 INPUT AND OUTPUT COMMANDS

`Print` ***{Expr,Expr;Expr...}*** Outputs the values of *Expr...* to the screen. A comma (,) between the expressions makes them print with no space between them and a semicolon (;) prints them with a tab space between them. If there are no expressions then a line feed is printed.

The first time `Print` is used the text will print at the top of the screen and then will print on subsequent lines until the screen is filled. When this happens the screen will scroll upward and the last line will always be used to print the text. See Sec. A.2 for more on this and a special feature for retrieving the text that scrolled off the screen.

The color of the text is according to the current default colors. If you desire to use different colors then use the `SetColor` and `GetColor` commands.

`XYstring` ***ExprN1,ExprN2,Expr3{;expr,expr;...}*** Outputs the result of *Expr3*, and so on at *ExprN1, ExprN2* position on screen. A comma (,) between expressions makes them print with no spaces between them and semicolon (;) prints them with tab separation.

The color of the text is according to the current default colors. If you desire to use different colors then use the `SetColor` command. See the `GetColor` and `GetXY` command.

`XYtext` ***ExprN1,ExprN2,Expr{,ExprS{,ExprN3{,ExprN4}}}*** Outputs the result of *Expr* at *ExprN1,ExprN2* position on screen. *ExprS* specifies the font name of the font to be assigned to the printed text. *ExprN3* specifies a size and *ExprN4* is a bitwise map of the style. *Expr* can be any valid expression that results in a number or string. If it is a number it will be converted to a string.

If either *ExprN1* or *ExprN2* is −1 then the current corresponding X or Y screen position will be used. *ExprS*, *ExprN3* and *ExprN4* are optional. If *ExprS* is not specified or is an empty string then the default font (Courier New) will be used. If *ExprN3* (Size) is not specified or is less than 1 then the default size will be 11. If *ExprN4* (Style) is not specified or is 0 then no style is applied.

ExprN3 is a number that specifies the size of the font. You can specify any number the system will apply the closest allowable size.

ExprN4 is a bitwise map of the desired styles to be applied. You can use bit wise OR (bOr or |) the values given in the Constants help page to apply multiple styles. So for example if you want bold and italic use *fs_Bold | fs_Italic*.

ExprS is a string that specifies the name of the font to be used. Different machines may have different fonts and if the name you specify is invalid the system will apply a default (not necessarily Courier New). To find out what names are available on your machine use the *Fonts* menu option from within the *Help* menu on the Editor Screen. This will bring up a dialog that shows all fonts available and what sizes are available for each. You can select a font and when you exit the dialog the font's name will be copied to the clipboard. You can then use *Ctlr+V* to insert the name into your program. Remember different machines may not have this font. So try to choose fonts that are universally available on most machines.

Many fonts do not have all the characters of the same width. These fonts will not print within a consistent width depending on the number of characters in the string, since different characters have different width. The font "Courier New" has all characters of the same width. Text written with a font that does not have the same width characters will be hard to line up from line to line.

The text will be printed with the currently set foreground and background colors. If you desire to use different colors then use the `SetColor` command. Also see the `GetColor` and `GetXY` command.

`Beep` ***{ExprN}*** Beeps the PC speaker *ExprN* times. If *ExprN* is not given only one beep will be sounded.

`Sound` ***ExprN1,ExprN2{,ExprN3}*** Makes the PC speaker create a sound at the frequency *ExprN1* for a duration of *ExprN2* milliseconds. The optional parameter *ExprN3* is either true or false. If it is true then the sound is made in the background and if it is false the sound will stop the system for the duration of the note. Your program will still pause until the note finishes but the Windows operating system will not be paused if *ExprN3* is true. The default for *ExprN3,* if you do not specify it, is true.

`Speaker` ***ExprN*** This will turn the speaker off if *ExprN* is zero (off) or on if *ExprN* is other than zero (on). The notes played with the `Sound` command will play and will take the time required but no sound will be heard if the speaker is turned off. The speaker is on by default upon the start of the program.

`PlaySong` ***ExprS*** This will play a song defined by the string *ExprS*. The string will contain notes and other specs as defined below:

- The notes are A, A#, B, C, C#, D, D#, E, F, F#, G, G#, also P which is a pause.
- You can use lower or upper case. The # must immediately follow a note or it will be ignored.
- Immediately after defining the note you can specify a number to define the duration of the note. This number can be any number greater than 0 and is usually 1, 2, 4, 8, 16, 32, or 64. But you can specify any number you desire. The number will be used to calculate the duration of the note by dividing into the defined tempo (see later). The formula is *Duration = Tempo/Number*. You do not need to define duration all the time. If a note does not have a duration defined after it then the duration last defined will be used. If you have not previously defined duration for a previous note then the duration will be 8 by default.
- You define a tempo for the song and can change it at any time by specifying the letter "T" followed by a number. The number will be in milliseconds. So if you say T1500 then the tempo will be 1 and half seconds. If you never define a tempo then it will be 1000 (i.e., 1 second) by default.
- You can define a scale for the song and can change it any time by specifying the letter "S" followed by a number. The number must be in the range 0 to 6. This means that there are 7 scales and scale 3 is the middle C scale. If you never define a scale it will be 4 by default. If you define a number greater than 6 it will be made to be 6.
- The letter "P" is taken to be a pause of the duration defined as for the notes (see above).
- If the speaker is off (see the command `Speaker` *ExprN*) then the song will play but no sound will be heard.

The following example will play a song with the scale being 4 to start with then it will be changed to 3. Also the tempo will start as 1000 and will be changed to 1500 afterward. The notes A and B will play at 8th and then the rest will be played at 16th

```
PlaySong "abc16dc#T1500dgS3abdg#"
```

The following will play the song "Jingle Bells" followed by "La Cucaracha." Notice how you have versatility in defining and playing the songs. Also notice how the scale is changed in the "La Cucaracha" song.

```
data Jingles;"T1000S4E8EP32E4P32E8EP32E4P32E8GP32C4"
data Jingles;"D16P32E2P16F8FP32F8F16P32F8EP32E8"
data Jingles;"E16P32G8GFDP32C2"
Cucaracha = "T2500S4C16CCP64F8P64A7C16CCP64F8P64"
Cucaracha = Cucaracha+"A7F16F8P64E16EP64D16DP64C4"
Cucaracha = Cucaracha+"P32S4C16CCP64E8P64G7C16CCP64"
Cucaracha = Cucaracha+"E8P64G7S5C8D16P64C16S4A#16"
Cucaracha = Cucaracha+"AGP64F8P4"
playsong Jingles[0]
playsong Jingles[1]
playsong Jingles[2]
Delay 1000 \playsong Cucaracha
```

`Input` ***Expr,Var***

`Input` ***Expr,VarA[ExprN{,ExprN...}]*** Prints the text resulting from *Expr* and waits for input that will be assigned to *Var.*

If *Expr* results in a blank string a > is printed as the prompt. If *Expr* results in a numeric it will be converted to a string. In the second format an array element is specified and will be assigned the input. The array element must be within the dimension of *VarA*.

`WaitKey` ***{ExprS,}Var*** Prints the text resulting from *ExprS* and waits for a key to be pressed. The text is printed just above the input area at the bottom of the Terminal Screen. When a key is pressed its ASCII code value is assigned to *Var*. If *ExprS* is not given then "Press Any Key" is printed as a prompt. The code assigned to *Var* is the ASCII code of the key pressed so if you press "a" then 97 will be assigned, while *Shift+a* (i.e., "A") will assign 65. Other key combinations like *Ctrl+a* or *Alt+a* will return non-ASCII codes if valid. Keys like *Up-arrow*, *Home*, and so on do not return any values. For these keys see the command `GetKeyE`. Also see the functions `Ascii()` and `Char()` which convert between the ASCII code and string characters.

`GetKey` ***Var*** Does not cause the program to pause and wait for a key, but if a key is pressed then its ASCII code value is assigned to *Var*, otherwise a 0 is stored. This is useful in loops that need to be exited if a key is pressed but without halting the loop until a key is pressed. The code assigned to *Var* is the ASCII code of the key pressed so if you press "a" then 97 will be assigned, while *Shift+a* (i.e., "A") will give 65. Other key combinations like *Ctrl+a* or *Alt+a* will return non-ASCII codes if valid. Keys like *Up-arrow, Home,* and so on do not return any values. For these keys see the command `GetKeyE`. Also see the functions `Ascii()` and `Char()` which convert between the ASCII code and string characters.

If you use this command within a loop you may get too many repetitions of the key due to the speed of the system not giving the user time to release the key before it is read many times as being a new key press. This can be counteracted by using the `Delay` command to delay between successive reads of the key (150 to 200 milliseconds might be sufficient). Or you can use looping to wait until the key is released (*Var* will be zero).

`GetKeyE` ***Var*** Does not cause the program to pause and wait for a key, but if a key is pressed then a code value corresponding to the key is assigned to *Var*, otherwise a 0 is stored. This is useful in loops that need to be exited if a key is pressed but without halting the loop until a key is pressed and also for detecting key presses of keys like up and down arrows and so on.

This command will return the code of the key pressed (not its ASCII code). The keys *Shift, Alt,* and *Ctrl* have their own codes if pressed on their own. However, if these keys are pressed in combination with another key then the code of the key alone is added to 1000 (*Shift*), 2000 (*Alt*), or 4000 (*Ctrl*). For example, if you press "a" alone you will get 65 inside *Var* (notice that this is the ASCII code for "A" not "a"). If you press *Shift+a* you will get 1065, *Alt+a* will give 2065, *Ctrl+a* will give 4065, *Alt+Shift+a* will give 3065, *Ctrl+Alt+a* will give 6065, and so on.

This command allows you to examine more keys than in the command `GetKey`. Keys like *Up-arrow* will return a value. The codes returned by this command must not be converted to characters using the `Char()` function since they are not ASCII codes. Rather, they are codes that represent the key inside the operating system (see the Constants help page within the IDE).

This command may not return a value for certain combinations of keys pressed if they correspond to key combinations that have a meaning for the operating system. So for example *Ctrl+Esc* will cause the Windows' "Start" menu to fire up and will not be possible to detect within your program. Also certain function keys and *Ctrl+* combinations have a meaning within RobotBASIC and will not be passed to your program. So for example the function key *F1* will bring up the help window of RobotBASIC and thus will not return a value for your program.

To find out what key code will be returned for a particular key you can experiment with the keys using the program below (also see the Constants help page within the IDE):

```
while true
  getkeyE k
  if k <> 0
    xystring 1,2,"Extended Code=",k;" "
 endif
wend
```

If you use this command within a loop you may get too many repetitions of the key due to the speed of the system not giving the user time to release the key before it is read many times as being a new key press. This can be counteracted by using the `Delay` command to delay between successive reads of the key (150 to 200 milliseconds might be sufficient). Or you can use looping to wait until the key is released (*Var* will be zero).

`ReadMouse` ***Var1,Var2{,Var3}*** Reads the current mouse position on the screen and sets *Var1*=x and *Var2*=y *Var3* is set to a number that indicates a variety of things as follows:

Two digit integer where the ones digit is:

1 if left mouse button is down
2 if right mouse button is down
3 if middle mouse button is down

The tens digit is:

10 if the *Shift* key is pressed
20 if the *Ctrl* key is pressed
30 if the *Alt* key is pressed
40 if the mouse button was double clicked

For example: *Var* will be 21 if the left mouse button was pressed while the *Ctrl* button was held down.

`SetMousePos` ***ExprN1,ExprN2*** Positions the mouse cursor within the Terminal Screen to any position specified by *ExprN1,ExprN2*.

`AddButton` ***ExprS,ExprN1,ExprN2{,ExprN3{,ExprN4}}*** Creates a push button in the Terminal Screen at position *ExprN1,ExprN2* (X,Y). The button will have the caption *ExprS* and will be of height *ExprN4* and width *ExprN3*. *ExprN3* and 4 are optional and if not given the button will be sized to fit the caption. The button will remain active until removed with the `RemoveButton` command below. The caption is important and should be unique for each button you create. The caption is used to identify the button using the commands `RemoveButton` and `GetButton`.

Once you create a group of buttons the interpreter will keep track of which button was pressed last. You can find out what button was pushed last with the `GetButton` command. If no button has been pushed since the last interrogation the value returned will be a blank string. The value returned from the `GetButton` command will be the caption of the last button that was pushed. See example below for details.

If you use the character '&' in *ExprS* before any letter then that letter will be displayed as an underlined letter in the caption of the button and pressing *Alt+* the letter will be the same as pushing the button. If you desire to have the '&' letter display as is then use a double &&. But remember the caption string *ExprS* will contain these letters and you must take them as part of the string when defining the *ExprS* for `RemoveButton` and when checking the returned string from `GetButton`.

`RemoveButton` ***ExprS*** Removes the button created with the `AddButton` command above. *ExprS* must be the same as the one used to create the button.

`GetButton` ***Var*** Assigns the variable *Var* the caption of the last button pushed. The caption returned is the string used to create the button. If no button has been pushed since the last interrogation the returned string will be null (zero length). Example:

```
for i=0 to 4
   AddButton "Test&"+i,300,20+i*40
next
while true
     GetButton Btn
     if Btn != "" then xystring 10,10,Btn
     if Btn == "Test&3" then RemoveButton "Test&3"
wend
```

C.7.2 SCREEN AND GRAPHICS COMMANDS

`ClearScr` ***{ExprN}*** Clears the screen with color given by *ExprN*. If *ExprN* is not given the default color is used (see list of colors in Sec. B.7.6). The specified color does not set the default color.

`ScrLimits` ***Var1,Var2*** Sets *Var1* to the maximum *x* coordinate of the screen, and *Var2* to the maximum *y* coordinate of the screen. This command allows you to find out the extent of the screen *x, y* coordinates. Zero is always the lower limit.

`SaveScr` ***{ExprN1{,ExprN2{,ExprN3{,ExprN4}}}}*** Saves a copy of a portion of the screen to memory. If you do not specify any parameters the entire screen will be saved. The expressions specify the coordinates of the top-left corner and coordinates of the bottom-right corner of the portion to be saved. *ExprN1* and *ExprN2* default to 0. *ExprN3* defaults to 800 and *ExprN4* defaults to 600.

This command is useful in animations and in drawing temporary objects on top of existing ones and then erasing them without having to redraw the original screen. See `RestoreScr` below for an example.

`RestoreScr` ***{ExprN1{,ExprN2}}*** Restores an already saved copy of a portion of the screen from memory. If you do not specify any parameter then the saved portion will be restored to the top-left corner of the screen. *ExprN1* and *ExprN2* default to 0. If you specify parameters then the previously saved rectangle will be restored over the area starting at *ExprN1, ExprN2* coordinate. The width and height are determined by the saved data specified in the `SaveScr` command. If a `SaveScr` command has not been previously issued then this command will have no effect (the buffer is *not* cleared between program runs).

This command and the `SaveScr` command are used to save and restore rectangular portions of the screen. Example:

```
//--create a screen some how then do the stuff below
SaveScr 100,100,200,200  //save the portion to be drawn over
rectangle 100,100,150,150,red,red  //draw over an area
RestoreScr 100,100       //restore the original stuff in the area
RestoreScr 10,10         //make it appear as if a portion of the
                         //screen moved over
SaveScr                  //save the entire screen
ClearScr                 //clear it
RestoreScr               //restores the entire screen
RestoreScr 10,10  //makes it appear asf if the entire screen
                  //moved over and down
```

`WriteScr` ***{ExprS}*** Saves the screen to a bitmap file on disk. If you do not specify *ExprS* the file name will default to"RobotBASICScreen.bmp". If you do specify *ExprS* do not include the extension, the system will automatically append ".bmp" to the string that results from *ExprS*.

`ReadScr` ***{ExprS}*** Restores the screen from a bitmap file on disk. If you do not specify *ExprS* the file name will default to "RobotBASICScreen.bmp". If you do specify *ExprS* do not include the extension, the system will automatically append ".bmp" to the string that results from *ExprS*. So if you have a file "Test.bmp" and you want to read it into the screen then say:

```
ReadScr "Test"
```

Notice the extension ".bmp" is not given. The interpreter will add this extension automatically, *do not include it*, if you do you will get an error.

`GetXY` ***Var1,Var2*** Reads the current pen position on the screen and sets *Var1*=x and *Var2*=y.

`GotoXY` ***ExprN1,ExprN2*** Sets the pen position to a point on the screen *ExprN1*=x, *ExprN2*=y.

`SetColor` ***ExprN1{,ExprN2}*** Sets the pen color to *ExprN1* and the background color to *ExprN2*.

`GetColor` ***Var1,Var2*** Sets *Var1*= pen color *Var2* = background color.

`ReadPixel` ***ExprN1,ExprN2,Var*** Reads the pixel color at the position x=ExprN1, y=ExprN2 and sets Var to that value (See color codes in Sec. B.7.6). (Also see FloodFill command below).

`SetPixel` ***ExprN1,ExprN2,ExprN3*** Sets the color of the pixel at *x=ExprN1*, *y=ExprN2* to the color *ExprN3*. (see color codes Sec. B.7.6) (Also see the FloodFill command below).

`LineWidth` ***ExprN*** Sets the pen width for drawing lines and other shapes.

`GetLineWidth` ***Var*** Sets *Var* to the current pen width for drawing lines and other shapes.

`LineTo` ***ExprN1,ExprN2{,ExprN3{,ExprN4}}*** Draws a line from the current pen position to *ExprN1*, *ExprN2*. If *ExprN3* is given then the pen width will be temporarily set to *ExprN3*. If *ExprN4* is given then pen color will be temporarily set to *ExprN4*.

ExprN3 and *ExprN4* will only affect the line drawn, not any subsequent lines or other drawings. If you desire to specify *ExprN4* you must also specify *ExprN3*.

`Line` ***ExprN1,ExprN2,ExprN3,ExprN4{,ExprN5{,ExprN6}}*** Draws a line from the point *ExprN1*,*ExprN2* to the point *ExprN3*,*ExprN4*. If *ExprN5* is given then the pen width will be temporarily set to *ExprN5*. If *ExprN6* is given then pen color will be temporarily set to *ExprN6*.

ExprN5 and *ExprN6* will only affect the line drawn, not any subsequent lines or other drawings. If you desire to specify *ExprN6* you must also specify *ExprN5*, but if *ExprN5* is less than 1 it will be ignored. Also if *ExprN6* is less than 0 it will be ignored. If you do not specify a width and color the current default width and color will be used.

`Rectangle` ***ExprN1,ExprN2,ExprN3,ExprN4{,ExprN5,ExprN6}*** Draws a rectangle defined by *x1=ExprN1*, *y1=ExprN2* and *x2=ExprN3*, *y2=ExprN4* filled with color given by *ExprN6* and bordered with color given by *ExprN5*. If *ExprN6* is not given then the default background color is used. If *ExprN5* is not given then the default pen color is used but if you want to specify *ExprN6* you must also give *ExprN5*. *x1, y1* are coordinates on the screen of the top-left corner, and *x2, y2* are of the bottom right corner of the rectangle.

`ERectangle` ***ExprN1,ExprN2,ExprN3,ExprN4{,ExprN5{,ExprN6}}*** The above command (`Rectangle`) will always draw the rectangle and fill it with the color specified or the default background color. This means the inside of the rectangle will be erased if there happens to be any previous drawings there. This command only draws the perimeter of the rectangle and does not fill the inside. Thus, any previous drawings inside the area of the rectangle will still be visible.

The rectangle is defined by X1=*ExprN1*,*ExprN2*=Y1 and X2=*ExprN3*,Y2=*ExprN4*. If *ExprN5* is given then the pen width will be temporarily set to *ExprN5*. If *ExprN6* is given then pen color will be temporarily set to *ExprN6*.

ExprN5 and *ExprN6* will only affect the current drawing, not any subsequent drawings. If you desire to specify *ExprN6* you must also specify *ExprN5*, but if *ExprN5* is less than 1 it will be ignored. Also, if *ExprN6* is less than 0 it will be ignored. For an example see the example given in the Pie command.

`Circle` ***ExprN1,ExprN2,ExprN3,ExprN4{,ExprN5,ExprN6}*** Draws a circle/ellipse inside a rectangle defined as above. If the rectangle is a square then it is a circle, otherwise it is an ellipse.

`Arc` ***ExprN1,ExprN2,ExprN3,ExprN4{,ExprN5{,ExprN6{,ExprN7{,ExprN8}}}}*** The above command (`Circle`) will always draw the circle/ellipse and fill it with the color specified or the default background color. This means the inside of the circle/ellipse will be erased if there happens to be any previous drawings there. This command only draws the perimeter of the circle/ellipse and does not fill the inside. Thus, any previous drawings inside the area of the circle/ellipse will still be visible. Additionally, this command allows you to draw a fraction of the arc of the circle or ellipse.

The circle/ellipse is defined by the bounding rectangle defined by X1=*ExprN1*, Y1=*ExprN2* and X2=*ExprN3*, Y2=*ExprN4*. If *ExprN7* is given then the pen width will be temporarily set to *ExprN7*. If *ExprN8* is given then pen color will be temporarily set to *ExprN8*.

ExprN5 and *ExprN6* are angles in radians that specify the start point of the arc and the length of the arc. The angles are defined counter-clock-wise from the right hand horizontal position (i.e. positive x-axis). If *ExprN6* is zero or an even multiple of Pi() (i.e. 0,360,720... degrees) the entire circle/ellipse will be drawn. *ExprN5* and 6 default to 0 if not given. *ExprN5* defines the angle (in radians) from the positive x-axis (counterclockwise) at which to start drawing the arc. *ExprN6* defines the length of the arc (angle inside the arc in radians). To specify angles in degrees use the function `DtoR()`.

ExprN7 and *ExprN8* will only affect the current drawing, not any subsequent drawings. If you desire to specify *ExprN8* you must also specify *ExprN5*,6,7, but if *ExprN7* is less than 1 it will be ignored. Also if *ExprN8* is less than 0 it will be ignored. For an example see the example given in the Pie command.

`Pie` ***ExprN1,ExprN2,ExprN3,ExprN4{,ExprN5{,ExprN6{,ExprN7{,ExprN8}}}}***
This command is very similar to the command Arc above. The difference is that radials from the center of the bounding rectangle to the start and end points of the arc will also be drawn. Also the inside will be filled with the default background color or the color *ExprN8* if given. This command allows for creating Pie graphs.

The circle/ellipse is defined by the bounding rectangle defined by X1=*ExprN1*, Y1=*ExprN2* and X2=*ExprN3*, Y2=*ExprN4*. If *ExprN7* is given then the pen color will be temporarily set to *ExprN7*. If *ExprN8* is given the pie is filled with the color *ExprN8*.

ExprN5 and *ExprN6* are angles in radians that specify the start point of the arc and the length of the arc. The angles are defined counter-clock-wise from the right hand horizontal position (i.e. positive x-axis). If *ExprN6* is zero or an even multiple of Pi() (i.e. 0,360,720... degrees) the entire circle/ellipse will be drawn. *ExprN5* and 6 default to 0 if not given. *ExprN5* defines the angle (in radians) from the positive x-axis (counter-clockwise) at which to start drawing the arc. ExprN6 defines the length of the arc (angle inside the arc in radians). To specify angles in degrees use the function `DtoR()`.

ExprN7 and *ExprN8* will only affect the current drawing, not any subsequent drawings. If you desire to specify *ExprN8* you must also specify *ExprN5*,6,7, but if *ExprN7* is less than 0 it will be ignored. Also if *ExprN8* is less than 0 it will be ignored.

Example:

```
n = 360/16
for i=0 to 15
  Pie 100,100,500,500,DtoR(i*n),DtoR(n),i,i
next
Arc 100,100,500,500,0,0,3,blue
ERectangle 100,100,500,500,5,lightgreen
```

`FloodFill` ***ExprN1,ExprN2{,ExprN3{,ExprN4}}*** Given a coordinate *ExprN1*, *ExprN2* the interpreter will start filling the area surrounding this coordinate with old color given by *ExprN4* and replacing it with new color given by *ExprN3*. It will do so as long as the pixels have the old color, but it will not convert any pixels with a different color from old color. So if you have a box that has color white surrounded by color blue, doing `FloodFill` *X,Y,Red,White* will fill the box with the new color red, but only the box since it will not flow into the blue areas. (See `ReadPixel` and `SetPixel` commands above.)

ExprN3 and *ExprN4* are optional. If *ExprN4* is not given then the color of the pixel at *ExprN1*, *ExprN2* position will be used as the old color. If *ExprN3* is not given then the current pen color will be used as the new color.

`DrawShape` ***ExprS,ExprN2,ExprN3{,ExprN4,ExprN5}*** This command will draw an image specified by the string in *ExprS*. *ExprN2* and *ExprN3* are screen position *x, y*. Optional *ExprN4* and *ExprN5* are to specify a scale factor (pixels) and color correspondingly. If scale is not given then 1 is assumed and if color is not given the default pen color is used. If you are to specify the color you must specify the scale. If color is a negative number then the color will be the background color. This is handy in appearing to erase the image. The data in *ExprS* indicates how to move (`LineTo`) from the *x, y* position (see Sec. C.7.6 for more details on the `DrawShape` command).

C.7.3 ARRAY COMMANDS

`Dim` ***VarA[ExprN{,ExprN...}]*** This will specify that *VarA* is an array of the dimensions *[ExprN{,ExprN...}]* the brackets are required and so is the comma between each dimension. You can use an array element anywhere as any variable (see Sec. B.7.4). The `Dim` statement establishes the maximum value for each dimension and the over all dimension of the array. The index of the dimension starts at 0 and ends at *ExprN-1*. *ExprN* must result in an integer value, otherwise an error will occur. Each element of the array can be any of the data types (string, float, or integer).

If you try to access any unassigned element an error will occur. If you try to access outside the specified range in the `Dim` statement an error will occur.

If you have a two-dimensional array `Dim` *N*[4,5] then there are 4 rows and 5 columns, that is, there are 4 rows with 5 elements in each row. If you have `Dim` *N*[6,7,8] then there are 6 rows where each element in the row constitutes a matrix in itself, where those matrices each have 7 rows and 8 columns.

This can go on for as long as you care, but remember that the row count is the first dimension and the second dimension is a count of elements in each row. Each element in the row can be a string, integer, float, or another matrix as described above.

```
Dim XY[2,4,5]
//means that XY is an array of 3D and that the
//first dimension goes from 0 to 1
//second dimension goes from 0 to 3
//third dimension goes from 0 to 4

//assigns element [1,3,2] as given
XY[1,3,2] = sin(pi(2.0/3))+4
print XY[1,3,2]  //prints element [1,3,2]
XY[2,4,5] = 9  //will cause an error outside the range

Print XY[0,3,1] //will cause an error because [0,3,1]
                // was not assigned
```

`Data` ***VarA;Expr{,Expr....}*** This command creates an array of one dimension named *VarA* and puts all the resulting values from the *Expr*'s into the array. *VarA* is the array name, the colon after the name is required to separate the name from the data. The data is separated by a comma. *Expr* can result in any value type (string, integer, or float). If you specify the same array name in two or more `Data` statements then the data is appended to the end of the array.

If you desire to erase the array, so as to start populating it at the first element, use `Dim` *VarA*[0] where *VarA* is the array name used in the `Data` statement. If you have previously dimensioned *VarA* before issuing the `Data` statements, then you must use the `Dim` *VarA*[0] before you use the Data statements. If you do not the data will not be loaded into the array.

```
data test;1,2,"some text",4.5,sin(3/4.6),length("test")
data test;66,44.5,log(3)
print test[7]           //print 44.5
print test[2]           //prints 'some text'
print test[0]+test[3]   //prints 5.5
dim test[0]    //effectively erases the array data
```

```
data test;"again",3 //repopulate the array
print test[0]       //prints 'again'
print test[7] //gives an error since there are not 7
              //elements
```

`MCopy` ***VarA1,VarA2*** *VarA1* must be an existing previously dimensioned array, or an array created by the `Data` command above. *VarA2* does not have to exist. If it does not exist then it will be created by the command and will be an exact copy of *VarA1*, including the dimension, and dimensional constraints. If it does exist and has been dimensioned, then the data in *VarA1* will be copied into *VarA2* row wise. That is each data element from *VarA1* will be copied into elements of *VarA2* until all the elements in the first row are filled, then it starts with the elements in the next row and so on until *VarA2* is filled or *VarA1* runs out of elements.

This command is useful for looking at the data of one array in different row-column dimensions. For example, the `Data` command can only create one-dimensional arrays with all the data in one row. But if you want to load the data into a two-dimensional array then dimension an array according to the desired dimensions and copy the array created by the `Data` command into it.

```
Data a;1,2,3,4,5,6,7,8,9
Print a[5]  //will print 6
Dim b[3,3]
MCopy  a,b
Print b[1,2]  //will print 6
```

`MWrite` ***VarA,ExprS*** Will write the contents of the array to the file specified in *ExprS*. When specifying the file name you can use directory structures, for example, "C:\RobotBASIC\Programs\MySimulation.sim".

The directory must exist. The file does not have to exist, but if it does it will be overwritten. Any error in writing to the file will cause an error to issue. The *VarA* array must be a valid previously dimensioned array or an array created with the `Data` command.

`MRead` ***VarA,ExprS*** This will create a new array *VarA* (if it already exists it will be erased first) and will populate its elements with the data from the file *ExprS*. The file must exist and it must be of the format written previously with the command `MWrite` (see above). The array will be dimensioned the same as the matrix that was used to write the file in the first place. You can find out the dimensions of *VarA* with the functions `MDim`(*VarA*) and `MaxDim`(*VarA*,*ExpN*) (see Sec. C.8). You will get an error if the file does not exist or if it is the wrong format. If the file is the correct format but there was an error in reading it an error will be issued.

`MPolygon` ***VarA{,ExprN}*** This command is used to draw multiple polygons on the screen with one command rather than use looping and the commands `LineTo`, `GotoXY`, and/or `Rectangle`. These polygons can be filled with a specified color or the default pen color. *VarA* must be a one-dimensional array created with `Dim` or `Data` commands. It must contain only numbers. If any element in the array contains non-numbers it will ignored and will not affect the pairing of elements, it would be as if it did not exist.

ExprN is optional. *ExprN* must result in a number. If it is not an integer it will be made into one by rounding. The data in *VarA* is a set of paired *x, y* coordinates. The command will execute a `LineTo` *x,y* or `GotoXY` *x,y* or `FloodFill` *x,y{,ExprN}* depending on the following logic:

- If the *x* and *y* value-pair are both positive then a `LineTo` *x,y* is executed.
- If the *x* is negative and *y* is positive then a `GotoXY` *−x,y* is executed.
- If the *y* is negative and *x* is positive then a `FlooFill` *x,−y,ExprN* is executed. If *ExprN* is not given then `FloodFill` *x,−y* is executed.(See the `FloodFill` command above.)
- If both *x* and *y* are negative they are both made positive and `LineTo` is executed.

If there are not enough pairs then drawing will occur only to the last pair. The array can contain any number of point pairs. The plotting will occur up to the last pair of points in the array. For example:

```
data p;-100,100,200,100,200,200,100,200,100,100,120,-120
data p;-500,100,600,100,600,200,500,200,500,100,520,-120
//---- you can also use MRead to read data from a file
MPolygon p //will draw the above and use default pen
           //color to do any filling
MPolygon p,blue //will draw the above and use blue color
                //to do any filling
```

C.7.4 ARRAY MATH COMMANDS

`MScale` ***VarA,ExprN*** This will multiply each element in the array *VarA* by the result of the numeric expression *ExprN*. *VarA* must exist. Also, all the elements must be numeric.

`MConstant` ***VarA,Expr*** his will fill each element in the array VarA with the result of the expression Expr. VarA must exist. Expr can be numeric or string.

`MDiagonal` ***VarA,Expr*** This will fill all the diagonal elements in the array *VarA* with the result of the expression *Expr*. *VarA* must exist. *Expr* can be numeric or string. If *Expr* is numeric then all the other elements will be zero. If *Expr* is a string then all the other elements will be blank. *VarA* must be two-dimensional.

`MAdd` ***VarA1,VarA2*** This will add elements of *VarA1* to *VarA2*. This is equivalent to saying *VarA2* = *VarA2+VarA1*. *VarA1* and *VarA2* must exist and must be of the same dimension with the same dimensional constraints. If any element is a string and the corresponding element to be added is a numeric then the result will be a string concatenation of the string with the numeric converted to a string. If the elements to be added are both strings then the result is a concatenation. Remember it is *VarA2+VarA1*. If one element is an integer while the other is a float then the result is a float.

`MSub` ***VarA1,VarA2*** This will subtract the elements of *VarA1* from *VarA2*. This is equivalent to saying *VarA2* = *VarA2*−*VarA1*. *VarA1* and *VarA2* must exist and must be of the same dimension and of the same dimensional constraints. If any element is a string then no operation will take place. If one element is an integer while the other is a float then the result is a float.

`MMultiply` ***VarA1,VarA2,VarA3*** This will multiply *VarA1* by *VarA2* and put the result in *VarA3*. This is equivalent to saying *VarA3* = *VarA1* × *VarA2*. **The order is important *VarA1* × *VarA2* is not equal to *VarA2* × *VarA1*.** *VarA3* does not have to exist, but if it does it will be erased and recreated. Both *VarA1* and *VarA2* must be two-dimensional. The number of Columns of *VarA1* must be the same as the number of rows of *VarA2*. That is if *VarA1* has the dimension [*R1,C1*] and *VarA2* has the dimension [*R2,C2*] then multiplication is possible only if *C1*=*R2*. The resulting array *VarA3* will have the dimension [*R1,C2*]. Also, the elements of *VarA1* and *VarA2* must be numeric.

`MInvert` ***VarA1,VarA2,Var*** This will calculate the inverse of array *VarA1* and assign it to *VarA2*, and also the determinant of *VarA1* will be assigned to *Var*. This is equivalent to saying *VarA2* = inverse(*VarA1*) and *Var* = det(*VarA1*). *VarA2* does not have to exist, but if it does it will be erased and recreated. *VarA1* must be two-dimensional and a square matrix. That is, the number of rows must equal the number of columns. All elements in *VarA1* must be numeric. If *VarA1* is not invertible then *Var* = 0 and elements of *VarA2* will be all zeros.

`MDeterminant` ***VarA1,Var*** This will calculate the determinant of array *VarA1* and put it in *Var*. This is equivalent to saying *Var* = det(*VarA1*). *VarA1* must be two-dimensional and a square matrix. That is, the number of rows must equal the number of columns. All elements in *VarA1* must be numeric.

`MTranspose` ***VarA1,VarA2*** This will transpose *VarA1* and put the result in *VarA2*. *VarA2* does not have to exist, but if it does it will be erased and recreated. *VarA1* must be two-dimensional. *VarA2* is the transpose of *VarA1*, that is, *VarA2*[*i,j*] = *VarA1*[*j,i*].

`MRegression` ***VarA,Var1,Var2*** This will perform a regression analysis (line fit) on the data in *VarA*. *VarA* must be two-dimensional and all data must be numeric. The first row contains the *x* values, and the second row contains the *y* values. There must be a *y* value for each *x* value. The line formula is $y=mx+b$ where *m* is the slope and *b* the *y*-axis intercept. *Var1* will be assigned the slope. That is, *Var1* = *m*. *Var2* will be assigned the intercept. That is, *Var2* = *b*.

`MSort` ***VarA1{,ExprN}*** This will sort the array *VarA*. *VarA* can be one- or two-dimensional. If *VarA* is two-dimensional then sorting will be done on the data in the row specified by *ExprN*. If *ExprN* is not specified then it will be done on the first row (row 0). (Remember row and column numbering start with 0.) The data in the other rows will be moved around to maintain the same association between the rows. That is, the columns are moved to fit in the correct sort order depending on the value in the row specified in

ExprN (or fisrt row if *ExprN* is not given). The elements in a row must be the same data type, but the elements in a column can be of different data types.

This can be helpful in creating databases, where the elements in a column are the different fields of the database and each column is a record.

```
data a;"Sam","Ted","Pam","Tom"  //names
data a;45   ,55   ,20   ,10      //ages
data a;30045,30067,30045,20022   //zip codes
dim b[3,4]    //make a 2-d array to hold data
mcopy a,b     //copy the data into it
msort b       //sort data by name
//a subroutine to display the data in a good format
gosub print_data
msort b,2     //sort data by zipcode
gosub print_data
msort b,1     //sort data by age
gosub print_data
msort b,0  //sort data by name again
gosub print_data
```

C.7.5 OTHER COMMANDS

`DebugOn`

`DebugOff`

`Debug` ***{Expr1,Expr2;Expr3...}*** Outputs the values of *Expr...* to the Debugging Screen. This only happens if the `DebugOn` command has been issued any prior time, or the *Debug On* button has been pressed on the Terminal Screen. The `DebugOff` command will turn debugging off again.

The Debug Screen will pause program execution and display the result of *Expr...* and wait for you to press the *Step* button to execute the rest of the program.

If `Debug{}` is in a loop it will be executed every time the command line is encountered. To stop any further execution use the *Debug Off* button on the Debug Screen. This is the same as issuing a `DebugOff` command, or you can close the Debug window using the close window icon on the top-right corner. This is the same as pressing the *Debug Off* button.

To remain in debug mode press the *Step* button. To eliminate any further program pausing and debugging press the *Debug Off* button, close the debug window or issue a `DebugOff` command within the program code. To turn on further debugging you must issue a `DebugOn` command again within the program flow, or press the *Debug On* button on the bottom-right side of the Terminal Screen. This button turns debugging on at any time during the program flow as if a `DebugOn` command was executed. Subsequently any `Debug` command will be executed. This can be useful to turn the debugging on at a certain stage in the program rather than having to step through until you get to a point of interest.

If you keep the `Debug{}` lines in the program but do not wish them executed next time you run the program, make sure that the command `DebugOff` is issued before any `Debug{}` is issued or that any prior `DebugOn` commands are commented out.

{Expr1,Expr2;Expr3...} are printed as described in the Print command above.

This combination of commands can help you step the program and view the values of variables while doing so. The *Clear* button on the Debug Screen clears any previous printed debug data. Every time you run the program from the start the Debug Screen is cleared.

You can swap windows back and forth between the Editor/Terminal/Debug windows.

`Delay` ***{ExprN}*** Causes a delay in milliseconds that is, 1000 = 1 second. If *ExprN* is not given 1000 will be assumed that is, 1 second.

`MicroDelay` ***{ExprN}*** Causes a delay in steps of 15 microseconds. So, for example, to get a delay of 1 second you would issue the statement `MicroDelay` (1.0e6)/15. Notice that you can use a floating-point number. Since the ticks are in intervals of 15 microseconds a value of 2 will make a delay of 30 microseconds. *ExprN* is optional and if it is not given or is less than 1 then it is assumed to be 1. This means that the minimum delay is 15 microseconds.

`Swap` ***Var1,Var2*** Swaps the values of *Var1* and *Var2* the variables can be of different types.

`GetError` ***VarN1{,VarS{,VarN2{,VarN3}}}*** Will fill the variable *VarN* with the last error number (-1 if no error), *VarS* will be filled with the description of the error (blank if no error), *VarN2* will be assigned the line number where the error occurred (−1 if no error) and *VarN3* will be assigned the character number in the line where the error occurred (−1 if no error). Issuing this command will retrieve the details of the last error to have occurred and then will clear the data. If an error occurs and you do not issue this command then another error occurs, issuing this command will get the details of the ***last*** error, the previous details will not be retrievable. If an error occurs and you have not issued an `OnError` *Label* (see Sec. C.6.10) statement then the error will halt the program and you won't be able to use this command.

C.7.6 DRAWSHAPE DETAILS

`DrawShape` ***ExprS,ExprN2,ExprN3{,ExprN4,ExprN5}***

- This command will draw an image specified by the string in *ExprS*.
- *ExprN2* and *ExprN3* are screen position *x*, *y*. Optional *ExprN4* and *ExprN5* are to specify a scale factor (pixels) and color, correspondingly.
- If scale is not given then 1 is assumed and if color is not given the default pen color is used. If you are to specify the color you must specify the scale.
- If color is a negative number then the color will be the background color. This is handy in appearing to erase the image.
- The data in *ExprS* indicates how to move (LineTo) from the *x*, *y* position. *ExprS* contains a list of letters "UDLRQAWS" or "udlrqaws" and "−0123456789". Where U=up, D=down, L=left, R=right, Q=diagonal to left and up, W=diagonal to right and up, A=diagonal to left and down, and S=diagonal to right and down. If lowercase letters are used then drawing and moving will take place. If uppercase letters are used then only moving will take place.
- Starting at the coordinate (*ExprN2*,*ExprN3*) the interpreter will *draw* a line to the next pixel (or *ExprN4* pixels) up/down, and so on if the letter used is lower case. If it is upper case it will *move* to the position instead of drawing a line, and the next letter will cause drawing or moving from that new position.

- The pen color will be according to *ExprN5* or as specified in the next paragraph. If a number, for example, 4 or 12 is given, it will be taken to indicate a change of color from the specified color in *ExprN5*. The default pen color is used if no *ExprN* is specified. The change will take effect until either another number is specified or -. If - is specified then the color will revert back to *ExprN5* or default color. This action will only take place if *ExprN5* is not negative (see above).
- When you specify numbers refer to the color codes in Sec. B.7.6. If you specify a number greater than the last color it will revert to the last color. Any other characters will be ignored. This command can be emulated with a combination of `GotoXY` and `LineTo` combined with `SetColor`.

C.8 Functions

See Sec. B.7.7 for details on how functions fit within the RoborBASIC language. Here the functions are grouped by functionality. For an alphabetic list see the end of this Sec. C.11.

Functions are not case sensitive. So `sin`(*Theta*), `SIN`(*Theta*), and `sIn`(*Theta*) are all the same function.

C.8.1 TRIGONOMETRIC FUNCTIONS

`Pi`({*ExprN*}) Returns the value of pie (p) (i.e., 3.141592654) multiplied by the result of *ExprN*. If *ExprN* is not given then it is assumed to be 1.

`Sin`(*ExprN*) Returns the sine of an angle. ExprN is the value of the angle in radians. If you want to specify degrees then use the conversion Sin(ExprN*Pi()/180).

`Cos`(*ExprN*) Returns the cosine of an angle. *ExprN* is the value of the angle in radians. If you want to specify degrees then use the conversion `Cos`(*ExprN**Pi()/180).

`Tan`(*ExprN*) Returns the tangent of an angle. *ExprN* is the value of the angle in radians. If you want to specify degrees then use the conversion Tan(*ExprN**`Pi()`/180). This function can cause an error if the angle is +/−`Pi()`/2 (i.e., +/−90°) since the result is infinity. If the angle is slightly more or less than 90° the result is valid but is an extremely large number.

`ASin`(*ExprN*) Returns the angle in radians whose sine is *ExprN*. If you want to get degrees then use the conversion 180*`ASin`(*ExprN*)/`Pi()`. This is the inverse of `Sin()`.

`ACos`(*ExprN*) Returns the angle in radians whose cosine is *ExprN*. If you want to get degrees then use the conversion 180*`ACos`(*ExprN*)/`Pi()`. This is the inverse of `Cos()`.

`ATan`(*ExprN*) Returns the angle in radians whose tangent is *ExprN*. If you want to get degrees then use the conversion 180*`ATan`(ExprN)/`Pi()`. This is the inverse of `Tan()`.

`ATan2`(*ExprN1,ExprN2*) Returns an angle in radians, given the *x* and *y* lengths. *ExprN1* = *x*, *ExprN2* = *y*. This gives the angle 0 to p and 0 to –pπ. Negative angles are clockwise from the x axis, and positive angles are counter clockwise from the x axis. So `ATan2`(1,1) will give 0.785398 which is 45°, while `ATan2`(1, –1) gives –0.78539816 which is –45°. If *x*=0 and *y*=0 the result will be 0.0.

C.8.2 CARTESIAN TO POLAR FUNCTIONS

`PolarR`(*ExprN1,ExprN2*) Returns the polar radius from the *x*, *y* coordinates (*x*=*ExprN1*, *y*=*ExprN2*) both must result in numbers (float or integer), otherwise an error will occur.

`PolarA`(*ExprN1,ExprN2*) Returns the polar angle from the *x*, *y* coordinates. This is effectively the same as `Atan2()` above. The angle returned is in radians. *ExprN1* = *x*, *ExprN2* = *y*. Both *ExprN1* and *ExpN2* must be numeric (integer, or float) or an error will occur.

C.8.3 POLAR TO CARTESIAN FUNCTIONS

`CartX`(*ExprN1,ExprN2*) This returns the Cartesian x coordinate given the polar *R*, *Theta*. *ExprN1*=*R*, *ExprN2*=*Theta* (in radians). Both *ExprN1* and *ExpN2* must be numeric (integer, or float) or an error will occur. This is effectively the inverse of the `PolarR()` and `PolarA()` above.

`CartY`(*ExprN1,ExprN2*) This returns the Cartesian *y* coordinate given the polar *R*, *Theta*. *ExprN1*=*R*, *ExprN2*=*Theta* (in radians). Both *ExprN1* and *ExpN2* must be numeric (integer, or float) or an error will occur. This is effectively the inverse of the `PolarR()` and `PolarA()` above.

C.8.4 LOGARITHMIC AND EXPONENTIAL FUNCTIONS

`NLog`(*ExprN*) Returns the log to base *e* (***e***=2.178281828) of *ExprN*. If *ExprN* is 0 or negative an error will occur.

`Log`(*ExprN*) Returns the log to base 10 of *ExprN*. If *ExprN* is 0 or negative an error will occur.

`Exp`(*ExprN*) Retruns *e* raised to the power *ExprN* (***e***=2.178281828).

`Exp10`(*ExprN*) Returns 10 raised to the power *ExprN*.

`SqRt`(ExprN) Returns the square root of *ExprN*. If *ExprN* is negative then an error will occur.

`CbRt`(*ExprN*) Returns the cube root of *ExprN*.

C.8.5 SIGN CONVERSION FUNCTIONS

Abs(*ExprN*) Returns the absolute value of *ExprN*. Returns *ExprN* as a positive number.

Sign(*ExprN*) Returns a −1 if *ExprN* is negative, 1 if positive, and 0 if zero.

C.8.6 FLOAT TO INTEGER CONVERSION FUNCTIONS

Round(*ExprN*) Rounds the float *ExprN* to an integer Round(3.4) => 3; Round(3.5) => 4; Round(3.6)=> 4.

RoundUp(*ExprN*) Rounds the float ExprN to an integer RoundUp(3.4) => 4; RoundUp(3.5) => 4; RoundUp(3.6) => 4.

RoundDn(*ExprN*) Rounds the float *ExprN* to an integer RoundDn(3.4) => 3; RoundDn(3.5) => 3; RoundDn(3.6) => 3.

Frac(*ExprN*) Returns the decimal fraction of the float *ExprN*. Frac(12.3456) => 0.3456.

Mod(*ExprN1,ExprN2*) Returns the remainder of dividing *ExprN1* by *ExprN2*. If *ExprN1* or *ExprN2* is not an integer it will be truncated [RoundDn()]. Mod(9,4) => 1.

C.8.7 NUMBER AND STRING CONVERSION FUNCTIONS

HexToInt(*Expr*) Returns the resulting integer number that is the equivalent to the hexadecimal value given as a string in *ExprS*. If *ExprS* is not a valid hexadecimal string or integer number then 0 will be returned. Examples:

```
print hextoint(10)       //prints 10
print hextoint("10")     //prints 10
print hextoint("0x10")   //prints 16
print hextoint("tt")     //prints 0
```

Hex(*ExprN*) Returns a string that contains the hexadecimal representation of the integer resulting from *ExprN*. If *ExprN* is not numeric an error is issued. If *ExprN* is not integer it will be converted to integer.

Bin(*ExprN*) Returns a string that contains the binary representation of the integer resulting from *ExprN*. If *ExprN* is not numeric an error is issued. If *ExprN* is not integer it will be converted to integer. The number of bits will depend on the value of *ExprN*. There will be no leading zeros. So Bin(7) will return 111 while Bin(3) will return 11 and Bin(4) gives 100. If the number must be a fixed number of bits with leading zeros then use the sRepeat() and Length() functions below. For example, to return a 10-bit binary of a number with leading zeros do sRepeat("0",10-length(bin(5))+bin(5) => 0000000101.

`IsNumber`(*Expr*) If *Expr* is a string that can be converted to a numeric or is a numeric returns 1, otherwise returns 0.

`IsString`(*Expr*) If *Expr* is a string (regardless of whether it can be converted to a number or not) returns 1, otherwise returns 0.

`ToNumber`(*Expr*) Returns the resulting string expression as a number if possible, otherwise it will return the string unconverted. If *Expr* results in a number then that number is returned.

`ToString`(*Expr*) If *Expr* results in a number it converts the number to a string and returns the value as a string. If *Expr* results in a string then the string is returned.

`Ascii`(*ExprS*) Returns the ASCII code of the first character in *ExprS*.

`Char`(*ExprN*) Returns a string of one character which is the character whose ASCII code is *ExprN*.

`Format`(*ExprN,ExprS*) Returns a string containing *ExprN* formatted according to *ExprS*. (See Sec. C.8.14)

C.8.8 STRING MANIPULATION FUNCTIONS

`Length`(*ExprS*) Returns the length of the resulting string from *ExprS*.

`Trim`(*ExprS*) Returns *ExprS* without leading or trailing spaces.

`LeftTrim`(*ExprS*) Returns *ExprS* without leading space.

`RightTrim`(*ExprS*) Returns *ExprS* without trailing spaces.

`NoSpaces`(*Exprs*) Returns *ExprS* without any spaces, even within the string not just leading and trailing spaces.

`Substring`(ExprS,ExprN1,ExprN2) Returns a string consisting of the characters from *ExprS* starting at character *ExprN1* up to and including the *ExprN2* character. If *ExprN1*<1 then first character is assumed. If *ExprN*> the length of the string then the last character is assumed. If *ExprN*<1 or *ExprN2*> the length of the string then the last character is assumed.

`Left`(*ExprS,ExprN*) Returns a string containing *ExprN* characters from *ExprS* starting from first character. This is equivalent to `Substring`(*ExprS,1,ExprN*). If *ExprN*< 1 then 1 is assumed. If *ExprN* > the length of the string then the length of the string is assumed.

`Right`(*ExprS,ExprN*) Returns a string containing *ExprN* characters from *ExprS* ending with the last character. This is equivalent to `Substring`(*ExprS*, `Length`(*ExprS*) + 1 – *ExprN*, *ExprN*). If *ExprN* < 1 then 1 is assumed. If *ExprN* > the length of the string then the length of the string is assumed.

`Extract`(ExprS1,ExprS2,ExprN) This will return a string that is the *ExprN*th part of *ExprS1* separated by the separator *ExprS2*. *ExprS1* is a string with data separated by characters specified in *ExprS2*. The `Extract()` function will return the data part that is the *ExprN*th part. If *ExprN* is greater than the number of parts `Extract()` will return the last part. If *ExprS1* does not contain the separator characters specified in *ExprS2* then *ExprS1* will be returned. If *ExprN* is not a number an error will be given. If it is not an integer it will be made into an integer. If it is less than 1 then it is converted to 1.

```
a = "test,5,2.4"
b = ","
print Extract(a,b,2)      //will print 5
print Extract(a,b,3)      //will print 2.4
print Extract(a,b,6)      //will print 2.4
b = ";"
print Extract(a,b,1)      //will print test,5,2.4
```

`InString`(*ExprS1,ExprS2*) Returns the position of the first occurrence of *ExprS2* within *ExprS1* if *ExprS2* does not occur inside *ExprS1* then 0 is returned. InString("testing","ting") => 4.

`Upper`(*ExprS*) Returns *ExprS* with all characters converted to upper case.

`Lower`(*ExprS*) Returns *ExprS* with all characters converted to lower case.

`Spaces`(*ExprN*) Returns a string of space characters *ExprN* long. If *ExprN* <0 then 0 is assumed.

`SRepeat`(*ExprS,ExprN*) Returns a string with *ExprS* repeated *ExprN* times. If *ExprN* <0 then 0 is assumed.

C.8.9 TIME AND DATE FUNCTIONS

`Time`(*ExprN*) Returns a string that has the time in the format "hh:mm:ss" (24 hour format) unless *ExprN=0*, then it returns "AM hh:mm:ss" or "PM hh:mm:ss" (12 hour format).

`Date`(*ExprN*) Returns a string "yyyy/mm/dd" unless *ExprN*=0 then returns "yyyy/mm/dd Day:Month" where Day is the day name and Month is the month name.

`Timer()` Returns a floating-point number that represents the time in milliseconds. You can save this value and then after the elapse of some time subtract the stored value from the new value to get the amount of time elapsed in millisecond (i.e., 1000 = 1 second). For Example:

```
StartTime = Timer()
while Timer() - StartTime < 5000
  // do stuff here that can be accomplished within 5 seconds
wend
print "done"
```

C.8.10 PROBABILITY FUNCTIONS

`Random(`*ExprN*`)` Returns a value between 0 and *ExprN* – 1 randomly. If *ExprN* is not an integer it will be rounded down.

`Factorial(`*ExprN*`)` Returns the factorial of *ExprN*. *ExprN* must be a number. If it is a float it will be converted to an integer. *ExprN* cannot be negative. Mathematically `Factorial`(*n*) = *n*!.

`nPr(`*ExprN1*,*ExprN2*`)` Returns the permutation of *ExprN1* and *ExprN2*. Both must be numeric and if not integer will be converted to integer. Neither can be negative. Also *ExprN2* must be less than or equal to *ExprN1*. The formula is `nPr`(*n*,*r*) = *n*!/(*n* – *r*)!.

`nCr(`*ExprN1*,*ExprN2*`)` Returns the combination of *ExprN1* and *ExprN2*. Both must be numeric and if not integer, will be converted to integer. Neither can be negative. Also *ExprN2* must be less than or equal to *ExprN1*. The formula is `nCr`(*n*,*r*) = *n*! /(*r*! *(*n* – *r*)!).

C.8.11 STATISTICAL FUNCTIONS

`Sum(`*VarA*`)` Returns the sum of all the elements of the array *VarA*. *VarA* must be one- or two-dimensional and contain only numerical data. If *VarA* is two-dimensional then the second row must contain the frequencies for the data in the first row. If there is no corresponding frequency for any data value in the first row then it will be considered 1.

`Average(`*VarA*`)` Returns the average of the elements of the array *VarA*. *VarA* must be one-or two-dimensional and contain only numeric data. If *VarA* is two-dimensional then the second row must contain the frequencies for the data in the first row. If there is no corresponding frequency for any data value in the first row then it will be considered 1.

`Median(`*VarA*`)` Returns the median value of the elements of the array *VarA*. *VarA* must be one-dimensional and contain only numeric data.

`Max(`*VarA*`)` Returns the largest element of the array *VarA*. *VarA* must be one- or two-dimensional and contain only numeric data. If *VarA* is two-dimensional then the second row will be ignored, as it has no bearing on the determination of the maximum value.

`Min(`*VarA*`)` Returns the smallest element of the array *VarA*. *VarA* must be one- or two-dimensional and contain only numeric data. If *VarA* is two-dimensional then the second row will be ignored, as it has no bearing on the determination of the minimum value.

`Range(`*VarA*`)` Returns the difference between the maximum and minimum elements of the array *VarA*. *VarA* must be one- or two-dimensional and contain only numeric data. If *VarA* is two-dimensional then the second row will be ignored, as it has no bearing on the determination of the range value.

`Count(`*VarA*`)` Returns the number of elements in the array *VarA*. *VarA* must be one- or two-dimensional and contain only numeric data. If *VarA* is two-dimensional then the first row will be ignored, as it has no bearing on the determination of the count. The count will be the sum of all the frequencies in the second row. If there are elements in the first row without a corresponding frequency then the frequency will be considered to be 1.

`Variance(`*VarA*`)` Returns the variance of the elements of the array *VarA*. *VarA* must be one- or two-dimensional and contain only numeric data. If *VarA* is two-dimensional then the second row must contain the frequencies for the data in the first row. If there is no corresponding frequency for any element in the first row then it is considered 1.

`StdDev(`*VarA*`)` Returns the standard deviation of the elements of the array *VarA*. *VarA* must be one- or two-dimensional and contain only numeric data. If *VarA* is two-dimensional then the second row must contain the frequencies of the elements in the first row. If there is no corresponding frequency for any element in the first row then it is considered 1.

`CorrCoef(`*VarA*`)` Returns the value of the correlation coefficient for the data in array *VarA*. *VarA* must be two-dimensional. The first row contains the *X* values, and the second row contains the *Y* values. There must be a corresponding *Y* value for each *X* value and all data must be numerical.

C.8.12 ARRAY FUNCTIONS

`MDim(`*VarA*`)` Returns the dimension of the array *VarA*. If *VarA* does not exist an error will occur. The dimension of the array starts at 1. For example if an array *MyData* has been created with the statement `Dim` *MyData*[4,5,6] then `MDim(`*MyData*`)` = 3.

This command can be useful if you read an array from a file with the `MRead` command (see above). The array read from a file will have the dimensions of the array that was used to write the file. It may be unknown to you as a programmer that this function and the `MaxDim` function can be used to determine the details of the matrix.

`MaxDim(`*VarA,ExprN*`)` Returns the limit of the *ExprN*th dimension. *ExprN* must be an integer.

This command can be useful if you read an array from a file with the `MRead` command (see the commands section). The array read from a file will have the dimensions of the array that was used to write the file. It may be unknown to you as a programmer and thus this function can be used with the `MDim` function to find out the details of the matrix. The dimension constraint has a minimum value of 1.

```
dim MyData[4,5,9]
gosub populate_mydata
mwrite MyData,"Test.Txt"
//later on in any program you read the data from a file
Mread NewData,"test.Txt"
print MDim(NewData)        //=== prints 3
for i = 1 to MDim(NewData)
   print MaxDim(NewData,i)  // will print 4 then 5 then 9
next
//===print out all the data elements
for i = 0 to MaxDim(NewData,1)-1
   For j = 0 to MaxDim(NewData,2)-1
      for k=0 to MaxDim(NewData,3)-1
          print NewData[i,j,k]
      next
   next
next
```

`MType`*(VarA[ExprN{,ExprN2,...}])* This function is used to find the type of an array element. You specify the array element in the normal way as described in the description for `Dim` or Data in the commands section. The returned values are:

102 = Ascii("f") → floating-point number
105 = Ascii("i") → integer number
115 = Ascii("s") → string
0 → no value, that is, has not been assigned a value

The values above are also defined in the constants section (Sec. B.7.6).

Float = floating-point number
Integer = integer number
String = string
Novalue = not a defined element

This function can be useful in iterating through arrays where you may need to determine if an element is a valid element before using it. If you try to use an invalid array element an error will occur. This function can be used to prevent this.

`MsgBox`(*VarA*) Shows a dialog box with text as specified in the one-dimensional array *VarA*. The dialog box has two buttons (*OK* and *Cancel*). The user can terminate the box by pressing either button. The value returned by the function is 1 (true) if the *OK* button (or *ENTER*) was used to close the box, or 0 (false) if the *Cancel* button (or *ESC*) was used. The dialog box can also be closed using the Windows methods and the returned value will be 0 (false).

The array *VarA* has to be a one-dimensional array. Each element will be displayed in the text on a line by itself. If the element is numeric it will be converted to string. If you need to have blank lines use a null string (""). The first element to not have an assigned value will be the end of the text. Use the Data command to create the array (or `Dim`). **The first element in the array will be used as a title (on the border frame of the box) to the dialog; it will not show inside the box with the rest of the text.**

The box will be centered on the screen and will be as wide as needed to display the longest line of text, but will not be wider than the screen. If the box is narrower than any line the line will wrap around. Also there is a scroll bar to scroll the text in the vertical direction. See the Commands help page for more user interaction commands Also see the `TextBox()` and `ErrMsg()` functions in the Functions Part II help page.

Example:

```
Data msg;"this is a test message box.",""
Data msg;"the next lines are numerical data displayed as text"
Data msg;-1,3,5.2,6.1e12
Data msg;"","this is the end."
n = MsgBox(msg)
if n
   print "OK"
else
   print "Cancel"
endif
```

C.8.13 OTHER FUNCTIONS

`Within(`*ExprN1,ExptN2,ExprN3*`)` Checks to see if *ExprN1* is within the range *ExprN2* to *ExprN3* inclusive. All the numbers can be floats or integers. If *ExprN2* is greater than *ExprN3* they are swapped to do the checking. The function returns 1 (true) if *ExprN1* >= *ExprN2* AND *ExprN1* <= *ExprN3* otherwise it returns 0 (false). The swapping is internal and does not affect the parameters.

`Evaluate(`*ExprS*`)` *ExprS* is a string that contains a valid expression (see Sec. B.7 for definition of expressions). Depending on the context you can use variables. But these must be existing variables for the evaluator to be able to evaluate the expression correctly. This can be useful when expressions are defined at runtime instead of design time. For example say you are writing a function plotter. Further, you want the user to define the function to be plotted at runtime. The user would enter a string in response to an `Input` command, for example, `sin`(*x*). The programmer can then loop through all values in the desired range of *x* and evaluate the expression given by the user as *y* = `Evaluate`(*ExprS*) where *ExprS* was given by the user as f(*x*).

```
Input "Enter f(x)",fx
for x = 0 to 10
      y = Evaluate(fx)
      print y
next
End
```

If an error is encountered while evaluating the function given by the user an error message is displayed indicating the error, and `Evaluate` will return the *ExprS* as it was, that is, if *ExprS* is not a valid expression the `Evaluate`(*ExprS*) function will display an error message and return the original input without evaluation. This allows the user to check if the return result is still the same as the input then execute any error handling within the program. The error message given by the interpreter is meaningful in that it indicates the type of error, but the line number will always be 0 because *ExprS* is just one line any way.

`KeyDown`(*ExprN*) The commands `WaitKey`, `GetKey` and `GetKeyE` allow you to obtain the code for the last key pressed. However, it may become necessary to test if a certain combinations of keys are pressed. Say you want to fire retrorockets horizontally and vertically and allow both at the same time. If you use the above commands the program will only be able to detect the last key pressed and thus if the user presses both the Up and Left Arrow keys together the commands will only report the code of the last key pressed. Also see the Constants help page in the IDE for handy constants that represent various key codes.

Using this function you can interrogate each key separately to see if it is pressed down or not. There are 255 keys on the keyboard and you can find out the state of each separately using this function. For instance, if you want to find out if the up arrow key is pressed and also at the same time the left arrow key then you would have a statement like

```
if KeyDown(kc_LArrow) && KeyDown(kc_RArrow) then GoSub DoSomething
```

Or you can use the numbers directly

```
if KeyDown(37) && KeyDown(38) then GoSub DoSomething
```

The function returns a non-zero number if the key is pressed and zero if it is not. *ExprN* is an integer value representing the code of the key you wish to query. The codes of the keys can be determined as described in the User Interface Commands help page under the command `GetKeyE`. There is a program segment given there that enables you to determine the code for any key on the keyboard. Most of the alphanumeric keys have a code that is the ASCII code of the UpperCase letter of the key. The shift, Control and Alt keys have the values 16, 17, and 18 respectively. So for example if you wish to find out if Ctrl+F2 are pressed simultaneously you would have:

```
if KeyDown(kc_Ctrl) && KeyDown(kc_F2) then GoSub DoSomething
```

The Function keys have the numbers 112 to 123 (F1 to F12). Remember the key code is specific for each key even if the key has multiple characters printed on it, it will have one code. So the 9key will have the code 57 which is actually the ASCII code for the numeral 9. The Left-Up-Right-Down arrow keys have the codes 37 to 40 in order (see the Constants help page in the IDE). You can also use this function to read the condition of the mouse buttons as if they were keyboard keys. See the Constants help page for their codes.

The function can be useful in gaming where a response to more than one key pressed together may be necessary. Also refer to the commands `GetKey`, and `GetKeyE` for a discussion on how to handle the speed of response when you do not wish to detect one press as multiple presses.

`VType(`***Var***`)` This function is used to find the type of a variable. You specify the variable name. The returned values are:

102 = Ascii("f")	→	floating-point number
105 = Ascii("i")	→	integer number
115 = Ascii("s")	→	string
0	→	no value, that is, has not been assigned a value

The values above are also defined in the constants section (Sec. B.7.6)

Float = floating point number
Integer = integer number
String = string
Novalue = not a defined variable

This function can be useful to determine the type of a variable or if it is an unassigned variable. You can check if a variable is valid before it is used. If you try to use an unassigned variable an error will occur. This function can be used to prevent this. Also see `IsString()`, `IsNumber()`, `ToString()`, `ToNumber()`, and so on.

C.8.14 FORMATTING CODES AND LOGIC

Specifier	Represents
0	*Digit place holder.* If the value being formatted has a digit in the position where the "0" appears in the format string, then that digit is copied to the output string. Otherwise, a "0" is displayed in that position in the output string.
#	*Digit placeholder.* If the value being formatted has a digit in the position where the "#" appears in the format string, then that digit is copied to the output string. Otherwise, nothing is stored in that position in the output string.
.	*Decimal point.* The first "." character in the format string determines the location of the decimal separator in the formatted value; any additional "." characters are ignored. The actual character used as the decimal separator in the output string is determined by the *DecimalSeparator* global variable. The default value of decimal separator is specified in the Number Format of the International section in the Windows Control Panel.
,	*Thousand separator.* If the format string contains one or more "," characters, the output will have thousand separators inserted between each group of three digits to the left of the decimal point. The placement and number of "," characters in the format string does not affect the output, except to indicate that thousand separators are wanted. The actual character used as the thousand separator in the output is determined by the *ThousandSeparator* global variable. The default value of *ThousandSeparator* is specified in the Number Format of the International section in the Windows Control Panel.

Specifier	Represents
E+	*Scientific notation.* If any of the strings "E+," "E–," "e+," or "e–" are contained in the format string, the number is formatted using scientific notation. A group of up to four "0" characters can immediately follow the "E+," "E–," "e+," or "e–" to determine the minimum number of digits in the exponent. The "E+" and "e+" formats cause a plus sign to be output for positive exponents and a minus sign to be output for negative exponents. The "E–" and "e–" formats output a sign character only for negative exponents.
'xx'/"xx"	Characters enclosed in single or double quotes are output asis, and do not affect formatting.
;	Separates sections for positive, negative, and zero numbers in the format string.
Notes:	The locations of the leftmost "0" before the decimal point in the format string and the rightmost "0" after the decimal point in the format string determine the range of digits that are always present in the output string. The number being formatted is always rounded to as many decimal places as there are digit placeholders ("0" or "#") to the right of the decimal point. If the format string contains no decimal point, the value being formatted is rounded to the nearest whole number. If the number being formatted has more digits to the left of the decimal separator than there are digit placeholders to the left of the "." character in the format string, the extra digits are displayed before the first digit placeholder. To allow different formats for positive, negative, and zero values, the format string can contain between one and three sections separated by semicolons. One section: The format string applies to all values. Two sections: The first section applies to positive values and zeros, and the second section applies to negative values. Three sections: The first section applies to positive values, the second applies to negative values, and the third applies to zeros. If the section for negative values or the section for zero values is empty, that is if there is nothing between the semicolons that delimit the section, the section for positive values is used instead.

C.9 The Robot Simulator Commands and Functions

The robot simulator is a set of commands and functions that allow for easy programming of a robot to move around a simulated environment. The environment is drawn on the screen using the drawing commands and functions (see Sec. C.7). The robot is then *located* in the environment and is made to programmatically move, turn, and sense around the screen.

C.9.1 GENERAL INFORMATION

The Robot comes with sensors to *feel* around, *look* for objects, and *sense* for lines on the ground. The view on the screen is as if you are looking at the robot and the environment from above. Objects drawn with `LineTo`, `Rectangle`, `Circle` and other commands will have colors and can be considered as furniture or other objects and obstacles. Additionally, you can draw lines on the floor and have the robot *sense* for them and follow them with the correct combination of commands.

To use the robot you must first `rLocate` it on the screen. Make sure the environment-drawing commands are done first. Once the robot is located you can issue commands to `rForward` and `rTurn` it. Also, the robot can `rSense`, `rLook`, and/or `rFeel` around the environment to avoid objects.

Since the environment is two-dimensional, you must specify any colors to be considered invisible to the robot and therefore nonobstacles. Any color on the screen that is not in the list of invisible colors will be considered an obstacle when encountered by the robot while moving around (except for the floor color). If you change the color of the floor, you must do so before locating the robot and then locate the robot on the new floor (see `rFloorColor`). You specify invisible colors with the `rInvisible` command.

The robot cannot be located or moved off the screen. The screen boundaries are the walls of a room. The robot will not move forward into an object. Any attempt to do so will cause an error. No sensor commands can sense beyond the walls.

The robot has a battery that discharges upon using the commands to move and sense. Moving depletes more charge than sensing. The battery level can be checked and can be recharged. The robot will refuse to move and will return nonsense values from the sensors if the battery is depleted. The default is to ignore the battery charge level but you can use a command to make the robot heed the charge level.

Throughout the next two sections there will be reference to the following items:

- *ExprN* implies that a numeric resulting expression is required.
- {*Expr*} or {*Var*} implies that it is optional and {*Expr*...} means many can be given.
- *Var* implies that a variable name must be given.
- If a *Var* is expected in any of the commands, then if *Var* exists it will be replaced with the result otherwise it will be created and assigned the result.

The robot simulator commands and functions *are not* case sensitive, so `rLocate`, `RLOCATE`, and `rlocate` are all the same.

C.9.2 SIMULATOR COMMANDS

In the following list the commands are arranged in order of functionality rather than alphabetically. See later for an alphabetic order.

NOTE: Commands with a * have additional functionality are explained in App. D.

`*rLocate` ***ExprN1,ExprN2,{ExprN3,{ExprN4, {ExprN5}}}*** Creates robot of specified radius (*ExprN4*) and color (*ExprN5*) at specified position (*ExprN1,ExprN2*) and heading (*ExprN3*).

If you need to specify *ExprN4*, you must also specify *ExprN3* and so on. *ExprN3/4/5* are optional. The heading defaults to 0 (north), size defaults to 20 pixels and color defaults to blue. The size is limited to between 5 and 50 pixels. The size value is the radius of the robot so a radius of 20 means a diameter of 40.

This command must be executed before any other robot functions or commands. It is the command that creates the robot, places it on the ground and switches it on. If you issue any commands/functions before this command an error will be issued and the program will halt.

The size of the robot should be considered in relation to the screen size. The screen is 800 × 600 pixels. So if you consider a room of 20 ft then each pixel is 1/3 in. So a robot of 20 pixels radius is 12 in in diameter. Which is a reasonable robot size, but you can specify a smaller or larger size.

If you locate the robot over a color other than the floor color, or if you change the floor color from the default screen color you must issue an `rFloorColor` command (see below) before you attempt any movement of the robot.

`rInvisible` *ExprN {,ExprN...}* This command sets the list of colors that will *not be* considered as objects when encountered by the robot while moving and also by many of the sensors as specified. You must pass at least one color and a maximum of 15 colors is allowed. The colors are according to the colors specified in the Constants help page within the IDE.

The first color in the list you specify is special in that it will be used as a default color for the `rPen` command and `rSense()` function if you do not specify a color when you issue the command or call the function (see below). The second color in the list is also special in that it will be the default color if you do not specify a color for the `rDFeel()` and `rDBumper()` functions.

Commands or functions that look or sense for colors will not detect the colors on the invisible colors list unless otherwise stated.

`rFloorColor` *ExprN* Sets the color that will be considered as the floor and will not affect sensors or the `rForward` command.

`*rForward` *ExprN* The robot moves in the direction it is heading *ExprN* pixels forward if *ExprN* is positive, or backwards if *ExprN* is negative. This command will cause an error if the robot tries to move into objects of any color not listed in the list of invisible colors (see the `rInvisible` command).

This command will cause an error if the battery is depleted and an `IgnoreCharge` *False* has been issued.

`*rTurn` *ExprN* This will cause the robot to turn *ExprN* degrees right (clockwise) if *ExprN* is positive and left (counterclockwise) if *ExprN* is negative.

This command will cause an error if the battery is depleted and an `IgnoreCharge` *False* has issued.

`rHeading` *ExprN* Sets the robot heading to *ExprN* (0–359). Zero is facing up on the screen (north), 90 is facing right (east), and so on.

`rSpeed` *ExprN* Sets the robot speed. The larger the number the slower the robot. *ExprN* must be >= 0.

`*rGps` *Var1,Var2* Sets *Var1* to the Robot's current *x* position and *Var2* to the *y* position, giving the robot's position on the screen. This command will *not* cause an error if the battery is depleted and an `IgnoreCharge` *False* has been issued, but the returned values will be 0.

`*rPen` *ExprN1 {,ExprN2}* The robot has a pen positioned at its center that can be put up or down using this command. When the pen is down the robot will draw a line using the color specified in *ExprN2*. If you do not specify *ExprN2* then the first color specified in the list of invisible colors (see `rInvisible` above) is used. If you have not specified a list of invisible colors then the pen will draw using the floor color, which means that you will not see the trace when it is over the floor color.

The width of the drawn line is determined by the current line width set by the drawing commands (see LineWidth *ExprN*, Sec. C.7).

If *ExprN1* is zero then the pen will be raised. If *ExprN1* is not zero then the pen will be lowered and drawing will take place. You can use the constants *Down* and *Up* so you can say:

```
rPen Down,Cyan  //---will set the pen down and
                //   use cyan for drawing
rPen Up         //---will set the pen up
```

NOTE:

- The line width is whatever was set by the last LineWidth command issued.
- The line color will be what is specified by *ExprN2*, or the first color in the invisible colors list.
- The color drawn by the pen will be considered an obstacle unless it is in the list of invisible colors.

`rCharge` *ExprN* Sets the batteries to percentage charge. *ExprN* is a percentage value 1 to 100 percent

`rIgnoreCharge` *ExprN* If *ExprN* is true (not zero) then the battery charge level will be ignored and the robot will continue to operate regardless of the battery charge level. This is the default state. If *ExprN* is false (zero) then the battery charge level will be taken into account. The robot will not operate when the battery is depleted and in some cases an error will occur as specified in the commands.

`rSensor` *ExprN1,ExprN2,Var1,Var2,Var3* The robot has a default set of sensors located at 90° and 45° to the left and right of the front and at the front. These sensors can look and return the color of the first object encountered and the distance to that object. *ExprN1* specifies the sensor number to use:

1 = 90° to the right
2 = 45° to the right

3 = Front
4 = 45° to the left
5 = 90° to the left

- *ExprN2* specifies the limit of the sensor range (pixels).
- *Var1* is set to the first color found directly in front of the sensor that is not on the list of invisible colors. If the range of the sensor or the walls of the room are reached (screen extent) without sensing an object then this color value will be −1, otherwise it will be the color of the object.
- *Var2* is set to the distance to the object found. If no object is found then the distance will be the maximum range of the sensor, or the distance to the wall if that was within the sensor range.
- *Var3* will be set to 1 (true) if an object is found or a wall is sensed within the sensor range. If the sensor range is reached without sensing a wall or an object then it is set to 0 (false). Colors on the invisible colors list will be ignored in the process.

This command will *not* cause an error if the battery is depleted and an `IgnoreCharge` *False* has been issued, but the values will be nonsense.

`rSensorA` ***ExprN1,ExprN2,Var1,Var2,Var3*** This is the same as the `rSensor` command, except *ExprN1* is not a sensor number but rather an angle (0–359) clockwise relative to the robot heading. A value of 90 is 90° to the right, 270 is 90° to the left. So this command can be used to do the same as above but instead of a limit of 5 sensors you have 360 of them. This command will *not* cause an error if the battery is depleted and an `IgnoreCharge` *False* has been issued, but the values will be nonsense.

`rSlip` ***{ExprN}*** This command defines a percentage for the slipping feature of the robot. Real robot motors tend to slip in a random way where going forward/backward can go less than expected. If the motors do not turn exactly the same amount while going forward a slight turning tendency may occur. Also, during turning if the motors do not turn as expected, less or more turning than commanded may occur.

Issuing this command will cause the robot to behave in this manner. If you do not pass *ExprN* the slip will occur 2 percent of the time. If you want to simulate a different percentage then issue the command with *ExprN*. *ExprN* resulting in 0 will turn the slipping off which is the default state. If *ExprN* is less than 0 then it will be made 0. If *ExprN* is greater than 100 it will be made 100. If *ExprN* does not result in a number an error will occur.

C.9.3 SIMULATOR FUNCTIONS

In the following if empty parenthesis "()" is specified then the function does not require a parameter but the () must still be typed. All functions will *not* cause an error if the battery is depleted and an `IgnoreCharge` *False* has been issued, but the values returned will be nonsense.

The functions are listed according to functionality rather than in an alphabetic order. See below for an alphabetical order.

Functions that look or sense for colors will not detect the colors on the invisible colors list (see `rInvisible` above) unless otherwise stated.

NOTE: Functions with a * have additional functionality as explained in App. D.

rChargeLevel() Returns the batteries charge percentage value.

rPoints() The robot keeps a count of the number of times its sensors and motors (`rTurn`, `rForward`) are used. This point value can be used as a measure of the efficiency of an algorithm to complete a task in the minimum possible points. This function returns the current count value.

rGpsX() Returns the robot's current *x* position.

rGpsY() Returns the robot's current *y* position.

***rCompass()** Returns the robot's current heading. 0–359 0= north.

***rLook()**
***rLook(*ExprN*)** Returns the first color seen directly in front of the robot. If the walls are seen before any object then −1 is returned. Colors on the invisible colors list are ignored.

If *ExprN* is given then it is an angle relative to the robot's heading. *ExprN* must be between −180 and +180. Positive is to the right of the robot, negative is to the left of the robot. The pivoting center is the center of the robot.

Specifying an angle places the camera at any angle relative to the robot's heading pivoted on the robot's center. With this sensor you can look at objects relative to the robot's heading 180° to the right (+180) up to 180° to the left (−180) at intervals of 1°.

***rRange()**
***rRange(*ExprN*)** Returns the distance to the first color seen directly in front of the robot. If the walls are seen before any object then the distance to the wall is returned. Colors on the invisible colors list are ignored.

If *ExprN* is given then it is an angle relative to the robot's heading. *ExprN* must be between −90 and +90. Positive is to the right of the robot, negative is to the left of the robot. Pivoting is about the front of the robot not the center of the robot. The ranger is fitted on the front point of the robot and pivots 90° left and right centered on this point.

Specifying an angle places the range finder at any angle relative to the robot's heading. With this sensor you can measure distances relative to the front of the robot from 90° to the right (+90) up to 90° to the left (−90) at intervals of 1°.

***rBeacon(*ExprN*)** Returns the distance to the specified color (*ExprN*) if it is in front of the robot, even if that color is blocked by other objects between the robot and the color. It returns 0 otherwise.

This is useful for detecting of a "flashing" beacon mounted above obstacles in the room. The command looks for any color specified. It *does not* ignore colors on the invisible colors list. The result can be treated as true/false (false = 0, true = otherwise) or you can use the number returned if it is *not zero* as the distance to the color specified.

`*rFeel()`
`*rDFeel(`*{ExprN}*`)` Returns a number 0 to 31 according to the following logic:

There are five infrared sensors around the robot. At 90° and 45° to the left and right of the robot and directly in the front. The sensors are able to feel any object within a robot's radius ahead of the sensor (ignoring invisible colors).

The number returned is a bitmap of the condition of the sensors. If any of the sensors feels something then its bit position is set to 1, otherwise 0. So there are 5 bits MSB...LSB 00000. The most significant bit (MSB) is the sensor 90° to the left, then 45° then 0° then 45° to the right, and the least significant (LSB) bit is 90° to the right. So if the number is 01110 (14) then the sensors in front and 45° right and left are feeling objects and the rest do not. Use bitwise operators (bAND, bOR, etc) to manipulate the result. Colors on the invisible colors list are not considered objects.

The difference between `rFeel()` and `rDFeel(`*{ExprN}*`)` is that `rDFeel(`*{ExprN}*`)` will draw radials out from the sensors to show the range of feel of the sensors using the color specified in *ExprN*. If you do not specify *ExprN* then the second color on the invisible colors list will be used. This is a useful feature for debugging the rFeel() results.

Do not use `rDFeel(`*{ExprN}*`)` unless you need to debug since it is significantly slower than `rFeel()`. You must specify *ExprN* or have a valid color (not a floor color) in the second position in the invisible colors list. The radial lines will be drawn using this *ExprN* color or the second color on the invisible colors list other wise the beams will not be visible.

`*rBumper()`
`*rDBumper(`*{ExprN}*`)` Returns a number 0 to 15 according to the following logic:

There are four bumpers around the robot. At the front covering an arc of 130° that is, from 65° to the left to 65° to the right. There is also a bumper at the back just like the one in front, that is, from 115° to 245°. To the left and to the right there are bumpers that cover an arc of 50°, that is, from 65° to 115° on the right and from −65° to −115° on the left. The bumpers will close (turn on) if there is any object within 2 pixels of the robot's perimeter (ignores invisible colors).

The number returned is a bit map of the condition of the bumpers. If any of the bumpers is closed then its bit position is set to 1, otherwise 0. So there 4 bits MSB...LSB 0000. The least significant bit is the back bumper, then the right, front and then the left one.

If the number is 1110 (14) then the right, front and left bumpers are closed while the rear one is not. Use bitwise operators (bAND, bOR, etc) to manipulate the result. floor/line/beacon colors are ignored.

The difference between `rBumper()` and `rDBumper(`*{ExprN}*`)` is that `rDBumper(`*{ExprN}*`)` will light up an LED where the bumper is touching objects using the *ExprN* color or the second color on the invisible colors list. If no color is specified with *ExprN* then the second color on the invisible colors list will be used. If that is not specified then the color will be the floor color and the beacon may not be visible. This is a useful feature for debugging the `rBumper()` result. Do not use `rDBumper(`*{ExprN}*`)` unless you are debugging since it is significantly slower than `rBumper()`.

`*rSense({ExprN})` Returns a number 0 to 7 according to the following logic:

There are three sensors around the robot, at 10° to the left and right of the robot and directly in the front. The sensors see only the color specified by *ExprN* or, if you don't specify *ExprN,* then the first color on the invisible colors list. If neither is specified then the floor color will be sensed. The sensors only look at the ground directly under the sensor situated directly at the perimeter of the robot. The number returned is a bitmap of the condition of the sensors. If any of the sensors sees the *ExprN* color then its bit position is set to 1, otherwise it is 0. There are 3 bits MSB.LSB 000. The most significant bit is the sensor 10° to the left, then ahead (0°) and then 10° to the right. If the number is 010 (2) then only the sensor in the front is on the line and the other two are outside of the line. Use bitwise operators (bAND, bOR, etc) to manipulate the result.

`rGround(ExprN)` The robot has a default of three sensors to look at the ground directly at the perimeter of the robot. *ExprN1* specifies the sensor number.

1 = 10° to the right
2 = In front
3 = 10° to the left.

It will return the color seen by the sensor. It will not ignore *any* color (i.e., the colors on the invisible colors list *are not* ignored). It is up to you what to make of the value. If you want to ignore any colors use programming logic to do so.

`rGroundA(ExprN)` This is the same as the `rGround` function, except *ExprN1* is not a sensor number but rather an angle (0–359) clockwise relative to the robot heading. A value of 90 is 90° to the right, 270 is 90° to the left. So this command can be used to do the same as above but instead of a limit of 3 sensors you have 360 of them.

C.9.4 SIMULATOR COMMANDS LISTED ALPHABETICALLY

```
rCharge ExprN
rFloorColor ExprN
*rForward ExprN
*rGps Var1,Var2
rHeading ExprN
rIgnoreCharge ExprN
rInvisible ExprN {,ExprN...}
*rLocate ExprN1,ExprN2,{ExprN3,{ExprN4, {ExprN5}}}
*rPen ExprN1 {,ExprN2}
rSensor ExprN1,ExprN2,Var1,Var2,Var3
rSensorA ExprN1,ExprN2,Var1,Var2,Var3
rSpeed ExprN
*rTurn ExprN
```

C.9.5 SIMULATOR FUNCTIONS LISTED ALPHABETICALLY

```
*rBeacon(ExprN)
*rBumper()
rChargeLevel()
*rCompass()
*rDBumper({ExprN})
```

```
*rDFeel({ExprN})
*rFeel()
rGpsX()
rGpsY()
rGround(ExprN)
rGroundA(ExprN)
*rLook({ExprN})
rPoints()
rRange({ExprN})
rSense({ExprN})
```

C.10 Commands and Functions Listed Alphabetically

C.10.1 COMMANDS

```
AddButton ExprS,ExprN1,ExprN2{,ExprN3{,ExprN4}}
Arc ExprN1,ExprN2,ExprN3,ExprN4{,ExprN5{,ExprN6{,ExprN7{,ExprN8}}}}
Beep {ExprN}
Circle ExprN1,ExprN2,ExprN3,ExprN4{,ExprN5,ExprN6}
ClearScr {ExprN}
Data VarA;Expr{,Expr....}
Debug {Expr1,Expr2;Expr3...}
DebugOff
DebugOn
Delay {ExprN}
Dim VarA[ExprN{,ExprN...}]
DrawShape ExprS,ExprN2,ExprN3{,ExprN4,ExprN5}
ERectangle ExprN1,ExprN2,ExprN3,ExprN4{,ExprN5{,ExprN6}}
FloodFill ExprN1,ExprN2{,ExprN3{,ExprN4}}
GetButton Var
GetColor Var1,Var2
GetError VarN1{,VarS{,VarN2{,VarN3}}}
GetLineWidth Var
GetXY Var1,Var2
GotoXy ExprN1,ExprN2
Input Expr,Var
Input Expr,VarA[ExprN{,ExprN...}]
Line ExprN1,ExprN2,ExprN3,ExprN4{,ExprN5{,ExprN6}}
LineTo ExprN1,ExprN2{,ExprN3{,ExprN4}}
LineWidth ExprN
MAdd VarA1,VarA2
MConstant VarA,Expr
MCopy VarA1,VarA2
MDeterminant VarA1,Var
MDiagonal VarA,Expr
MicroDelay {ExprN}
MInvert VarA1,VarA2,Var
MMultiply VarA1,VarA2,VarA3
MPolygon VarA{,ExprN}
MRead VarA,ExprS
MRegression VarA,Var1,Var2
MScale VarA,ExprN
MSort VarA1{,ExprN}
MSub VarA1,VarA2
```

```
MTranspose VarA1,VarA2
MWrite VarA,ExprS
Pie ExprN1,ExprN2,ExprN3,ExprN4{,ExprN5{,ExprN6{,ExprN7{,ExprN8}}}}
Print {Expr,Expr;Expr...}
ReadMouse Var1,Var2{,Var3}
ReadPixel ExprN1,ExprN2,Var
ReadScr {ExprS}
Rectangle ExprN1,ExprN2,ExprN3,ExprN4{,ExprN5,ExprN6}
RemoveButton ExprS
RestoreScr {ExprN1{,ExprN2}}
SaveScr {ExprN1{,ExprN2{,ExprN3{,ExprN4}}}}
ScrLimits Var1,Var2
SetColor ExprN1{,ExprN2}
SetPixel ExprN1,ExprN2,ExprN3
Sound ExprN1,ExprN2 {,ExprN3}
Speaker ExprN
Swap Var1,Var2
GetKey Var
GetKeyE Var
OnError Label
OnError Expr
OnError
WaitKey {ExprS,}
WriteScr {ExprS}
XYstring ExprN1,ExprN2,Expr3{;expr,expr;...}
XYtext ExprN1,ExprN2,Expr{,ExprS{,ExprN3{,ExprN4}}}
```

C.10.2 FUNCTIONS

```
Abs(ExprN)
ACos(ExprN)
Ascii(ExprS)
ASin(ExprN)
ATan(ExprN)
ATan2(ExprN1,ExprN2)
Average(VarA)
Bin(ExprN)
CartX(ExprN1,ExprN2)
CartX(ExprN1,ExprN2)
CbRt(ExprN)
Char(ExprN)
CorrCoef(VarA)
Cos(ExprN)
Count(VarA)
Date(ExprN)
Evaluate(ExprS)
Exp(ExprN)
Exp10(ExprN)
Extract(ExprS1,ExprS,ExprN)
Factorial(ExprN)
Format(ExprN,ExprS)
Frac(ExprN)
Hex(ExprN)
InString(ExprS1,ExprS2)
IsNumber(Expr)
IsString(Expr)
KeyDown(ExprN)
```

```
Left(ExprS,ExprN)
LeftTrim(ExprS)
Length(ExprS)
Log(ExprN)
Lower(ExprS)
Max(VarA)
MaxDim(VarA,ExprN)
MDim(VarA)
Median(VarA)
Min(VarA)
Mod(ExprN1,ExprN2)
MsgBox(VarA)
MType(VarA[ExprN{,ExprN,...}])
nCr(ExprN1,ExprN2)
NLog(ExprN)
NoSpaces(Exprs)
nPr(ExprN1,ExprN2)
Pi(ExprN)
PolarA(ExprN1,ExprN2)
PolarR(ExprN1,ExprN2)
Random(ExprN)
Range(VarA)
Right(ExprS,ExprN)
RightTrim(ExprS)
Round(ExprN)
RoundDn(ExprN)
RoundUP(ExprN)
Sign(ExprN)
Sin(ExprN)
Spaces(ExprN)
Sqrt(ExprN)
SRepeat(ExprS,ExprN)
StdDev(VarA)
Substring(ExprS,ExprN1,ExprN2)
Sum(VarA)
Tan(ExprN)
Time(ExprN)
ToNumber(Expr)
ToString(Expr)
Trim(ExprS)
Upper(ExprS)
Variance(VarA)
VType(Var)
Within(ExprN1,ExprN2,ExprN3
```

APPENDIX D

PORTS AND SERIAL INPUT/OUTPUT

D.1 General Information

The following commands allow for using direct port input/output (I/O) where you can read/write a byte from/to a particular port on the PC. There is a set of special commands for I/O to the parallel port as a special case for ease of use. There is also a set of commands that allow for serial I/O using either actual comm ports on the PC or virtual comm ports created by protocols such as Bluetooth devices or USB devices. These devices will create a virtual port number that can be used for all intents and purposes as if it were a physical serial comm port.

The robot simulator has an extension that allows for the use of serial I/O to communicate to and from a real robot, effectively enabling programming of a real robot using the RobotBASIC language. The protocol extension will be explained in its own section below.

To facilitate serial comm port setup there are a set of constants that are listed in the Constants help page within the IDE. The I/O commands *are not* case sensitive.

Throughout this section there will be reference to the following items:

- *ExprN* implies that a numeric resulting expression is required.
- *ExprS* implies that a string resulting expression is required.

- *Expr* implies that any type expression is required.
- *{Expr}* or *{Var}* implies that it is optional and *{Expr...}* means many can be given.
- *Var* implies that a variable name must be provided. If *VarS* is specified then it will be filled with a string if *VarN* is specified then it will be filled with a number.
- If a *Var* is expected in any of the commands, then if *Var* exists it will be replaced with the result otherwise it will be created and assigned the result.

NOTE: Always refer to the help pages within the RobotBASIC IDE for the most up-to-date information regarding material mentioned in this appendix.

D.2 Serial I/O Commands

`SetCommPort` ***ExprN1 {,ExprN2 {,ExprN2 {,ExprN3 {,ExprN4 {,ExprN5}}}}}***
This command sets the comm port number and parameters and activates the port for communications.

- *ExprN1* is the port number 1 to 5000.
- *ExprN2* is the baud rate 0 to 14 (see the Constants help page for details of codes). Defaults to 6, that is, 9600 baud.
- *ExprN3* is the number of data bits in the transmitted byte can be 4, 5, 6, 7, or 8. Defaults to 8.
- *ExprN4* is the parity check 0 to 4 (see Sec. B.7.6). Defaults to 0 (no parity).
- *ExprN5* is the number of stop bits in the transmission (see the Constants help page). Ranges from 0 to 2. Defaults to 1, that is, 1 stop bit.
- *ExprN6* is the flow-control protocol (see the Constants help page). Ranges from 0 to 2. Defaults to 0, that is, none.

To deactivate the communication port issue the command `SetCommPort 0`

`SerOut` ***Expr {,Expr {; Expr ...}}*** This command is similar to the Print command. The output of the command is sent to the serial port in place of the Terminal Screen.

Outputs the values of *Expr....* to the serial port. A comma (,) between the expressions makes them print with no space between them. A semicolon (;) prints them with a tab space between them. You must specify at least one expression. The result of all the expressions is put together as specified by the commas and semicolons into one string and the string is outputted to the serial port as specified by the `SetCommPort` command. **The total length of the resulting string will be truncated to 4095 bytes if it exceeds this size.**

`SerIn` ***VarS*** This command reads the data that is currently in the serial communications buffer all in one go and puts it as a string in the variable *VarS*. You can use string manipulation commands and functions as well as conversion functions to make use of the data. The command will not cause a time out. If there are no bytes in the input buffer the string returned will be null. Use the `CheckSerBuffer` command to determine how many bytes (if any) of data are waiting on the buffer.

`SerBytesIn` *ExprN,VarS,VarN* This command reads *ExprN* bytes from the serial input buffer and puts them in to the string variable *VarS*. *VarN* is set to the number of bytes actually read before a time out occurred. The command will wait until *ExprN* bytes are read. If a time-out occurs before all the characters are retrieved then what ever characters have been read are returned inside *VarS* and *VarN* is set to the number actually read. See `SetTimeOut` and `GetTimeOut` commands. The maximum amount for *ExprN* is 4095 bytes and the minimum is 1. *ExprN* will be set to the closest limit if it is outside the range 1 to 4095.

`SetTimeOut` *{ExprN}* This command sets the time-out for the serial port reading commands. *ExprN* is the number of milliseconds to wait before a time out occurs. *ExprN* is optional and if not given or is less than one then 5000 is assumed (i.e., 5 seconds). 5000 is also the default on program startup.

`GetTimeOut` *VarN* This command reads the current setting for the time-out for the serial port and puts the value in *VarN*. The number returned is in milliseconds.

`CheckSerBuffer` *VarN* This command assigns the number of characters in the serial input buffer to the variable *VarN*. The buffer can be read with the commands above.

`ClearSerBuffer` *{ExprN}* This command clears the input and output serial buffers. If *ExprN* is not given or is 0 then both buffers are cleared. A value of 1 will clear the input buffer. A value of 2 will clear the output buffer. Any other values are ignored.

D.3 Parallel Ports I/O Commands

These commands read and write byte values to the parallel port specified. These commands assume that your system has a bidirectional parallel port (ECP standard). If your system does not have this then only output is possible and no input can occur. You can check if your system has this capability and set it using the BIOS setup while starting your PC. If you have more than one parallel port you can specify which one to use by specifying its base address using the `SetPPortNumber` command.

`PPortOut` *ExprN1* Outputs the byte value *ExprN1* to the parallel port If *ExprN1* is greater than 255 only the lower byte will be outputted.

`PPortIn` *VarN* Reads the data byte at the parallel port.

`SetPPortNumber` *{ExprN}* This command sets the base address of the parallel port to be used by all the parallel port commands including the virtual ones below. The system defaults to address 888 (hex 0x378), which is usually the first and only parallel port on many systems. If your system has more than one port and you wish to use other than the default first port use this command to specify the port number (in decimal). You can find the addresses of available ports from the BIOS setup upon starting your computer, or you can use the Windows system to do the same. Beware the reported addresses are usually given in hexadecimal format. You need to convert them to integer to use it for *ExprN* in this command [see function `HextToInt()`]. *ExprN* is optional and if it is not given or it is less than one then the system will use port number 888 (hex 0x378).

WARNING! If you specify the wrong address the commands will not function and you may even damage your system.

WARNING! Use these commands with care. Badly designed hardware connected to the parallel port can damage the port. If you are going to experiment with electronics and these commands do so at your own risk and know what you are doing. Failure to use the correct design will damage your port if not your whole computer.

D.4 Virtual Parallel Port I/O Protocol

This protocol is included as a convenience for extending the single byte parallel port to 4 input bytes and 4 output bytes. The protocol assumes there is multiplexing hardware connected to the parallel port. You can write your own protocol using the `InPort` and `OutPort` commands, but this protocol is provided for convenience and speed.

This protocol makes use of the control port on the parallel port to put the port number on the multiplexer hardware (3 bits) and uses the fourth bit to clock the multiplexer and read the data or output the data depending on what port number is in use.

The result is that one parallel port with only 8 I/O pins can be expanded to 4 × 8 Input and 4 × 8 Output pins. The hardware will not be discussed here (see Chap. 17 for a suggested design).

On the parallel port connector, pins 2 to 9 are the data pins. Pins 1, 14, and 16 are used to set the address on the multiplexer. Where LSB (least significant bit) is pin 1. Pin 17 will be used to clock the multiplexer (low is HiZ, high is Clocked) assuming a rising edge trigger.

Addresses 0 to 3 will be input ports (to the PC) and 4 to 7 will be output ports (from the PC).

`VPPortOut` *ExprN1,ExprN2* Outputs the byte value *ExprN2* to the *virtual parallel port* number *ExprN1*. If *ExprN2* is greater than 255 only the lower byte will be outputted. *ExprN1* is limited to 1 to 4. If it is outside these limits the nearest limit will be set.

`VPPortIn` *ExprN1,VarN* Reads the byte value at the *virtual parallel port* number *ExprN1* and then puts it in the variable *VarN*. *ExprN1* is limited to 1 to 4. If it is outside these limits the nearest limit will be set.

WARNING! Use these commands with care. Badly designed hardware connected to the parallel port can damage the port. If you are going to experiment with electronics and these commands do so at your own risk and know what you are doing. Failure to do the correct design will damage your port if not your whole computer.

D.5 General Ports I/O Commands

`OutPort` ***ExprN1,ExprN2*** Outputs the byte value *ExprN2* to the PC port number *ExprN1*. If *ExprN2* is greater than 255 only the lower byte will be outputted. *ExprN1* can be any valid port number.

WARNING! Using this command incorrectly can damage your system if you write or read from an incorrect port number.

`InPort` ***ExprN1,VarN*** Reads the byte value at the PC port number *ExprN1* and then put it in the variable *VarN* as a number. *ExprN1* can be any valid port number.

WARNING! Using this command incorrectly can damage your system if you write or read from an incorrect port number.

D.6 Robot Simulator Serial I/O Protocol

This protocol uses serial communication (can be Bluetooth virtual serial port) to communicate between RobotBASIC and a real robot with the ability to send and receive serial data. You can create your own protocol using the above serial communications commands. This protocol is provided for convenience.

You can write programs normally using the simulator commands and functions. However if you signal RobotBASIC to use the serial port protocol the same program will run, but instead of simulating the robot on the screen it will send and receive data back and forth between a real robot and RobotBASIC. This allows you to test your simulated algorithms on a real physical robot.

The microcontroller on the real robot receives data representing the commands (like `rForward`, etc.), responds to the commands and returns data to RobotBASIC to tell it that it received the command and acted upon it and that the current state of sensors is as per the data sent.

You can write programs using the commands and function in the robot simulator, but the commands will not cause the simulated screen robot to move. Rather, a set of 2 bytes is sent via the serial port and then RobotBASIC will wait for a set of 5 bytes to be sent back. These 5 bytes are then stored and interpreted to provide data to be returned by the command or function.

To tell RobotBASIC to use the serial communications protocol rather than the simulator, use the command `rCommPort` (see below) to make all subsequent commands and functions use the protocol in place of the simulator.

The commands and functions will cause RobotBASIC to send 2 bytes where the first byte is a code for the command and the next byte is the parameter for the command. The receiving microcontroller on the physical robot can make use of these bytes and must respond within the time-out limit (see `SetTimeOut` above) by sending 5 bytes to be received by RobotBASIC. The first 3 bytes are the status of the bumpers, infrared sensors, and line sensors (in this order). The last 2 bytes are the returned values relating to the function/command. The order is MSB first. The first byte is the most significant byte (MSB) of the number. (`rGPS` is the exception to the 5-byte rule, see below).

The values received in the first 3 bytes of the 5 bytes are used to update the status of the bumper, infrared, and line sensors. These values will be used by the functions `rBumper()`, `rFeel()`, and `rSense()`. This means that these functions will still work in the same way as in the simulated situation but instead of returning the values read by the simulated robot off the screen they will return the real physical values received by the protocol from real sensors on the physical robot. The values can be used in exactly the same manner as before (as described in the robot simulator help page).

`rCommPort` *ExprN1 {,ExprN2 {,ExprN2 {,ExprN3 {,ExprN4 {,ExprN5}}}}}*
This command is exactly the same as the `SetCommPort` command (Sec. D.2). It sets the comm port number and parameters and opens the comm port for communications. However, it also signals a flag that makes the simulator use the comm port specified in place of the screen. To return to using the screen, issue the command again with 0 as the comm port number.

`rLocate` *ExprN1,ExprN2* (code 3) You still need to issue this command before using the protocol just as in the simulator. This command initializes the real robot and starts the command process. No use is made of the second (and the others if you happen to be using them) parameters but they need to still be there because the command is the same as in the simulated one but does not use the numbers when the protocol is active. This means that you do not need to change your simulator program to make it work with the protocol. The protocol sends 2 bytes 03 and *ExprN1*. It will expect to receive 5 bytes where the first three are as described above. The last two are not used.

`rForward` *ExprN* (code 6 or 7) This command will send 2 bytes 6 (or 7) and Abs(*ExprN*). The code 6 is for going forward and 7 is for going backward. *ExprN* is the distance required if it is positive then code 6 is sent and if it is negative then it is made positive and code 7 is sent. The received data contains 5 bytes but only the first 3 are used. They are in the order described above for the status of the sensors.

`rTurn` *ExprN* (code 12 or 13) This command will send 2 bytes 12 (or 13) and Abs(*ExprN*). The code 12 is for turning right and 13 is for turning left. *ExprN* is the required degrees, if positive then code 12 is sent and if negative then the number is made positive and code 13 is sent. If you specify a number greater than 180 then the turn is made into *ExprN* – 360 and if it is less than –180 (i.e., more negative than –180, e.g., –190) then it is made into 360 + *ExprN*. So if your simulator command says `rTurn` 190 then the value sent to the real robot will be 170 with code 13 (i.e., left turn of 170). The received data contains 5 bytes but only the first 3 are used. They are in the order described above for the status of the sensors.

`rCompass` () (code 24) This function will send 2 bytes 24 and 0. The code 24 is for the compass. The received data contains 5 bytes. The first 3 used are in the order described above for the status of the sensors. The last 2 bytes are the value returned by the compass on the real robot representing the heading (MSB first). This means that this function will return a number formed as a 16-bit number from the fourth byte of the received 5 bytes as the MSB of the number and the fifth byte as the LSB.

`rLook` ({*ExprN*}) (code 48 or 49) Will send 2 bytes 48 (or 49) and Abs(*ExprN*). The code 48 is for the camera sensors reading *ExprN* degrees to the right and code 49 is to the left. If *ExprN* is negative it will be made positive and code 49 will be sent. The received data contains 5 bytes. The first 3 used are in the order described above for the status of the sensors. The last 2 bytes are the value returned representing the color seen by the sensor. (MSB first). This means that this function will return a number formed as a 16-bit number from the fourth byte of the received 5 bytes as the MSB of the number and the fifth byte as the LSB.

`rBeacon` (*ExprN*) (code 96) Will send 2 bytes 96 and *ExprN*. The code 96 is for beacon sensors reading and *ExprN* is the required color. The received data contains 5 bytes. The first 3 used are in the order described above for the status of the sensors. The last 2 bytes are the value returned by the beacon representing the distance measured to the found beacon or 0 if the beacon is not found. (MSB first). This means that this function will return a number formed as a 16-bit number from the fourth byte of the received 5 bytes as the MSB of the number and the fifth byte as the LSByte.

`rRange` ({*ExprN*}) (code 192 or 193) Will send 2 bytes 192 (or 193) and Abs(*ExprN*). The code 192 is for range sensor reading at *ExprN* degrees to the right and 193 is to the left. If *ExprN* is negative it is made positive and code 193 is sent. The received data contains 5 bytes. The first 3 used are in the order described above for the status of the sensors. The last 2 bytes are the value returned by the rangers representing the distance measured. (MSB first). This means that this function will return a number formed as a 16-bit number from the fourth byte of the received 5 bytes as the MSB of the number and the fifth byte as the LSB.

`rPen` *ExprN* (code 129) This command will send 2 bytes 129 and *ExprN*. The code 129 is for pen control. *ExprN* is either *up* (0) or *down* (not zero, can be any number other than 0 usually it is 1). The received data contains 5 bytes. The first 3 are in the order described above for the status of the sensors. The last 2 bytes are not used and are set to 0.

`rSpeed` ***ExprN* (code 36)** This command will send 2 bytes 36 and *ExprN*. The code 36 is for setting a speed on the physical robot and *ExprN* is the required speed (0–255). The received data contains 5 bytes. The first 3 used are in the order described above for the status of the sensors. The last 2 bytes are not use and are set to 0.

`rGPS` ***VarNX,VarNY* (code 66)** This command will send 2 bytes 66 and 0. The code 66 is for global positioning system (GPS) sensors reading and *VarNX* will be set to the *x* position and *VarNY* to the *y* position. The received data contains 5 bytes. The first 2 are the *x* position and the second two are the *y* position. The fifth is not used and is set to 0. (MSB first).

This command is an exception to the 5 bytes received in all the other commands. The 5 bytes received after issuing this command are not used to update the bumpers, infrared, and line sensors status. Instead the first 2 bytes are used to calculate the *x* position and the second two bytes the *y* position. The last byte is not used.

INDEX

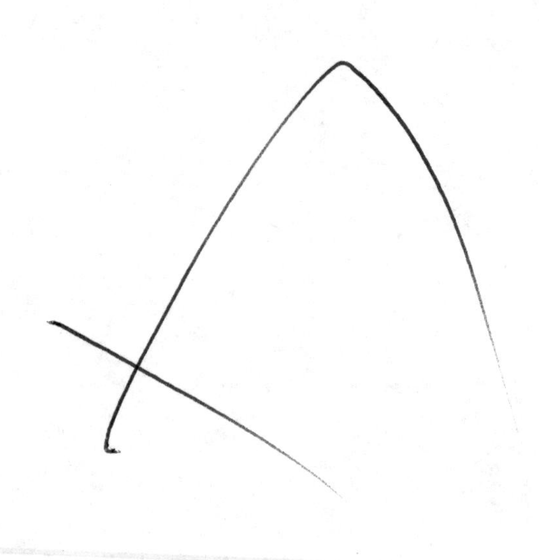